Differential Equations with Boundary Value Problems

EIGHTH EDITION

Dennis G. Zill
Loyola Marymount University

Warren S. Wright
Loyola Marymount University

Prepared by

Dennis G. Zill
Loyola Marymount University

Warren S. Wright
Loyola Marymount University

Carol D. Wright

BROOKS/COLE
CENGAGE Learning·

Australia • Brazil • Japan • Korea • Mexico • Singapore • Spain • United Kingdom • United States

BROOKS/COLE
CENGAGE Learning·

For product information and technology assistance, contact us at
**Cengage Learning Customer & Sales Support,
1-800-354-9706**

For permission to use material from this text or product, submit all requests online at **www.cengage.com/permissions**
Further permissions questions can be emailed to
permissionrequest@cengage.com

ISBN-13: 978-1-133-49195-8
ISBN-10: 1-133-49195-2

Brooks/Cole
20 Channel Center Street
Boston, MA 02210
USA

Cengage Learning is a leading provider of customized learning solutions with office locations around the globe, including Singapore, the United Kingdom, Australia, Mexico, Brazil, and Japan. Locate your local office at: **www.cengage.com/global**

Cengage Learning products are represented in Canada by Nelson Education, Ltd.

To learn more about Brooks/Cole, visit **www.cengage.com/brookscole**

Purchase any of our products at your local college store or at our preferred online store **www.cengagebrain.com**

Printed in the United States of America
1 2 3 4 5 6 7 16 15 14 13 12

Table of Contents

SRSM Introduction

This *Student Resource and Solutions Manual* is meant to accompany the text *A First Course in Differential Equations with Modeling Applications*, Ninth Edition. It is intended to serve several purposes:

- to provide a list of the terminology, concepts, and skills associated with each section in the text

- to provide further elaboration on some of the material discussed in the section

- to review some of the mathematics pertinent to the material being presented (such as integration, synthetic division, and partial fractions)

- to furnish extra examples of the material presented in the section

- to illustrate the use of the computer algebra systems *Mathematica* and *Maple* in dealing with the material in the section

- to provide solutions for every third problem in the exercise sets, excluding Discussion Problems, Mathematical Models, Computer Lab Assignments, Project Problems, and Contributed Problems

- to provide hints and suggestions for selected problems throughout the entire set of exercises in each section

- to provide examples illustrating how to solve some of the Computer Lab problems.

1 INTRODUCTION TO DIFFERENTIAL EQUATIONS

Section 1.1 Definitions and Terminology

The terminology and concepts listed below provide an outline of the main ideas encountered in this section. These can be useful when preparing for a quiz or test.

Terminology and Concepts

- ordinary DE

- partial DE

- order of a DE

- normal form of a DE

- linear DE

- explicit and implicit form of a solution of a DE

- interval of definition of a solution

- trivial solution of a DE

- solution curve

- family of solutions of a DE

- n-parameter family of solutions of an nth-order DE

- particular solution of a DE

- singular solution of a DE

- system of DEs

- general solution of a DE

1

The basic skills listed below summarize the more mechanical types of problems encountered in the exercise set for this section.

Basic Skills

- determine the order of a DE

- determine if a DE is linear or nonlinear

- verify that a given function is a solution of a DE

- determine the interval of definition of a solution of a DE

- use an implicit solution to determine an explict solution (this is not always possible)

- identify constant solutions of certain DEs

- verify that a pair of functions is a solution of a system of two DEs

Basic Properties of the Derivative

Definition: The derivative of a function $y = f(x)$ with respect to the independent variable x is

$$f'(x) = \lim_{h \to 0} \frac{f(x+h) - f(x)}{h},$$

whenever this limit exists.

The derivative is also denoted by y' and dy/dx. When the derivative is evaluated at $x = x_0$, denoted by $f'(x_0)$, the resulting value represents the *rate of change* of the function f at x_0. Geometrically, this is the slope of the tangent line to the graph of $y = f(x)$ at the point $(x_0, f(x_0))$. If the function f is differentiable at x_0 then the graph of f at the point $(x_0, f(x_0))$ does not have a sharp corner or vertical tangent line at that point.

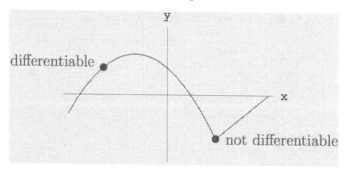

In DEs the independent variable frequently represents time and is denoted by t. In this case, if the position at time t of an object moving on a straight line is given by $s = s(t)$, then the velocity of the object at time t is the first derivative, $v(t) = s'(t)$, and the acceleration is the second derivative of $s(t)$, $a(t) = s''(t)$.

If the first derivative of a function is positive on an interval, then the function is **increasing** on that interval. Correspondingly, when the derivative is negative on an interval, the function is

decreasing on the interval. Relative extrema (maxima or minima) of a function occur where the derivative is either 0 or does not exist. The second derivative of a function can be used to determine concavity. The graph of a function f is **concave up** on an interval if $f''(x) > 0$ on the interval and **concave down** on an interval if $f''(x) < 0$ on the interval.

Hyperbolic Functions   It is unfortunate that many current texts and calculus instructors downplay the hyperbolic functions. These functions are important in applied mathematics and every serious student of differential equations should know their definitions, graphs, and some of their properties. In the text, solutions of DEs will be expressed in terms of hyperbolic functions when appropriate. The six hyperbolic functions are denoted by sinh, cosh, tanh, coth, sech and csch. If x denotes a real variable, the **hyperbolic sine** and **cosine** are defined in terms of exponential functions.

Definitions: $\qquad\qquad \sinh x = \dfrac{e^x - e^{-x}}{2} \qquad$ and $\qquad \cosh x = \dfrac{e^x + e^{-x}}{2} \qquad\qquad$ (1)

The remaining four hyperbolic functions are defined in a manner analogous to trigonometry. For example, the hyperbolic tangent and cotangent are, respectively,

$$\tanh x = \frac{\sinh x}{\cosh x} \qquad \text{and} \qquad \coth x = \frac{\cosh x}{\sinh x}.$$

The graphs of the hyperbolic sine and cosine functions, like the graphs of natural exponential, natural logarithmic, and trigonometric functions, should be memorized.

Graphs:

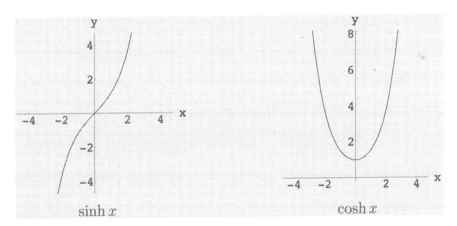

$\sinh x$ $\qquad\qquad\qquad\qquad\qquad\qquad\qquad$ $\cosh x$

Some Properties   As can be seen from both the graphs and the definitions in (1),

$$\sinh 0 = 0 \qquad \text{and} \qquad \cosh 0 = 1. \qquad\qquad (2)$$

Also, $\cosh x$ has no real zeros and, unlike the trigonometric functions, no hyperbolic function is periodic. However, a number of other properties similar to properties for trigonometric functions do hold. As you examine the hyperbolic identities (3) through (8) below, determine whether or not the given identity differs from the analogous identity for trigonometric functions.

$$\cosh^2 x - \sinh^2 x = 1 \tag{3}$$

$$\sinh(x_1 \pm x_2) = \sinh x_1 \cosh x_2 \pm \cosh x_1 \sinh x_2 \tag{4}$$

$$\cosh(x_1 \pm x_2) = \cosh x_1 \cosh x_2 \pm \sinh x_1 \sinh x_2 \tag{5}$$

$$\sinh 2x = 2 \sinh x \cosh x \tag{6}$$

$$\cosh 2x = \cosh^2 x + \sinh^2 x \tag{7}$$

The derivatives of $\sinh x$ and $\cosh x$, given below, can be obtained using the definitions in (1).

$$\frac{d}{dx} \sinh x = \cosh x \qquad \text{and} \qquad \frac{d}{dx} \cosh x = \sinh x \tag{8}$$

Use of Computers   To define a function, such as $f(x) = e^{x^2} + \sin 2x$, use the syntax

Clear[f]   (*Mathematica*)
f[x_]:= Exp[x^2]+Sin[2x]
f[x]

```
f:=x->exp(x^2)+sin(2*x);                                  (Maple)
or
f:=unapply(exp(x^2)+sin(2*x), x);
```

If you are not familiar with *Maple* the **unapply** function probably seems like an awkward way to define a function. We will see later that it is a useful command when solving DEs.

Piecewise-defined functions are, as the name suggests, defined in pieces. For example, the function

$$f(x) = \begin{cases} x^2 + 1, & x \le 2 \\ \sin x, & x > 2 \end{cases}$$

is defined in a CAS using:

Clear[f]   (*Mathematica*)
f[x_]:= x^2 + 1/; x <= 2
f[x_]:= Sin[x]/; x > 2

```
f:=x->piecewise(x<=2, x^2+1, x>2,sin(x));        (Maple)
```

The syntax for the first derivative, dy/dx, of $y = f(x)$ is:

D[f[x], x]   or   **f'[x]**   (*Mathematica*)

```
diff(f(x), x);
```
(*Maple*)

The second derivative, d^2y/dx^2, of $y = f(x)$ is:

$$\mathbf{D[f[x], \{x,2\}]} \quad \text{or} \quad \mathbf{f''[x]}$$
(*Mathematica*)

```
diff(f(x), x$2);  or  diff(f(x),x,x);
```
(*Maple*)

1.1 | Definitions and Terminology

3. Fourth order; linear

6. Second order; nonlinear because of R^2

9. Writing the boundary-value problem in the form $x(dy/dx) + y^2 = 1$, we see that it is nonlinear in y because of y^2. However, writing it in the form $(y^2 - 1)(dx/dy) + x = 0$, we see that it is linear in x.

12. From $y = \frac{6}{5} - \frac{6}{5}e^{-20t}$ we obtain $dy/dt = 24e^{-20t}$, so that

$$\frac{dy}{dt} + 20y = 24e^{-20t} + 20\left(\frac{6}{5} - \frac{6}{5}e^{-20t}\right) = 24.$$

15. The domain of the function, found by solving $x + 2 \geq 0$, is $[-2, \infty)$. From $y' = 1 + 2(x + 2)^{-1/2}$ we have

$$\begin{aligned}
(y - x)y' &= (y - x)[1 + (2(x + 2)^{-1/2}] \\
&= y - x + 2(y - x)(x + 2)^{-1/2} \\
&= y - x + 2[x + 4(x + 2)^{1/2} - x](x + 2)^{-1/2} \\
&= y - x + 8(x + 2)^{1/2}(x + 2)^{-1/2} = y - x + 8.
\end{aligned}$$

An interval of definition for the solution of the differential equation is $(-2, \infty)$ because y' is not defined at $x = -2$.

18. The function is $y = 1/\sqrt{1 - \sin x}$, whose domain is obtained from $1 - \sin x \neq 0$ or $\sin x \neq 1$. Thus, the domain is $\{x \mid x \neq \pi/2 + 2n\pi\}$. From $y' = -\frac{1}{2}(1 - \sin x)^{-3/2}(-\cos x)$ we have

$$2y' = (1 - \sin x)^{-3/2}\cos x = [(1 - \sin x)^{-1/2}]^3\cos x = y^3\cos x.$$

An interval of definition for the solution of the differential equation is $(\pi/2, 5\pi/2)$. Another interval is $(5\pi/2, 9\pi/2)$ and so on.

21. Differentiating $P = c_1 e^t / \left(1 + c_1 e^t\right)$ we obtain

$$\frac{dP}{dt} = \frac{\left(1 + c_1 e^t\right) c_1 e^t - c_1 e^t \cdot c_1 e^t}{\left(1 + c_1 e^t\right)^2} = \frac{c_1 e^t}{1 + c_1 e^t} \frac{\left[\left(1 + c_1 e^t\right) - c_1 e^t\right]}{1 + c_1 e^t}$$

$$= \frac{c_1 e^t}{1 + c_1 e^t} \left[1 - \frac{c_1 e^t}{1 + c_1 e^t}\right] = P(1 - P).$$

24. From $y = c_1 x^{-1} + c_2 x + c_3 x \ln x + 4x^2$ we obtain

$$\frac{dy}{dx} = -c_1 x^{-2} + c_2 + c_3 + c_3 \ln x + 8x,$$

$$\frac{d^2 y}{dx^2} = 2c_1 x^{-3} + c_3 x^{-1} + 8,$$

and

$$\frac{d^3 y}{dx^3} = -6c_1 x^{-4} - c_3 x^{-2},$$

so that

$$x^3 \frac{d^3 y}{dx^3} + 2x^2 \frac{d^2 y}{dx^2} - x \frac{dy}{dx} + y = (-6c_1 + 4c_1 + c_1 + c_1)x^{-1} + (-c_3 + 2c_3 - c_2 - c_3 + c_2)x$$

$$+ (-c_3 + c_3)x \ln x + (16 - 8 + 4)x^2$$

$$= 12x^2.$$

27. From $y = e^{mx}$ we obtain $y' = me^{mx}$. Then $y' + 2y = 0$ implies

$$me^{mx} + 2e^{mx} = (m + 2)e^{mx} = 0.$$

Since $e^{mx} > 0$ for all x, $m = -2$. Thus $y = e^{-2x}$ is a solution.

30. From $y = e^{mx}$ we obtain $y' = me^{mx}$ and $y'' = m^2 e^{mx}$. Then $2y'' + 7y' - 4y = 0$ implies

$$2m^2 e^{mx} + 7me^{mx} - 4e^{mx} = (2m - 1)(m + 4)e^{mx} = 0.$$

Since $e^{mx} > 0$ for all x, $m = \frac{1}{2}$ and $m = -4$. Thus $y = e^{x/2}$ and $y = e^{-4x}$ are solutions.

In Problems 33 and 36 we substitute $y = c$ into the differential equations and use $y' = 0$ and $y'' = 0$.

33. Solving $5c = 10$ we see that $y = 2$ is a constant solution.

36. Solving $6c = 10$ we see that $y = 5/3$ is a constant solution.

1.2 | Initial-Value Problems

The terminology and concepts listed below provide an outline of the main ideas encountered in this section. These can be useful when preparing for a quiz or test.

Terminology and Concepts

- IVPs
- initial conditions
- how the interval of definition of a solution of an IVP may differ from the domain of the function in the solution
- existence and uniqueness of solutions of IVPs

The basic skills listed below summarize the more mechanical types of problems encountered in the exercise set for this section.

Basic Skills

- given a family of solutions of a DE, find a particular solution satisfying given initial conditions
- determine a region in the plane for which a given DE will have a unique solution whose graph passes through a point in the region
- match graphs of solution curves with given initial conditions

Existence and Uniqueness Theorem 1.2.1 is local in nature, so the interval I_0 may not be very large. The words "existence" and "uniqueness" are often not clearly understood. Use of the notation such as $y = y(x)$ and saying that a solution "exists" on the interval I_0 does *not* necessarily mean that we can actually find (meaning display) an explicit or implicit solution of the DE defined on I_0. We know in theory that a solution $y = y(x)$ exists but we may have to approximate it on I_0. The word "uniqueness" basically means we cannot have *two* (or more) solutions $y = y_1(x)$ and $y = y_2(x)$ defined on I_0 such that their solution curves both pass through the initial point (x_0, y_0). In other words, on I_0 we cannot have either of the following situations:

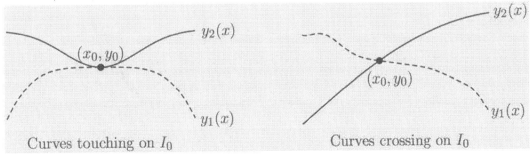

Curves touching on I_0 Curves crossing on I_0

Since any point on the graph of a solution curve may be considered as an initial condition of the DE resulting in that curve, no two solution curves can intersect.

3. Letting $x = 2$ and solving $1/3 = 1/(4 + c)$ we get $c = -1$. The solution is $y = 1/(x^2 - 1)$. This solution is defined on the interval $(1, \infty)$.

6. Letting $x = 1/2$ and solving $-4 = 1/(1/4 + c)$ we get $c = -1/2$. The solution is $y = 1/(x^2 - 1/2) = 2/(2x^2 - 1)$. This solution is defined on the interval $(-1/\sqrt{2}, 1/\sqrt{2})$.

9. From the initial conditions we obtain

$$\frac{\sqrt{3}}{2} c_1 + \frac{1}{2} c_2 = \frac{1}{2}$$

$$-\frac{1}{2} c_1 + \frac{\sqrt{3}}{2} c_2 = 0.$$

Solving, we find $c_1 = \sqrt{3}/4$ and $c_2 = 1/4$. The solution of the initial-value problem is

$$x = (\sqrt{3}/4) \cos t + (1/4) \sin t.$$

12. From the initial conditions we obtain

$$e c_1 + e^{-1} c_2 = 0$$
$$e c_1 - e^{-1} c_2 = e.$$

Solving, we find $c_1 = \frac{1}{2}$ and $c_2 = -\frac{1}{2} e^2$. The solution of the initial-value problem is

$$y = \frac{1}{2} e^x - \frac{1}{2} e^2 e^{-x} = \frac{1}{2} e^x - \frac{1}{2} e^{2-x}.$$

15. Two solutions are $y = 0$ and $y = x^3$.

18. For $f(x, y) = \sqrt{xy}$ we have $\partial f/\partial y = \frac{1}{2} \sqrt{x/y}$. Thus, the differential equation will have a unique solution in any region where $x > 0$ and $y > 0$ or where $x < 0$ and $y < 0$.

21. For $f(x, y) = x^2/(4 - y^2)$ we have $\partial f/\partial y = 2x^2 y/(4 - y^2)^2$. Thus the differential equation will have a unique solution in any region where $y < -2$, $-2 < y < 2$, or $y > 2$.

24. For $f(x, y) = (y + x)/(y - x)$ we have $\partial f/\partial y = -2x/(y - x)^2$. Thus the differential equation will have a unique solution in any region where $y < x$ or where $y > x$.

27. Since $(2, -3)$ is not in either of the regions defined by $y < -3$ or $y > 3$, there is no guarantee of a unique solution through $(2, -3)$.

30. (a) Since $\dfrac{d}{dx} \tan(x + c) = \sec^2(x + c) = 1 + \tan^2(x + c)$, we see that $y = \tan(x + c)$ satisfies the differential equation.

(b) Solving $y(0) = \tan c = 0$ we obtain $c = 0$ and $y = \tan x$. Since $\tan x$ is discontinuous at $x = \pm \pi/2$, the solution is not defined on $(-2, 2)$ because it contains $\pm \pi/2$.

(c) The largest interval on which the solution can exist is $(-\pi/2, \pi/2)$.

33. (a) Differentiating $3x^2 - y^2 = c$ we get $6x - 2yy' = 0$ or $yy' = 3x$.

(b) Solving $3x^2 - y^2 = 3$ for y we get

$$y = \phi_1(x) = \sqrt{3(x^2 - 1)}, \qquad 1 < x < \infty,$$

$$y = \phi_2(x) = -\sqrt{3(x^2 - 1)}, \qquad 1 < x < \infty,$$

$$y = \phi_3(x) = \sqrt{3(x^2 - 1)}, \qquad -\infty < x < -1,$$

$$y = \phi_4(x) = -\sqrt{3(x^2 - 1)}, \qquad -\infty < x < -1.$$

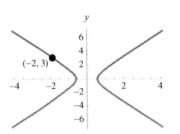

(c) Only $y = \phi_3(x)$ satisfies $y(-2) = 3$.

36. The graph satisfies the conditions in (e).

In Problems 39 and 42 $y = c_1 \cos 2x + c_2 \sin 2x$ is a two parameter family of solutions of the second-order differential equation $y'' + 4y = 0$. In some of the problems we will use the fact that $y' = -2c_1 \sin 2x + 2c_2 \cos 2x$.

39. From the boundary conditions $y(0) = 0$ and $y\left(\dfrac{\pi}{4}\right) = 3$ we obtain

$$y(0) = c_1 = 0$$

$$y\left(\frac{\pi}{4}\right) = c_1 \cos\left(\frac{\pi}{2}\right) + c_2 \sin\left(\frac{\pi}{2}\right) = c_2 = 3.$$

Thus, $c_1 = 0$, $c_2 = 3$, and the solution of the boundary-value problem is $y = 3 \sin 2x$.

42. From the boundary conditions $y(0) = 1$ and $y'(\pi) = 5$ we obtain

$$y(0) = c_1 = 1$$

$$y'(\pi) = 2c_2 = 5.$$

Thus, $c_1 = 1$, $c_2 = \dfrac{5}{2}$, and the solution of the boundary-value problem is $y = \cos 2x + \dfrac{5}{2} \sin 2x$.

1.3 Differential Equations as Mathematical Models

The terminology and concepts listed below provide an outline of the main ideas encountered in this section. These can be useful when preparing for a quiz or test.

Terminology and Concepts

- construction of mathematical models
- level of resolution of a model
- use of assumptions and empirical laws
- steps of the modeling process
- model for population growth
- model for radioactive decay
- Newton's law of cooling/warming
- model for the spread of a disease
- model for the draining of a tank
- models for first- and second-order chemical reactions
- models to describe mixtures of differing concentrations of elements in liquid solutions
- models for electrical series circuits
- models for the motion of falling bodies involving Newton's laws of motion
- approaches to DEs: analytical, qualitative, numerical

The basic skills listed below summarize the more mechanical types of problems encountered in the exercise set for this section.

Basic Skills

- set up DEs, sometimes with initial conditions, that can be used to model the situations listed above

Mixtures   In setting up DEs such as

$$\frac{dA}{dt} + \frac{1}{100} A = 6 \qquad \text{[(8) in the text]}$$

it is helpful to keep in mind the units. The derivative, dA/dt, is the rate at which the amount of salt (pounds) is changing in time (minutes) and so is measured in pounds per minute. In the derivation of (8) in the text observe that the unit "gal" cancels out in both R_{in} and R_{out} to give "lb/min." Also, note carefully that even though the tank contains 300 gallons of brine, the initial condition associated with (8) is *not* $A(0) = 300$ since $A(t)$, which represents the amount of salt at time t, is measured in pounds, whereas 300 is a measure of the number of gallons of fluid in the tank. Indeed, an initial condition for $A(t)$ is not specified. See the text: Problem 9 in Exercises 1.3.

Population Dynamics

3. Let b be the rate of births and d the rate of deaths. Then $b = k_1 P$ and $d = k_2 P^2$. Since $dP/dt = b - d$, the differential equation is $dP/dt = k_1 P - k_2 P^2$.

Newton's Law of cooling/Warming

6. By inspecting the graph in the text we take T_m to be $T_m(t) = 80 - 30 \cos \pi t/12$. Then the temperature of the body at time t is determined by the differential equation

$$\frac{dT}{dt} = k \left[T - \left(80 - 30 \cos \frac{\pi}{12} t \right) \right], \quad t > 0.$$

Mixtures

9. The rate at which salt is leaving the tank is

$$R_{out} \ (3 \ \text{gal/min}) \left(\frac{A}{300} \ \text{lb/gal} \right) = \frac{A}{100} \ \text{lb/min}.$$

Thus $dA/dt = -A/100$ (where the minus sign is used since the amount of salt is decreasing). The initial amount is $A(0) = 50$.

12. The rate at which salt is entering the tank is

$$R_{in} = (c_{in} \ \text{lb/gal})(r_{in} \ \text{gal/min}) = c_{in} r_{in} \ \text{lb/min}.$$

Now let $A(t)$ denote the number of pounds of salt and $N(t)$ the number of gallons of brine in the tank at time t. The concentration of salt in the tank as well as in the outflow is $c(t) = x(t)/N(t)$. But the number of gallons of brine in the tank remains steady, is increased, or is decreased depending on whether $r_{in} = r_{out}$, $r_{in} > r_{out}$, or $r_{in} < r_{out}$. In any case, the number of gallons of brine in the tank at time t is $N(t) = N_0 + (r_{in} - r_{out})t$. The output rate of salt is then

$$R_{out} = \left(\frac{A}{N_0 + (r_{in} - r_{out})t} \ \text{lb/gal} \right) (r_{out} \ \text{gal/min}) = r_{out} \frac{A}{N_0 + (r_{in} - r_{out})t} \ \text{lb/min}.$$

The differential equation for the amount of salt, $dA/dt = R_{in} - R_{out}$, is

$$\frac{dA}{dt} = c_{in} r_{in} - r_{out} \frac{A}{N_0 + (r_{in} - r_{out})t} \quad \text{or} \quad \frac{dA}{dt} + \frac{r_{out}}{N_0 + (r_{in} - r_{out})t} A = c_{in} r_{in}.$$

Series Circuits

15. Since $i = dq/dt$ and $L\,d^2q/dt^2 + R\,dq/dt = E(t)$, we obtain $L\,di/dt + Ri = E(t)$.

Newton's Second Law and Archimedes' Principle

18. Since the barrel in Figure 1.3.17(b) in the text is submerged an additional y feet below its equilibrium position the number of cubic feet in the additional submerged portion is the volume of the circular cylinder: $\pi\times$(radius)$^2\times$height or $\pi(s/2)^2y$. Then we have from Archimedes' principle

upward force of water on barrel = weight of water displaced

$$= (62.4) \times \text{(volume of water displaced)}$$
$$= (62.4)\pi(s/2)^2y = 15.6\pi s^2 y.$$

It then follows from Newton's second law that

$$\frac{w}{g}\frac{d^2y}{dt^2} = -15.6\pi s^2 y \qquad \text{or} \qquad \frac{d^2y}{dt^2} + \frac{15.6\pi s^2 g}{w}y = 0,$$

where $g = 32$ and w is the weight of the barrel in pounds.

Newton's Second Law and Rocket Motion

21. Since the positive direction is taken to be upward, and the acceleration due to gravity g is positive, (14) in Section 1.3 becomes

$$m\frac{dv}{dt} = -mg - kv + R.$$

This equation, however, only applies if m is constant. Since in this case m includes the variable amount of fuel we must use (17) in Exercises 1.3:

$$F = \frac{d}{dt}(mv) = m\frac{dv}{dt} + v\frac{dm}{dt}.$$

Thus, replacing $m\,dv/dt$ with $m\,dv/dt + v\,dm/dt$, we have

$$m\frac{dv}{dt} + v\frac{dm}{dt} = -mg - kv + R \qquad \text{or} \qquad m\frac{dv}{dt} + v\frac{dm}{dt} + kv = -mg + R.$$

Newton's Second Law and the Law of Universal Gravitation

24. The gravitational force on m is $F = -kM_r m/r^2$. Since $M_r = 4\pi\delta r^3/3$ and $M = 4\pi\delta R^3/3$ we have $M_r = r^3 M/R^3$ and

$$F = -k\frac{M_r m}{r^2} = -k\frac{r^3 Mm/R^3}{r^2} = -k\frac{mM}{R^3}r.$$

Now from $F = ma = d^2r/dt^2$ we have

$$m\frac{d^2r}{dt^2} = -k\frac{mM}{R^3}r \quad \text{or} \quad \frac{d^2r}{dt^2} = -\frac{kM}{R^3}r.$$

Additional Mathematical Models

27. The differential equation is $x'(t) = r - kx(t)$ where $k > 0$.

1.R   Chapter 1 in Review

3. $\dfrac{d}{dx}(c_1 \cos kx + c_2 \sin kx) = -kc_1 \sin kx + kc_2 \cos kx;$

$\dfrac{d^2}{dx^2}(c_1 \cos kx + c_2 \sin kx) = -k^2 c_1 \cos kx - k^2 c_2 \sin kx = -k^2(c_1 \cos kx + c_2 \sin kx);$

$\dfrac{d^2 y}{dx^2} = -k^2 y \quad \text{or} \quad \dfrac{d^2 y}{dx^2} + k^2 y = 0$

6. $y' = -c_1 e^x \sin x + c_1 e^x \cos x + c_2 e^x \cos x + c_2 e^x \sin x;$

$y'' = -c_1 e^x \cos x - c_1 e^x \sin x - c_1 e^x \sin x + c_1 e^x \cos x - c_2 e^x \sin x + c_2 e^x \cos x + c_2 e^x \cos x + c_2 e^x \sin x$

$= -2c_1 e^x \sin x + 2c_2 e^x \cos x;$

$y'' - 2y' = -2c_1 e^x \cos x - 2c_2 e^x \sin x = -2y; \qquad y'' - 2y' + 2y = 0$

9. b   **(12.)** a, b, d

15. The slope of the tangent line at (x, y) is y', so the differential equation is $y' = x^2 + y^2$.

18. (a) Differentiating $y^2 - 2y = x^2 - x + c$ we obtain $2yy' - 2y' = 2x - 1$ or $(2y - 2)y' = 2x - 1$.

(b) Setting $x = 0$ and $y = 1$ in the solution we have $1 - 2 = 0 - 0 + c$ or $c = -1$. Thus, a solution of the initial-value problem is $y^2 - 2y = x^2 - x - 1$.

(c) Solving $y^2 - 2y - (x^2 - x - 1) = 0$ by the quadratic formula we get

$$y = \frac{2 \pm \sqrt{4 + 4(x^2 - x - 1)}}{2} = 1 \pm \sqrt{x^2 - x} = 1 \pm \sqrt{x(x-1)}.$$

Since $x(x-1) \geq 0$ for $x \leq 0$ or $x \geq 1$, we see that neither $y = 1 + \sqrt{x(x-1)}$ nor $y = 1 - \sqrt{x(x-1)}$ is differentiable at $x = 0$. Thus, both functions are solutions of the differential equation, but neither is a solution of the initial-value problem.

21. (a)

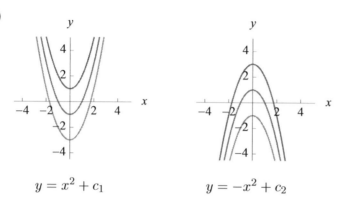

$$y = x^2 + c_1 \qquad\qquad y = -x^2 + c_2$$

(b) When $y = x^2 + c_1$, $y' = 2x$ and $(y')^2 = 4x^2$. When $y = -x^2 + c_2$, $y' = -2x$ and $(y')^2 = 4x^2$.

(c) Pasting together x^2, $x \geq 0$, and $-x^2$, $x \leq 0$, we get

$$f(x) = \begin{cases} -x^2, & x \leq 0 \\ x^2, & x > 0. \end{cases}$$

24. Differentiating $y = x \sin x + (\cos x) \ln(\cos x)$ we get

$$y' = x \cos x + \sin x + \cos x \left(\frac{-\sin x}{\cos x} \right) - (\sin x) \ln(\cos x)$$

$$= x \cos x + \sin x - \sin x - (\sin x) \ln(\cos x)$$

$$= x \cos x - (\sin x) \ln(\cos x)$$

and,

$$y'' = -x \sin x + \cos x - \sin x \left(\frac{-\sin x}{\cos x} \right) - (\cos x) \ln(\cos x)$$

$$= -x \sin x + \cos x + \frac{\sin^2 x}{\cos x} - (\cos x) \ln(\cos x)$$

$$= -x \sin x + \cos x + \frac{1 - \cos^2 x}{\cos x} - (\cos x) \ln(\cos x)$$

$$= -x \sin x + \cos x + \sec x - \cos x - (\cos x) \ln(\cos x)$$

$$= -x \sin x + \sec x - (\cos x) \ln(\cos x).$$

Thus

$$y'' + y = -x\sin x + \sec x - (\cos x)\ln(\cos x) + x\sin x + (\cos x)\ln(\cos x) = \sec x.$$

To obtain an interval of definition we note that the domain of $\ln x$ is $(0, \infty)$, so we must have $\cos x > 0$. Thus, an interval of definition is $(-\pi/2, \pi/2)$.

27. Using implicit differentiation on $x^3 y^3 = x^3 + 1$ we have

$$3x^3 y^2 y' + 3x^2 y^3 = 3x^2$$
$$xy^2 y' + y^3 = 1$$
$$xy' + y = \frac{1}{y^2}.$$

30. Using implicit differentiation on $y = e^{xy}$ we have

$$y' = e^{xy}\left(xy' + y\right)$$
$$\left(1 - xe^x y\right)y' = ye^{xy}.$$

Since $y = e^{xy}$ we have

$$(1 - xy)y' = y \cdot y \qquad \text{or} \qquad (1 - xy)y' = y^2.$$

33. The initial conditions imply

$$c_1 e^3 + c_2 e^{-1} - 2 = 4$$
$$3c_1 e^3 - c_2 e^{-1} - 2 = -2,$$

so $c_1 = \frac{3}{2}e^{-3}$ and $c_2 = \frac{9}{2}e$. Thus $y = \frac{3}{2}e^{3x-3} + \frac{9}{2}e^{-x+1} - 2x$.

36. Figure 1.3.3 in the text can be used for reference in this problem. The differential equation is

$$\frac{dh}{dt} = -\frac{cA_0}{A_w}\sqrt{2gh}.$$

Using $A_0 = \pi(1/24)^2 = \pi/576$, $A_w = \pi(2)^2 = 4\pi$, and $g = 32$, this becomes

$$\frac{dh}{dt} = -\frac{c\pi/576}{4\pi}\sqrt{64h} = \frac{c}{288}\sqrt{h}.$$

2 | FIRST-ORDER

DIFFERENTIAL EQUATIONS

2.1 | Solution Curves Without a Solution

The terminology and concepts listed below provide an outline of the main ideas encountered in this section. These can be useful when preparing for a quiz or test.

Terminology and Concepts

- lineal element
- direction field (slope field)
- autonomous first-order DE
- critical (stationary, equilibrium) point
- equilibrium solution
- phase portrait (phase line)
- asymptotically stable critical point (attractor)
- unstable critical point (repeller)
- semi-stable critical point

The basic skills listed below summarize the more mechanical types of problems encountered in the exercise set for this section.

Basic Skills

- sketch a direction field of a first-order DE by hand
- sketch a solution curve through a given point in a direction field
- find the critical points of an autonomous first-order DE
- draw the phase portrait of an autonomous first-order DE
- classify critical points as asymptotically stable (attractor), unstable (repeller) or semi-stable
- use a phase portrait to sketch solution curves of autonomous first-order DEs
- determine the behavior of solutions of autonomous first-order DEs as $t \to \pm\infty$

16

Signs of Derivatives The first derivative and its algebraic sign over an interval plays an important role in the qualitative analysis in Section 2.1. If $y = f(x)$ is a differentiable function and if $dy/dx > 0$ on an interval then f is increasing on the interval, whereas if $dy/dx < 0$ on an interval then f is decreasing on the interval. Similarly, if $y = f(x)$ possesses a second derivative, then the sign of d^2y/dx^2 determines the concavity of the graph of f. If $d^2y/dx^2 > 0$ on an interval then the graph of f is concave upward on the interval, whereas if $d^2y/dx^2 < 0$ on an interval, then the graph of f is concave downward.

Use of Computers Both *Mathematica* and *Maple* can be used to obtain direction fields of first-order DEs. In *Mathematica*, a graphics package containing the command **PlotVectorField** must first be loaded. (It is important that this be done only once during any session of *Mathematica* and that it be done before the command is first used. If this protocol is not followed, you should save your current file, quit *Mathematica*, and restart the program.) **PlotVectorField** is actually used to graph the vector field corresponding to a vector function so it must be adapted to the case of a first-order DE having the form $dy/dx = f(x, y)$. The routine below plots the direction field for the DE $dy/dx = 0.2xy$ where $-5 < x < 5$ and $-5 < y < 5$.

```
<<Graphics`                                              (Mathematica)
PlotVectorField[{1, 2x y}, {x, -5, 5}, {y, -5, 5},
    Frame->True, Axes->True, AxesLabel->{"x", "y"},
    Plotpoints->15]
```

This example will plot a direction field in which the lineal elements (arrows) have their base at a grid point. The number of grid points in both horizontal and vertical directions is specified by the **PlotPoints** option. The output is shown below in Figure 2.1. To obtain a direction field with linear elements all the same length, the graphics options **ScaleFunction** and **ScaleFactor** can be added as shown below.

```
PlotVectorField[{1, 2x y},                               (Mathematica)
    {x, -5, 5}, {y, -5, 5}, Frame->True, Axes->True,
    AxesLabel->{"x", "y"}, ScaleFunction->(0.5&),
    ScaleFactor->0.5, PlotPoints->9]
```

The output for this command is shown below in Figure 2.2.

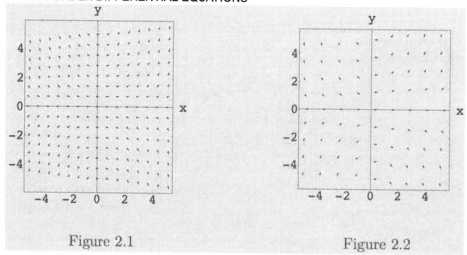

Figure 2.1   Figure 2.2

In *Maple* the command `fieldplot` is used to obtain a direction field. The routine below plots the direction field for the DE $dy/dx = 0.2xy$.

```
with(plots);                                    (Maple)
fieldplot([1,0.2*x*y], x=-5..5, y=-5..5);
```

To obtain a direction field with lineal elements all having the same length it is necessary to scale the input by the length of each vector as shown below.

```
with(plots);                                    (Maple)
fieldplot([1/sqrt(1+(0.2*x*y)^2), 0.2*x*y/sqrt(1+(0.2*x*y)^2)],
    x=-5..5, y=-5..5);
```

2.1.1 DIRECTION FIELDS

In Problem 3 the graph corresponding to the initial condition in Part (a) *is red,* Part (b) *is green,* Part (c) *is blue, and* Part (d) *is brown. The pictures are obtain using Mathematica with*

VectorPlot[{1, f[x, y]}, {x, lhs, rhs}, {y, down, up}, . . .].

3.

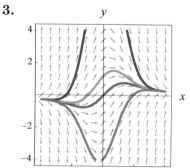

6.

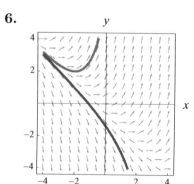

In Problems 9 and 12 the graph corresponding to the initial condition in Part (a) is red, and Part (b) is blue. The pictures are obtain using Mathematica, as mentioned before Problem 1.

9.

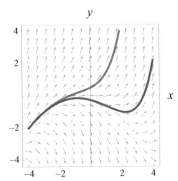

12.

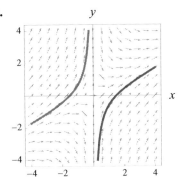

15. (a) The isoclines have the form $y = -x + c$, which are straight lines with slope -1.

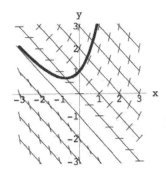

(b) The isoclines have the form $x^2 + y^2 = c$, which are circles centered at the origin.

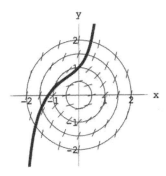

2.1.2   AUTONOMOUS FIRST-ORDER DEs

*In Problems 21–30 graphs of typical solutions are shown. However, in some of the solutions, even though the upper and lower graphs either actually bend up or down, they display as straight line segments. This is a peculiarity of the Mathematica graphing routine and may be due to the fact that the **NDSolve** function was used rather than **DSolve**. NDSolve uses a numerical routine (see Section 2.6 in the text), and involves sampling x-coordinates where the corresponding y-coordinates are approximated. It may be that the routine involved breaks down as the graph becomes nearly vertical, forcing the x-coordinates on the graph to becomes closer and closer together.*

21. Solving $y^2 - 3y = y(y-3) = 0$ we obtain the critical points 0 and 3. From the phase portrait we see that 0 is asymptotically stable (attractor) and 3 is unstable (repeller).

24. Solving $10 + 3y - y^2 = (5-y)(2+y) = 0$ we obtain the critical points -2 and 5. From the phase portrait we see that 5 is asymptotically stable (attractor) and -2 is unstable (repeller).

27. Solving $y \ln(y+2) = 0$ we obtain the critical points -1 and 0. From the phase portrait we see that -1 is asymptotically stable (attractor) and 0 is unstable (repeller).

30. The critical points are approximately at -2, 2, 0.5, and 1.7. Since $f(y) > 0$ for $y < -2.2$ and $0.5 < y < 1.7$, the graph of the solution is increasing on $(-\infty, -2.2)$ and $(0.5, 1.7)$. Since $f(y) < 0$ for $-2.2 < y < 0.5$ and $y > 1.7$, the graph is decreasing on $(-2.2, 0.5)$ and $(1.7, \infty)$.

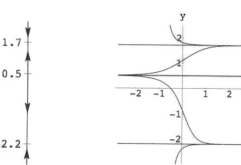

2.2 | Separable Equations

The terminology and concepts listed below provide an outline of the main ideas encountered in this section. These can be useful when preparing for a quiz or test.

Terminology and Concepts

- separable DE

- solutions with nonelementary functions

The basic skills listed below summarize the more mechanical types of problems encountered in the exercise set for this section.

Basic Skills

- determine whether or not a DE is separable

- use integration to solve a separable DE

- when possible, solve an equation for the dependent variable to find an explicit solution of a DE

Integration

The method of separation of variables is the first of several techniques for solving DEs that require integration. Recall from calculus that an indefinite integral (or antiderivative) of a function $f(x)$ is another function $F(x)$ such that $F'(x) = f(x)$. The most general indefinite integral of f is written:

$$\int f(x)dx = F(x) + c, \tag{1}$$

where $F'(x) = f(x)$ and c is an arbitrary constant. Of the many formulas given in typical tables of integrals there are only a few that appear as a matter of course in typical texts in DEs. If $u = g(x)$ is a differentiable function, then its differential is defined as $du = g'(x)dx$. Each of the following results can be verified by differentiation.

$$\int u^n du = \frac{u^{n+1}}{n+1} + c, \quad n \neq 1 \tag{2}$$

$$\int u^{-1} du = \int \frac{1}{u} du = \ln|u| + c \tag{3}$$

$$\int e^u du = e^u + c \tag{4}$$

$$\int \sin u \, du = -\cos u + c \tag{5}$$

$$\int \cos u \, du = \sin u + c \tag{6}$$

$$\int \sec^2 u \, du = \tan u + c \tag{7}$$

$$\int \csc^2 u \, du = -\cot u + c \tag{8}$$

$$\int \frac{1}{a^2 + u^2} \, du = \frac{1}{a} \tan^{-1}\left(\frac{u}{a}\right) + c \tag{9}$$

For example, to evaluate $\int \sqrt{5x - 2} \, dx$ we identify $u = 5x - 2$, $du = 5 \, dx$, and write

$$\int (5x - 2)^{1/2} \, dx = \frac{1}{5} \int \overbrace{(5x - 2)^{1/2}}^{u^{1/2}} \underbrace{5 \, dx}_{du} = \frac{1}{5}\frac{2}{3}(5x - 2)^{3/2} + c = \frac{2}{15}(5x - 2)^{3/2} + c.$$

You are encouraged to always check your answer. In the last example, in view of the Chain Rule we have

$$\frac{d}{dx}\left[\frac{2}{15}(5x - 2)^{3/2} + c\right] = \frac{2}{15}\frac{3}{2}(5x - 2)^{1/2} \cdot 5 = (5x - 2)^{1/2}.$$

A formula very similar to (9) uses inverse hyperbolic functions

$$\int \frac{1}{a^2 - u^2} \, du = \frac{1}{a} \tanh^{-1}\left(\frac{u}{a}\right) + c, \quad |u| < a. \tag{10}$$

Every inverse hyperbolic function is a natural logarithm. Hence, if the integrand on the left-hand side of (10) is expanded into partial fractions, then an alternative form of (10) is found to be

$$\int \frac{1}{a^2 - u^2} \, du = \frac{1}{2a} \ln\left|\frac{a + u}{a - u}\right| + c, \quad |u| \neq a. \tag{11}$$

Formulas (10) and (11) are useful in some of the problems in Exercises 3.2.

Integration by Parts   Integration by parts is a method for integrating certain products of functions as well as some inverse functions such as $\ln x$ (the inverse of e^x) and $\sin^{-1} x$. If $u = f(x)$ and $v = g(x)$, then the differentials are $du = f'(x) \, dx$ and $dv = g'(x) \, dx$, respectively. Integration by parts is a two-step procedure that begins with an integration: Integrate the function that you choose to call dv, and then differentiate the other function (by default after dv is chosen, the function remaining is u):

$$\int u \, \boxed{dv} = \boxed{u}\,\boxed{v} - \int v \, \boxed{du} . \tag{12}$$

$$\text{(differentiate / integrate)}$$

When choosing which portion of the integrand to designate dv you should take account of the fact that this function will need to be integrated, first to obtain v, and then integrated again to obtain

$\int v \, du$. Thus, you will normally want to choose dv both so that its integral is known to you and so that the integral of dv is no more complicated than v itself. For example, in $\int x^3 \sin x \, dx$ you want to let $dv = \sin x \, dx$ because $\sin x$ is easily integrated and $\int \sin x \, dx = -\cos x + c$ is no more complicated than $\sin x$. [If you chose $dv = x^3 \, dx$, $\int x^3 \, dx = \frac{1}{4}x^4 + c$ is more complicated than x^3.] There are exceptions however. For example, when finding $\int x^3 \ln x \, dx$, you should let $u = \ln x$ and $dv = x^3 \, dx$.

Some Important Integrals   The following integrals require integration by parts.

(a) If $n = 1, 2, 3, \ldots$, the indefinite integral of the form $\int x^n e^{ax} dx$, a a constant, demands integration by parts n times. For example, to evaluate $\int x^2 e^{5x} dx$ we must use (12) twice. It is also important to remember that for integrals of this type, integration by parts always starts by integrating the exponential function. For example,

$$\overset{\text{differentiate}}{\int x \boxed{e^{-2x}} dx = \boxed{x}\left[\left(-\tfrac{1}{2}e^{-2x}\right)\right] - \int \boxed{1}\cdot\left(-\tfrac{1}{2}e^{-2x}\right) dx = -\tfrac{1}{2}xe^{-2x} - \tfrac{1}{4}e^{-2x} + c.}$$

$$\text{integrate}$$

(b) An integral $\int x^\alpha \ln x \, dx$, $\alpha \neq -1$, begins by integrating x^α. For example,

$$\int x^3 \ln x \, dx = \frac{1}{4}x^4 \ln x - \frac{1}{4}\int x^4 \frac{1}{x} \, dx = \frac{1}{4}x^4 \ln x - \frac{1}{4}\int x^3 \, dx = \frac{1}{4}x^4 \ln x - \frac{1}{16}x^4 + c.$$

The special case when $\alpha = -1$, does not require integration by parts. By writing

$$\int x^{-1} \ln x \, dx = \int \ln x \, \frac{dx}{x}$$

and identifying $u = \ln x$, $du = dx/x$, the integral is recognized as the form given in (2) with $n = 1$:

$$\int x^{-1} \ln x \, dx = \int \overset{u}{\overbrace{\ln x}} \,\overset{du}{\overbrace{\frac{dx}{x}}} = \frac{1}{2}(\ln x)^2 + c.$$

Verify this last result by differentiating the right-hand side.

(c) You might recall that integrals of the type $\int e^{\alpha x} \sin \beta x \, dx$ or $\int e^{\alpha x} \cos \beta x \, dx$ require integration by parts twice. The unusual feature of these integrals is that in the second application of integration by parts the original integral is recovered on the right-hand side. We then solve for the original integral. It is hard to make a mistake here since we can begin the process by integrating either the exponential function or the trigonometric function. But it is important to remember that if the first integraton by parts is begun by integrating, say, the exponential function, then the second integration by parts must also begin by integrating the exponential function. For example, differentiate

$$\int \boxed{e^{2x}} \sin 3x \, dx = \boxed{\frac{1}{2}e^{2x}} \; \boxed{\sin 3x} - \int \frac{1}{2}e^{2x} \boxed{(3\cos 3x)} \; dx$$

integrate

differentiate

$$= \frac{1}{2}e^{2x} \sin 3x - \frac{3}{2} \int \boxed{e^{2x}} \cos 3x \, dx = \frac{1}{2}e^{2x} \sin 3x - \frac{3}{2}\left[\boxed{\frac{1}{2}e^{2x}} \; \boxed{\cos 3x} - \int \frac{1}{2}e^{2x} \boxed{(-3\sin 3x)} \; dx \right]$$

integrate

$$= \frac{1}{2}e^{2x} \sin 3x - \frac{3}{4}e^{2x} \cos 3x - \frac{9}{4} \int e^{2x} \sin 3x \, dx.$$

If the original integral is denoted by the symbol I then the last result is equivalent to

$$I = \frac{1}{2}e^{2x} \sin 3x - \frac{3}{4}e^{2x} \cos 3x - \frac{9}{4}I.$$

Solving for I then gives $\frac{13}{4}I = \frac{1}{2}e^{2x} \sin 3x - \frac{3}{4}e^{2x} \cos 3x$ or

$$I = \int e^{2x} \sin 3x \, dx = \frac{2}{13}e^{2x} \sin 3x - \frac{3}{13}e^{2x} \cos 3x + c.$$

See **Use of Computers** later in this section for *Mathematica* and *Maple* implementations of both indefinite and definite integration.

Partial Fractions   The following discussion illustrates a procedure for integrating certain rational functions $P(x)/Q(x)$, where we shall assume for this discussion that the degree of $P(x)$ is less than the degree of $Q(x)$. (If the degree $P(x) \geq$ degree $Q(x)$, then start the process with long division.) This method, known as **partial fractions**, consists of expanding such a rational function into simpler component fractions, and then evaluating the integral term-by-term. We begin with an example. When the terms in the sum

$$\frac{2}{x+5} + \frac{1}{x+1} \tag{13}$$

are combined by means of a common denominator, we obtain the single rational expression

$$\frac{3x+7}{(x+5)(x+1)}. \tag{14}$$

Now suppose that we are faced with the problem of evaluating the integral $\displaystyle\int \frac{3x+7}{(x+5)(x+1)} \, dx$. Of course, the solution is obvious. We use the equality of (13) and (14) to write:

$$\int \frac{3x+7}{(x+5)(x+1)} \, dx = \int \frac{2}{x+5} \, dx + \int \frac{1}{x+1} \, dx = 2\ln|x+5| + \ln|x+1| + c.$$

In the discussion that follows, we review the assumptions and the algebra of four cases of partial fraction decomposition.

CASE 1   Nonrepeated Linear Factors

We state the following fact from algebra without proof. If

$$\frac{P(x)}{Q(x)} = \frac{P(x)}{(a_1 x + b_1)(a_2 x + b_2) \cdots (a_n x + b_n)}$$

where all the factors $a_i x + b_i$, $i = 2, \ldots, n$, in the denominator are distinct and the degree of $P(x)$ is less than n, then unique real constants $C_1, C_2, \ldots, C_n$ exist such that

$$\frac{P(x)}{Q(x)} = \frac{C_1}{a_1 x + b_1} + \frac{C_2}{a_2 x + b_2} + \cdots + \frac{C_n}{a_n x + b_n}.$$

For example, to expand $(2x + 1)/(x - 1)(x + 3)$ into individual partial fractions we make the assumption that the integrand can be written as

$$\frac{2x + 1}{(x - 1)(x + 3)} = \frac{A}{x - 1} + \frac{B}{x + 3}.$$

Combining the terms of the right-hand member of the equation over a common denominator gives

$$\frac{2x + 1}{(x - 1)(x + 3)} = \frac{A(x + 3) + B(x - 1)}{(x - 1)(x + 3)}.$$

Since the denominators are identical, the numerators are also identical

$$2x + 1 = A(x + 3) + B(x - 1) = (A + B)x + (3A - B), \tag{15}$$

and the coefficients of the powers of x are the same:

$$2 = A + B \quad \text{and} \quad 1 = 3A - B.$$

These simultaneous equations can now be solved for A and B. The results are $A = \frac{3}{4}$ and $B = \frac{5}{4}$. Therefore,

$$\frac{2x + 1}{(x - 1)(x + 3)} = \frac{\frac{3}{4}}{x - 1} + \frac{\frac{5}{4}}{x + 3}.$$

Note: The numbers A and B in the preceding example can be determined in an alternative manner. Since (15) is an identity, that is, the equality is true for every value of x, it holds for $x = 1$ and $x = -3$, *the zeros of the denominator*. Setting $x = 1$ in (15) gives $3 = 4A$, from which it follows immediately that $A = \frac{3}{4}$. Similarly, by setting $x = -3$ in (15), we obtain $-5 = (-4)B$ or $B = \frac{5}{4}$. In the *Remarks* at the end of Section 7.2 in the text an equivalent and simpler method for evaluating the coefficients in a partial fraction decomposition is given. It is important to be aware that this technique, called the **cover-up method**, works only when the denomintor is a product of linear factors, none of which are repeated.

CASE II   Repeated Linear Factors

If

$$\frac{P(x)}{Q(x)} = \frac{P(x)}{(ax + b)^n}$$

where $n > 1$ and the degree of $P(x)$ is less than n, then unique real constants $C_1, C_2, \ldots, C_n$ can be found such that

$$\frac{P(x)}{(ax + b)^n} = \frac{C_1}{ax + b} + \frac{C_2}{(ax + b)^2} + \cdots + \frac{C_n}{(ax + b)^n}.$$

Combining the Cases: When the denominator $Q(x)$ contains distinct as well as repeated inear factors, we combine the two cases. For example, to expand $(6x - 1)/x^3(2x - 1)$ into partial fractions we write

$$\frac{6x - 1}{x^3(2x - 1)} = \frac{A}{x} + \frac{B}{x^2} + \frac{C}{x^3} + \frac{D}{2x - 1}$$

since x^3 means that the linear term x is used as a factor x three times. Putting the right-hand side over a common denominator and equating numerators gives

$$6x - 1 = Ax^2(2x - 1) + Bx(2x - 1) + C(2x - 1) + Dx^3 \tag{16}$$

$$= (2A + D)x^3 + (-A + 2B)x^2 + (-B + 2C)x - C. \tag{17}$$

If we set $x = 0$ and $x = \frac{1}{2}$ in (16), we find $C = 1$ and $D = 16$, respectively. Now since the denominator of the original expression has only two distinct zeros, we can find A and B by equating the coefficients of x^3 and x^2 in

$$6x - 1 = 0x^3 + 0x^2 + 6x - 1 = (2A + 16)x^3 + (-A + 2B)x^2 + (-B + 2)x - 1.$$

Thus, $A = -8$ and $B = -4$. Therefore,

$$\frac{6x - 1}{x^3(2x - 1)} = -\frac{8}{x} - \frac{4}{x^2} + \frac{1}{x^3} + \frac{16}{2x - 1}.$$

CASE III   Nonrepeated Quadratic Factors

Suppose the denominator of the rational function $P(x)/Q(x)$ can be expressed as a product of distinct quadratic factors $a_i x^2 + b_i x + c_i$, $i = 1, 2, \ldots, n$, that are irreducible (not factorable) over the real numbers. If the degree of $P(x)$ is less than $2n$, we can find unique real constants $A_1, A_2, \ldots, A_n, B_1, B_2, \ldots, B_n$ such that

$$\frac{P(x)}{(a_1 x^2 + b_1 x + c_1)(a_2 x^2 + b_2 x + c_2) \cdots (a_n x^2 + b_n x + c_n)}$$

$$= \frac{A_1 x + B_1}{a_1 x^2 + b_1 x + c} + \frac{A_2 x + B_2}{a_2 x^2 + b_2 x + c_2} + \cdots + \frac{A_n x + B_n}{a_n x^2 + b_n x + c_n}.$$

For example, to expand $4x/(x^2+1)(x^2+2x+3)$ into partial fractions we first write

$$\frac{4x}{(x^2+1)(x^2+2x+3)} = \frac{Ax+B}{x^2+1} + \frac{Cx+D}{x^2+2x+3}$$

and, after putting the right-hand side over a common denominator, equate numerators:

$$4x = (Ax+B)(x^2+2x+3) + (Cx+D)(x^2+1)$$

$$= (A+B)x^3 + (2A+B+D)x^2 + (3A+2B+C)x + (3B+D).$$

Since the denominator of the original fraction has no real zeros, we compare coefficients of powers of x:

$$0 = A + C$$

$$0 = 2A + B + D$$

$$4 = 3A + 2B + C$$

$$0 = 3B + D.$$

Solving the equations yields $A = 1$, $B = 1$, $C = -1$, and $D = -3$. Therefore,

$$\frac{4x}{(x^2+1)(x^2+2x+3)} = \frac{x+1}{x^2+1} - \frac{x+3}{x^2+2x+3}.$$

CASE IV Repeated Quadratic Factors

In the case when the denominator contains a repeated irreducible quadratic polynomial ax^2+bx+c,

$$\frac{P(x)}{Q(x)} = \frac{P(x)}{(ax^2+bx+c)^n},$$

then if $P(x)$ is of degree less than $2n$, we can find real constants $A_1, A_2, \ldots, A_n, B_1, B_2, \ldots, B_n$ such that

$$\frac{P(x)}{(ax^2+bx+c)^n} = \frac{A_1x+B_1}{ax^2+bx+c} + \frac{A_2x+B_2}{(ax^2+bx+c)^2} + \cdots + \frac{A_nx+B_n}{(ax^2+bx+c)^n}.$$

Combining the Cases: The denominator $Q(x)$ in the fraction $(x+3)/x^2(x-5)(x^2+1)^2$ contains a distinct linear factor $x-5$, repeated linear factors x^2, and repeated irreducible quadratic factors $(x^2+1)^2$. To expand into partial fractions, we write

$$\frac{x+3}{x^2(x-5)(x^2+1)^2} = \underbrace{\frac{A}{x} + \frac{B}{x^2}}_{\textit{repeated linear}} + \underbrace{\frac{C}{x-5}}_{\textit{linear}} + \underbrace{\frac{Dx+E}{x^2+1} + \frac{Fx+G}{(x^2+1)^2}}_{\textit{repeated quadratic}}.$$

Important Note: Although we have used it as a technique of integration, the ability to expand a rational expression $P(x)/Q(x)$ into partial fractions is also used in another context in the study of DEs; *it is frequently essential* when computing the inverse Laplace transform in Chapter 7. See

Section 7.2 in this manual for the syntax used in *Mathematica* and *Maple* to obtain a partial fraction decomposition of a rational function.

Use of Computers   We show below how *Mathematica* and *Maple* can be used to graph a function, graph an equation, find (and name) the solution of an equation, and find the indefinite integral (or antiderivative) and definite integral of a function.

Graphing a Function: To graph a function, such as $f(x) = \sqrt{4 - x^2}$, use the syntax

> **Clear[f]**   (*Mathematica*)
> **f[x_]:= Sqrt[4 - x^2]**
> **f[x]**
> **Plot[f[x], {x, -3, 3}]**

> ```
> f:=x->sqrt(4-x^2);                                     (Maple)
> plot(f(x), x = -3..3);
> ```

Because $\sqrt{4 - x^2}$ is not a real number for $x < -2$ or $x > 2$, *Mathematica* returns some error messages when it tries to evaluate the function at points in these intervals; it does however then plot the graph. *Maple*, on the other hand, appears to ignore such problems and simply plots the graph. Since the graph of $f(x)$ is the upper half of the circle $x^2 + y^2 = 4$, you would expect the graph to look like a semicircle. Due, however, to the fact that both programs choose the widths and heights of their plots without regard to the scales on the horizontal and vertical axes, the semicircle may look more like a semi-ellipse. To insure that one unit on each axis has the same length on the computer screen, a graphics option is used as shown below.

> **Plot[f[x], {x, -2, 2}, AspectRatio—>2/4]**   (*Mathematica*)

> ```
> plot(f(x), x=-3..3, scaling=CONSTRAINED);              (Maple)
> ```

(The **AspectRatio** option in the **Plot** command is the height of the plot divided by its width.)

Both programs provide a large number of graphics options that can be used to control the appearance of plots. Probably the most commonly used of these is an option that specifies the extent of the vertical axis to be shown. This may be necessary to see details of the graph that are obscured if the CAS (as opposed to the user) is allowed to choose the range of the graph shown. Examples are shown below with pictures displayed for the *Mathematica* output. The pictures generated by *Maple* are similar.

> **Plot[(x^2 - 1)Exp[2x^2], {x, -2, 2},**   (*Mathematica*)
> **PlotRange—>{-2, 2}]**

```
plot((x^2-1)*exp(2*x^2), x = -2..2,          (Maple)
     y = -2..2));
```

without **PlotRange**   with **PlotRange**

Finally, you may want to superimpose the graphs of two or more functions on the same set of coordinate axes. The commands below show the graphs of $f(x) = x^3 - 3x^2 + x + 1$ and $g(x) = 1 - 5x/4$, the tangent line to the graph of $f(x)$ at $x = 3/2$.

Clear[f, g]   (Mathematica)
f[x_]:= x^3 - 3x^2 + x + 1
g[x_]:= 1 - 5x/4
{f[x], g[x]}
Plot[{f(x), g(x)}, {x, -2, 3}]

```
f:=x->x^3-3*x^2+x+1;          (Maple)
g:=x->1-5*x/4;
plot({f(x),g(x)}, x = -2..3);
```

Graphing an Equation: In addition to graphing functions, a CAS can be used to graph equations that implicitly define functions. Equations such as this frequently arise when solving a DE using separation of variables. For example, the solution of $(1 + e^y)y' = x \cos x + \sin x$, $y(0) = 0$, obtained using separation of variables, is $y + e^y = x \sin x + 1$. We cannot solve this equation for y in terms of elementary functions, but we can graph the equation using **ContourPlot** in *Mathematica* and **contourplot** in *Maple* as shown below.

f[x_, y_]:= y + Exp[y] - x Sin[x] - 1   (Mathematica)
f[x, y]
ContourPlot[f[x, y], {x, -15, 15}, {y, -10, 5}, PlotPoints->50,
   Contours->{0}, ContourShading->False, Frame->False,
   Axes->True, AxesOrigin->{0, 0}, AxesLabel->{"x", "y"}]

Many of the graphics options used above, such as **Frame** and **Axes** are self-explanatory. The default setting for **PlotPoints** is 15 and may result in a graph consisting of straight line segments and sharp corners. The **Contours** options is set to 0 in this case because the equation being graphed is $f(x, y) = y + e^y - x\sin x - 1 = 0$. Alternatively, the function could have been specified as $f(x, y) = y + e^x - x\sin x$ and the **Contours** option set to 1.

In *Maple* we effect the same result using the commands

```
f:=(x,y)->y+exp(y)-x*sin(x)-1;              (Maple)
with(plots);
contourplot(f(x,y), x=-15..15, y=-10..5,
    contours=[0], grid=[80,80]);
```

The `grid` option is set to numbers that are large enough to insure a fairly smooth-looking graph.

Solving an Equation: Depending on the nature of an equation, there are a number of commands in both *Mathematica* and *Maple* that can be used to find solutions of the equation. If the equation involves only polynomial or rational functions, use **Solve** or **NSolve** in *Mathematica* and **solve** in *Maple*.

$$\text{Solve}[2\text{x}^2 - 3\text{x} - 4 == 0, \text{x}] \qquad (Mathematica)$$

or

$$\text{NSolve}[2\text{x}^2 - 3\text{x} - 4 == 0, \text{x}]$$

and

$$\text{solve}(2*\text{x}^2 - 3*\text{x} - 4 = 0, \text{ x}); \qquad (Maple)$$

Both **Solve** and **solve** will give exact solutions when possible. To obtain decimal versions of the solutions use **NSolve** in *Mathematica* and insert a decimal point in one of the numbers in the equation (`2*x^2-3*x-4.=0`) in *Maple*. (This will also work with the **Solve** command in *Mathematica* as an alternative to **NSolve**.)

To solve equations involving more complicated expressions such as $e^x = 3\cos x + 1$ use **FindRoot** in *Mathematica* and **fsolve** in *Maple*. In both programs, either Newton's method or a variation of it is used, so some form of an initial guess should be provided. This can be obtained from graphs and will enable you to select the solution you want in the event there is more than one.

$$\text{Plot}[\{\text{Exp}[\text{x}], 3\text{Cos}[\text{x}]+1\}, \{\text{x}, -5, 5\}] \qquad (Mathematica)$$
$$\text{FindRoot}[\text{Exp}[\text{x}]==3\text{Cos}[\text{x}]+1, \{\text{x}, 1\}]$$

```
plot([exp(x), 3*cos(x)+1], x=-5..5, y=-3..10);  (Maple)
fsolve(exp(x), 3*cos(x)+1, x=0..2);
```

As shown above, in *Maple*, a range of values can be specified. Other options are available, but this is generally the most effective way to find the solution you are looking for. If there is no solution within this range, no solution will be given. If there are multiple solutions within the range, only one will be given, unless you are solving a polynomial equation, in which case all solutions within the range will be given.

Naming the Solution of an Equation: If you intend to use the solution of an equation in a subsequent computation in your CAS, then it is convenient to have a name for this value so that you do not have to read it from the screen and then type it in. The routines below show how to do this, assigning the name a to the solution of $e^{3x-x^2}\cos x = 0$.

> **Clear**[f]   (*Mathematica*)
> f[x_]:= **Exp**[x - x^2] + **Cos**[x]
> f[x]
> **Plot**[f[x], {x, -5, 5}]
> sol = **FindRoot**[f[x]== 0, {x, 2}]
> a = x/.sol

```
f:=x->exp(x-x^2)+cos(x);                               (Maple)
plot(f(x), x=-5..5);
a:=fsolve(f(x)=0, x=1..3);
```

Integrals: *Mathematica* and *Maple* can find both antiderivatives (indefinite integrals) of many functions and definite integrals of most functions. To find an antiderivative of $f(x)$ use

> **Integrate**[f[x], x]   (*Mathematica*)

```
int(f(x), x);                                          (Maple)
```

If you want to further work with the antiderivative in the CAS it can be helpful to define it as a function that can itself be evaluated at a point, graphed, or differentiated. The syntax to accomplish this is

> g[x_]:= **Evaluate**[**Integrate**[f(x), x]]   (*Mathematica*)
> g[x]

```
g:=unapply(int(f(x),x),x);                             (Maple)
```

To find the definite integral, $\int_a^b f(x)\,dx$, of f we use

$$\textbf{Integrate[f(x), \{x, a, b\}]} \qquad\qquad \textit{(Mathematica)}$$

$$\texttt{int(f(x),x=a..b);} \qquad\qquad \textit{(Maple)}$$

The comands **Integrate** and `int` will try to find exact values of a definite integral by evaluating the antiderivative at the endpoints of the interval. If the CAS is unable to find an antiderivative of $f(x)$ (for example, when $f(x) = \sin(\sin x)$, or if you simply want a numerical approximation of the definite integral, use

$$\textbf{NIntegrate[f(x), \{x, a, b\}]} \qquad\qquad \textit{(Mathematica)}$$

$$\texttt{evalf(int(f(x),x=a..b));} \qquad\qquad \textit{(Maple)}$$

In this section and ones following we will encounter an expression of the form $\ln|g(y)| = f(x) + c$. *To solve for* $g(y)$ *we exponentiate both sides of the equation. This yields* $|g(y)| = e^{f(x)+c} = e^c e^{f(x)}$ *which implies* $g(y) = \pm e^c e^{f(x)}$. *Letting* $c_1 = \pm e^c$ *we obtain* $g(y) = c_1 e^{f(x)}$.

3. From $dy = -e^{-3x}\,dx$ we obtain $y = \frac{1}{3}e^{-3x} + c$.

6. From $\frac{1}{y^2}\,dy = -2x\,dx$ we obtain $-\frac{1}{y} = -x^2 + c$ or $y = \frac{1}{x^2 + c_1}$.

9. From $\left(y + 2 + \frac{1}{y}\right)dy = x^2 \ln x\,dx$ we obtain $\frac{y^2}{2} + 2y + \ln|y| = \frac{x^3}{3}\ln|x| - \frac{1}{9}x^3 + c$.

12. From $2y\,dy = -\frac{\sin 3x}{\cos^3 3x}\,dx$ or $2y\,dy = -\tan 3x \sec^2 3x\,dx$ we obtain $y^2 = -\frac{1}{6}\sec^2 3x + c$.

15. From $\frac{1}{S}\,dS = k\,dr$ we obtain $S = ce^{kr}$.

18. From $\frac{1}{N}\,dN = \left(te^{t+2} - 1\right)dt$ we obtain $\ln|N| = te^{t+2} - e^{t+2} - t + c$ or $N = c_1 e^{te^{t+2} - e^{t+2} - t}$.

21. From $x\,dx = \frac{1}{\sqrt{1 - y^2}}\,dy$ we obtain $\frac{1}{2}x^2 = \sin^{-1} y + c$ or $y = \sin\left(\frac{x^2}{2} + c_1\right)$.

24. From $\frac{1}{y^2 - 1}\,dy = \frac{1}{x^2 - 1}\,dx$ or $\frac{1}{2}\left(\frac{1}{y - 1} - \frac{1}{y + 1}\right)dy = \frac{1}{2}\left(\frac{1}{x - 1} - \frac{1}{x + 1}\right)dx$ we obtain

$\ln|y - 1| - \ln|y + 1| = \ln|x - 1| - \ln|x + 1| + \ln c$ or $\frac{y - 1}{y + 1} = \frac{c(x - 1)}{x + 1}$. Using $y(2) = 2$ we find

$c = 1$. A solution of the initial-value problem is $\frac{y - 1}{y + 1} = \frac{x - 1}{x + 1}$ or $y = x$.

27. Separating variables and integrating we obtain

$$\frac{dx}{\sqrt{1-x^2}} - \frac{dy}{\sqrt{1-y^2}} = 0 \quad \text{and} \quad \sin^{-1}x - \sin^{-1}y = c.$$

Setting $x = 0$ and $y = \sqrt{3}/2$ we obtain $c = -\pi/3$. Thus, an implicit solution of the initial-value problem is $\sin^{-1}x - \sin^{-1}y = -\pi/3$. Solving for y and using an addition formula from trigonometry, we get

$$y = \sin\left(\sin^{-1}x + \frac{\pi}{3}\right) = x\cos\frac{\pi}{3} + \sqrt{1-x^2}\sin\frac{\pi}{3} = \frac{x}{2} + \frac{\sqrt{3}\sqrt{1-x^2}}{2}.$$

30. Separating variables, integrating from -2 to x, and using t as a dummy variable of integration gives

$$\int_{-2}^{x} \frac{1}{y^2}\frac{dy}{dt}\,dt = \int_{-2}^{x} \sin t^2 dt$$

$$-y(t)^{-1}\Big|_{-2}^{x} = \int_{-2}^{x} \sin t^2 dt$$

$$-y(x)^{-1} + y(-2)^{-1} = \int_{-2}^{x} \sin t^2 dt$$

$$-y(x)^{-1} = -y(-2)^{-1} + \int_{-2}^{x} \sin t^2 dt$$

$$y(x)^{-1} = 3 - \int_{-2}^{x} \sin t^2 dt.$$

Thus

$$y(x) = \frac{1}{3 - \int_{-2}^{x} \sin t^2 dt}.$$

33. Writing the differential equation as $e^x dx = e^{-y}dy$ and integrating we have $e^x = -e^{-y} + c$. Using $y(0) = 0$ we find that $c = 2$ so that $y = -\ln(2 - e^x)$.

To find the interval of definition of this solution we note that $2 - e^x > 0$ so x must be in $(-\infty, \ln 2)$.

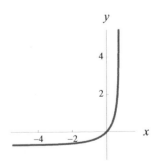

36. Separating variables, we have

$$\frac{dy}{y^2 - y} = \frac{dx}{x} \qquad \text{or} \qquad \int \frac{dy}{y(y-1)} = \ln|x| + c.$$

Using partial fractions, we obtain

$$\int \left(\frac{1}{y-1} - \frac{1}{y} \right) dy = \ln|x| + c$$

$$\ln|y-1| - \ln|y| = \ln|x| + c$$

$$\ln \left| \frac{y-1}{xy} \right| = c$$

$$\frac{y-1}{xy} = e^c = c_1.$$

Solving for y we get $y = 1/(1 - c_1 x)$. We note by inspection that $y = 0$ is a singular solution of the differential equation.

(a) Setting $x = 0$ and $y = 1$ we have $1 = 1/(1-0)$, which is true for all values of c_1. Thus, solutions passing through $(0, 1)$ are $y = 1/(1 - c_1 x)$.

(b) Setting $x = 0$ and $y = 0$ in $y = 1/(1 - c_1 x)$ we get $0 = 1$. Thus, the only solution passing through $(0, 0)$ is $y = 0$.

(c) Setting $x = \frac{1}{2}$ and $y = \frac{1}{2}$ we have $\frac{1}{2} = 1/(1 - \frac{1}{2}c_1)$, so $c_1 = -2$ and $y = 1/(1 + 2x)$.

(d) Setting $x = 2$ and $y = \frac{1}{4}$ we have $\frac{1}{4} = 1/(1 - 2c_1)$, so $c_1 = -\frac{3}{2}$ and $y = 1/(1 + \frac{3}{2}x) = 2/(2 + 3x)$.

39. The singular solution $y = 1$ satisfies the initial-value problem.

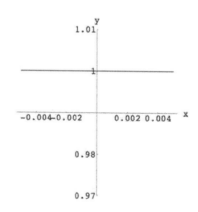

42. Separating variables we obtain $\dfrac{dy}{(y-1)^2 + 0.01} = dx.$ Then

$$10\tan^{-1} 10(y-1) = x+c \quad \text{and} \quad y = 1+\frac{1}{10}\tan\frac{x+c}{10}.$$

Setting $x = 0$ and $y = 1$ we obtain $c = 0$. The solution is

$$y = 1 + \frac{1}{10}\tan\frac{x}{10}.$$

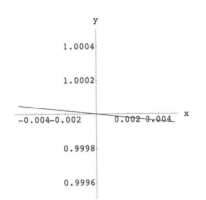

Alternatively, we can use the fact that

$$\int \frac{dy}{(y-1)^2 - 0.01} = -\frac{1}{0.1}\tanh^{-1}\frac{y-1}{0.1} = -10\tanh^{-1} 10(y-1).$$

We use the inverse hyperbolic tangent because $|y-1| < 0.1$ or $0.9 < y < 1.1$. This follows from the initial condition $y(0) = 1$. Solving the above equation for y we get $y = 1 + 0.1\tanh(x/10)$.

45. We separate variable and rationalize the denominator:

$$dy = \frac{1}{1+\sin x} \cdot \frac{1-\sin x}{1-\sin x}\,dx = \frac{1-\sin x}{1-\sin^2 x}\,dx = \frac{1-\sin x}{cos^2 x}\,dx$$

$$= (\sec^2 x - \tan x\sec x)\,dx.$$

Integrating, we have $y = \tan x - \sec x + C$.

48. Separating variables and integrating we have

$$\int \frac{dy}{y^{2/3}(1 - y^{1/3})} = \int dx$$

$$\int \frac{y^{2/3}}{1 - y^{1/3}}\,dy = x + c_1$$

$$-3\ln\left|1 - y^{1/3}\right| = x + c_1$$

$$\ln\left|1 - y^{1/3}\right| = -\frac{x}{3} + c_2$$

$$\left|1 - y^{1/3}\right| = c_3 e^{-x/3}$$

$$1 - y^{1/3} = c_4 e^{-x/3}$$

$$y^{1/3} = 1 + c_5 e^{-x/3}$$

$$y = \left(1 + c_5 e^{-x/3}\right)^3.$$

2.3 | Linear Equations

The terminology and concepts listed below provide an outline of the main ideas encountered in this section. These can be useful when preparing for a quiz or test.

Terminology and Concepts

- linear first-order DE

- homogeneous linear DE

- variation of parameters as applied to a linear first-order DE
- integrating factor
- general solution of a linear first-order DE
- singular points of a linear first-order DE
- transient term
- special functions

The basic skills listed below summarize the more mechanical types of problems encountered in the exercise set for this section.

Basic Skills

- find the integrating factor for a linear first-order DE
- use the integrating factor to find the general solution of a linear first-order DE
- find transient terms in the general solution of a linear first-order DE
- solve a linear first-order DE satisfying an initial condition
- solve a linear first-order IVP with a piecewise-defined coefficient
- determine an interval of definition of the general solution of a linear first-order DE

An Identity   In precalculus mathematics you learned that the exponential statement $e^N = M$ is equivalent to the logarithmic expression $N = \ln M$. If we substitute this last expression into the first, we obtain the important identity

$$e^{\ln M} = M,$$

which is very useful in this section when computing the integrating factor $e^{\int P(x)\,dx}$. For example, to determine the integrating factor for $(5+x^2)\dfrac{dy}{dx} - 2xy = x$ we first put the DE into standard form $\dfrac{dy}{dx} - \dfrac{2x}{5+x^2}\,y = \dfrac{x}{5+x^2}$ and identify $P(x) = -2x/(5+x^2)$:

$$e^{-\int \frac{2x}{5+x^2}\,dx} = e^{-\ln|5+x^2|} = e^{\ln|5+x^2|^{-1}} = (5+x^2)^{-1} = \frac{1}{5+x^2}.$$

We were able to discard the absolute value sign in $|5 + x^2|$ because $5 + x^2 > 0$ for all x.

It may take some time to get used to the fact that once the DE has been put into standard form and the integrating factor determined, multiplying both sides of the standard form by the integrating factor recasts the left-hand side of the equation into the derivative of the integrating factor and the dependent variable. Above, we saw that the integrating factor for $\dfrac{dy}{dx} - \dfrac{2x}{5 + x^2} y = \dfrac{x}{5 + x^2}$ is $\dfrac{1}{5 + x^2}$. Multiplying the DE by this factor gives

$$\frac{1}{5 + x^2} \frac{dy}{dx} - \frac{2x}{(5 + x^2)^2} y = \frac{x}{(5 + x^2)^2}. \tag{1}$$

The left-hand side of the last DE can then be written compactly as the derivative of a product of the integrating factor and $y(x)$:

$$\frac{d}{dx} \left[\frac{1}{5 + x^2} y \right] = \frac{x}{(5 + x^2)^2}. \tag{2}$$

Until you become comfortable with this notion we suggest that you actually multiply the differential equation $dy/dx + P(x)y = f(x)$ by the integrating factor and then verify using the product rule that

$$e^{\int P(x)\,dx} \frac{dy}{dx} + P(x) e^{\int P(x)\,dx} y = e^{\int P(x)\,dx} f(x) \qquad \text{(as in (1))}$$

and

$$\frac{d}{dx} \left[e^{\int P(x)\,dx} y \right] = e^{\int P(x)\,dx} f(x) \qquad \text{(as in (2))}$$

are formally equivalent. After that, once you have computed the integrating factor, you can jump to the last form and proceed with integration. In the case of (2) the integral of the left-hand side is

$$\int \frac{d}{dx} \left[\frac{1}{5 + x^2} y \right] dx = \frac{1}{5 + x^2} y.$$

The integral of the right-hand side is

$$\int \frac{x}{(5 + x^2)^2} dx = \frac{1}{2} \int (5 + x^2)^{-2}(2x\,dx) = -\frac{1}{2}(5 + x^2)^{-1} + c.$$

Hence the solution of the DE on $(-\infty, \infty)$ is

$$\frac{1}{5 + x^2} y = -\frac{1}{2}(5 + x^2)^{-1} + c \qquad \text{or} \qquad y = -\frac{1}{2} + c(5 + x^2).$$

Use of Computers As we have seen in Section 2.2, a CAS can help you find the solution of a DE by performing some of the more complicated integrations required. In addition to this, a CAS can also frequently find directly the solution of a DE. For example, to solve the DE $xy' - xy = e^x$ use

$$\textbf{Dsolve}[\textbf{x y}\,'[\textbf{x}] - \textbf{x y}[\textbf{x}] == \textbf{Exp}[\textbf{x}], \textbf{y}[\textbf{x}], \textbf{x}] \qquad (\textit{Mathematica})$$

```
dsolve(x*diff(y(x),x)-x*y(x)=exp(x),y(x));      (Maple)
```

To solve an IVP such as $xy' - xy = e^x$, $y(1) = 2$, use

$$\textbf{Dsolve}[\{\textbf{x y}\,'[\textbf{x}] - \textbf{x y}[\textbf{x}] == \textbf{Exp}[\textbf{x}], \textbf{y}[\textbf{1}] == \textbf{2}\}, \textbf{y}[\textbf{x}], \textbf{x}] \qquad (\textit{Mathematica})$$

```
dsolve({x*diff(y(x),x)-x*y(x)=exp(x),y(1)=2},      (Maple)
    y(x));
```

The output is

$$\{\{y[x] \to e^{x-1}(2 + e \ln x)\}\} \qquad (\textit{Mathematica})$$

$$y[x] = \left(\ln x + \frac{2}{e}\right) e^x \qquad (\textit{Maple})$$

In neither case can the output be used subsequently in the CAS to easily obtain numerical values or graphs of the function which is the solution, $y(x)$, of the IVP. The following routines can be used to assign a specific function name to the solution. This function can then be evaluated at a number, differentiated, graphed, etc.

$$\textbf{Clear}[\textbf{y}] \qquad (\textit{Mathematica})$$
```
sol=Flatten[Dsolve[{x^2 y'[x] - x y[x]==y[x]^2, y[1]==2}, y[x], x]];
y[x_]:=Evaluate[y[x]/.sol]
y[x]
```

The semicolon at the end of line of input in *Mathematica* is quite different than in *Maple*, where it is required at the end of every line of input. In *Mathematica* it simply suppresses the display of the output. For example, typing **a=2** in *Mathematica* will set a equal to 2 and display 2 on the monitor, whereas typing **a=2;** will set a equal to 2, but will not display the 2.

```
sol:=dsolve({x^2*diff(y(x),x)-x*y(x)=y(x)^2,      (Maple)
    y(1)=2}, y(x));
y:=unapply(eval(y(x),sol), x);
```

Special Functions: As discussed in the text, a number of nonelementary functions are defined by integrals. Others, such as the Bessel functions discussed in Chapter 6 of the text, can be defined using infinite series. Virtually all special functions that have been tabulated in standard handbooks are built into *Mathematica* and *Maple*. Below is a table listing all of the special functions used in the text, together with the syntax for each function in both *Mathematica* and *Maple*.

Name	Notation	*Mathematica*	*Maple*
Bessel function of the first kind	$J_n(x)$	**BesselJ[n,x]**	BesselJ(n,x)
Bessel function of the second kind	$Y_n(x)$	**BesselY[n,x]**	BesselY(n,x)
Chebyshev polynomial	$T_n(x)$	**ChebyshevT[n,x]**	ChebyshevT(n,x)
Complementary error function	$\mathrm{erfc}(x)$	**Erfc[x]**	erfc(x)
Dirac delta function	$\delta(t-a)$	**DiracDelta[t-a]**	Dirac(t-a)
Error function	$\mathrm{erf}(x)$	**Erf[x]**	erf(x)
Fresnel sine integral	$S(x)$	**FresnelS[x]**	FresnelS(x)
Gamma	$\Gamma(x)$	**Gamma[x]**	GAMMA(x)
Heaviside (unit step function)	$(t-a)$	**UnitStep[t-a]**	Heaviside(x)
Hermite polynomial	$H_n(x)$	**HermiteH[n,x]**	HermiteH(n,x)
Imaginary error function	$\mathrm{erfi}(x)$	**Erfi[x]**	erfi(x)
Laguerre polynomial	$L_n(x)$	**LaguerreL[n,x]**	orthopoly[L](n,x)
Legendre polynomial	$P_n(x)$	**LegendreP[n,x]**	orthopoly[P](n,x)
Modified Bessel function of the first kind	$I(x)$	**BesselI[n,x]**	BesselI(n,x)
Modified Bessel function of the second kind	$K(x)$	**BesselK[n,x]**	BesselK(n,x)
Sine integral	$Si(x)$	**SinIntegral[x]**	Si(x)

3. For $y' + y = e^{3x}$ an integrating factor is $e^{\int dx} = e^x$ so that $\dfrac{d}{dx}[e^x y] = e^{4x}$ and $y = \frac{1}{4}e^{3x} + ce^{-x}$ for $-\infty < x < \infty$. The transient term is ce^{-x}.

6. For $y' + 2xy = x^3$ an integrating factor is $e^{\int 2x\,dx} = e^{x^2}$ so that $\dfrac{d}{dx}\left[e^{x^2}y\right] = x^3 e^{x^2}$ and $y = \frac{1}{2}x^2 - \frac{1}{2} + ce^{-x^2}$ for $-\infty < x < \infty$. The transient term is ce^{-x^2}.

9. For $y' - \dfrac{1}{x}y = x\sin x$ an integrating factor is $e^{-\int(1/x)dx} = \dfrac{1}{x}$ so that $\dfrac{d}{dx}\left[\dfrac{1}{x}y\right] = \sin x$ and $y = cx - x\cos x$ for $0 < x < \infty$. There is no transient term.

12. For $y' - \dfrac{x}{(1+x)}y = x$ an integrating factor is $e^{-\int[x/(1+x)]dx} = (x+1)e^{-x}$ so that $\dfrac{d}{dx}\left[(x+1)e^{-x}y\right] = x(x+1)e^{-x}$ and $y = -x - \dfrac{2x+3}{x+1} + \dfrac{ce^x}{x+1}$ for $-1 < x < \infty$. There is no transient term.

15. For $\dfrac{dx}{dy} - \dfrac{4}{y}x = 4y^5$ an integrating factor is $e^{-\int(4/y)dy} = e^{\ln y^{-4}} = y^{-4}$ so that $\dfrac{d}{dy}\left[y^{-4}x\right] = 4y$ and $x = 2y^6 + cy^4$ for $0 < y < \infty$. There is no transient term.

18. For $y' + (\cot x)y = \sec^2 x \csc x$ an integrating factor is $e^{\int \cot x\,dx} = e^{\ln|\sin x|} = \sin x$ so that $\dfrac{d}{dx}[(\sin x)\,y] = \sec^2 x$ and $y = \sec x + c\csc x$ for $0 < x < \pi/2$. There is no transient term.

21. For $\dfrac{dr}{d\theta} + r\sec\theta = \cos\theta$ an integrating factor is $e^{\int \sec\theta\,d\theta} = e^{\ln|\sec x + \tan x|} = \sec\theta + \tan\theta$ so that $\dfrac{d}{d\theta}[(\sec\theta + \tan\theta)r] = 1 + \sin\theta$ and $(\sec\theta + \tan\theta)r = \theta - \cos\theta + c$ for $-\pi/2 < \theta < \pi/2$.

24. For $y' + \dfrac{2}{x^2 - 1}y = \dfrac{x+1}{x-1}$ an integrating factor is $e^{\int[2/(x^2-1)]dx} = \dfrac{x-1}{x+1}$ so that $\dfrac{d}{dx}\left[\dfrac{x-1}{x+1}y\right] = 1$ and $(x-1)y = x(x+1) + c(x+1)$ for $-1 < x < 1$.

27. For $y' + \dfrac{1}{x}y = \dfrac{1}{x}e^x$ an integrating factor is $e^{\int(1/x)dx} = x$ so that $\dfrac{d}{dx}[xy] = e^x$ and $y = \dfrac{1}{x}e^x + \dfrac{c}{x}$ for $0 < x < \infty$. If $y(1) = 2$ then $c = 2 - e$ and $y = \dfrac{1}{x}e^x + \dfrac{2-e}{x}$. The solution is defined on $I = (0, \infty)$.

30. For $\dfrac{dT}{dt} - kT = -T_m k$ an integrating factor is $e^{\int(-k)dt} = e^{-kt}$ so that $\dfrac{d}{dt}[e^{-kt}T] = -T_m k e^{-kt}$ and $T = T_m + ce^{kt}$ for $-\infty < t < \infty$. If $T(0) = T_0$ then $c = T_0 - T_m$ and $T = T_m + (T_0 - T_m)e^{kt}$. The solution is defined on $I = (-\infty, \infty)$.

33. For $y' + \dfrac{1}{x+1}\,y = \dfrac{\ln x}{x+1}$ an integrating factor is $e^{\int[1/(x+1)]dx} = x+1$ so that $\dfrac{d}{dx}[(x+1)y] = \ln x$

and

$$y = \frac{x}{x+1}\ln x - \frac{x}{x+1} + \frac{c}{x+1}$$

for $0 < x < \infty$. If $y(1) = 10$ then $c = 21$ and $y = \dfrac{x}{x+1}\ln x - \dfrac{x}{x+1} + \dfrac{21}{x+1}$. The solution is

defined on $I = (0, \infty)$.

36. For $y' + (\tan x)y = \cos^2 x$ an integrating factor is $e^{\int \tan x\, dx} = e^{\ln|\sec x|} = \sec x$ so that

$\dfrac{d}{dx}[(\sec x)\,y] = \cos x$ and $y = \sin x \cos x + c \cos x$ for $-\pi/2 < x < \pi/2$. If $y(0) = -1$ then

$c = -1$ and $y = \sin x \cos x - \cos x$. The solution is defined on $I = (-\pi/2,\, \pi/2)$.

39. For $y' + 2xy = f(x)$ an integrating factor is e^{x^2} so that

$$ye^{x^2} = \begin{cases} \frac{1}{2}e^{x^2} + c_1, & 0 \le x < 1 \\ c_2, & x \ge 1. \end{cases}$$

If $y(0) = 2$ then $c_1 = 3/2$ and for continuity we must have
$c_2 = \frac{1}{2}e + \frac{3}{2}$ so that

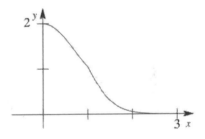

$$y = \begin{cases} \frac{1}{2} + \frac{3}{2}e^{-x^2}, & 0 \le x < 1 \\ \left(\frac{1}{2}e + \frac{3}{2}\right)e^{-x^2}, & x \ge 1. \end{cases}$$

42. For $y' + e^x y = 1$ an integrating factor is e^{e^x}. Thus

$$\frac{d}{dx}\left[e^{e^x} y\right] = e^{e^x} \quad \text{and} \quad e^{e^x} y = \int_0^x e^{e^t}\, dt + c.$$

From $y(0) = 1$ we get $c = e$, so $y = e^{-e^x}\int_0^x e^{e^t}\, dt + e^{1-e^x}$.

When $y' + e^x y = 0$ we can separate variables and integrate:

$$\frac{dy}{y} = -e^x\, dx \quad \text{and} \quad \ln|y| = -e^x + c.$$

Thus $y = c_1 e^{-e^x}$. From $y(0) = 1$ we get $c_1 = e$, so $y = e^{1-e^x}$.

When $y' + e^x y = e^x$ we can see by inspection that $y = 1$ is a solution.

2.4 | Exact Equations

The terminology and concepts listed below provide an outline of the main ideas encountered in this section. These can be useful when preparing for a quiz or test.

Terminology and Concepts

- differential of a function f of two variables

- exact DE

- criterion for exactness of $M(x, y)\, dx + N(x, y)\, dy = 0$

- use of an integrating factor to change a nonexact DE into one that is exact

The basic skills listed below summarize the more mechanical types of problems encountered in the exercise set for this section.

Basic Skills

- determine whether a DE is exact

- solve an exact DE

- find and use an integrating factor to make a DE exact

Partial Differentiation

In this section you will have to use **partial differentiation** and a reverse process which could be called **partial integration** or **partial antidifferentiation**.

Suppose $f(x, y)$ is a function of two variables defined in a region R of the plane. The mechanics of computing the partial derivative of f with respect to x, written $\partial f / \partial x$, can be summarized simply as:

To compute $\dfrac{\partial f}{\partial x}$, use the rules of ordinary differentiation with respect to x while treating y as a constant.

For example, if $f(x, y) = 3x^4 y^2 + e^{5y} + 4\cos xy^2$, then by treating y as a constant we obtain

$$\frac{\partial f}{\partial x} = y^2 \frac{\partial}{\partial x} 3x^4 + \frac{\partial}{\partial x} e^{5y} + 4 \frac{\partial}{\partial x} \cos xy^2 = y^2 \cdot 12x^3 + 0 + 4(-y^2 \sin xy^2)$$

$$= 12x^3 y^2 - 4y^2 \sin xy^2.$$

On the other hand, to compute the partial derivative of f with respect to y, written $\partial f/\partial y$, use the rules of ordinary differentiation, except treat x as a constant. For example, for the function f in the preceding example, we have

$$\frac{\partial f}{\partial y} = 3x^4 \frac{\partial}{\partial y} y^2 + \frac{\partial}{\partial y} e^{5y} + 4 \frac{\partial}{\partial y} \cos xy^2 = 3x^4 \cdot 2y + 5e^{5y} + 4(-x \cdot 2y \sin xy^2)$$

$$= 6x^4 y + 5e^{5y} - 8xy \sin xy^2.$$

Now suppose that we are given the partial derivative of a function f, say $\partial f/\partial x = M(x, y)$. We can recover f up to an "additive constant" by partial integration with respect to x:

$$f(x, y) = \int M(x, y)\, dx.$$

In the foregoing integral we use the rules of ordinary indefinite integration with respect to x while treating y as a constant. If $\partial f/\partial y = N(x, y)$, then by partial integration with respect to y we have

$$f(x, y) = \int N(x, y)\, dy,$$

where x is treated as a constant. For example, if $\partial f/\partial x = x + y$, then

$$f(x, y) = \int (x + y)\, dx = \frac{1}{2}x^2 + xy + g(y).$$

Here any function $g(y)$ is a "constant" of integration since

$$\frac{\partial f}{\partial x} = \frac{1}{2} \frac{\partial}{\partial x} x^2 + y \frac{\partial}{\partial x} x + \frac{\partial}{\partial x} g(y) = \frac{1}{2} \cdot 2x + y \cdot 1 + 0 = x + y.$$

As another example, suppose that $\partial f/\partial y = 10e^{-x^2 y}$, then, with x held constant,

$$f(x, y) = \int 10e^{-x^2 y}\, dy = \frac{10}{-x^2} \int e^{-x^2 y}(-x^2\, dy) = -\frac{10}{x^2} e^{-x^2 y} + h(x),$$

where this time the "constant" of integration is $h(x)$. You are encouraged to verify the last result by partial differentiation with respect to y.

Use of Computers To obtain the partial derivative of $f(x, y) = xe^y + x^2 y^3$ with respect to y in a CAS use

(Mathematica)

```
Clear[f]
f[x_, y_]:= x Exp[y] + x^2 y^3
f[x, y]
D[f[x, y], y]
```

(Maple)

```
f:=(x,y)->x*exp(y)+x^2*y^3;
diff(f(x,y),y);    or    D[2](f);
```

In *Maple*, `diff(f(x,y),y);` returns the expression $xe^y + 3x^2y^2$, which cannot be directly evaluated at a point. On the other hand, `D[2](f);` returns the function $(x, y) \to xe^y + 3x^2y^2$. In this case, the 2 in `D[2]` signifies the partial derivative with respect to the second variable, y, in the definition of f.

3. Let $M = 5x + 4y$ and $N = 4x - 8y^3$ so that $M_y = 4 = N_x$. From $f_x = 5x + 4y$ we obtain $f = \frac{5}{2}x^2 + 4xy + h(y)$, $h'(y) = -8y^3$, and $h(y) = -2y^4$. A solution is $\frac{5}{2}x^2 + 4xy - 2y^4 = c$.

6. Let $M = 4x^3 - 3y\sin 3x - y/x^2$ and $N = 2y - 1/x + \cos 3x$ so that $M_y = -3\sin 3x - 1/x^2$ and $N_x = 1/x^2 - 3\sin 3x$. The equation is not exact.

9. Let $M = y^3 - y^2\sin x - x$ and $N = 3xy^2 + 2y\cos x$ so that $M_y = 3y^2 - 2y\sin x = N_x$. From $f_x = y^3 - y^2\sin x - x$ we obtain $f = xy^3 + y^2\cos x - \frac{1}{2}x^2 + h(y)$, $h'(y) = 0$, and $h(y) = 0$. A solution is $xy^3 + y^2\cos x - \frac{1}{2}x^2 = c$.

12. Let $M = 3x^2y + e^y$ and $N = x^3 + xe^y - 2y$ so that $M_y = 3x^2 + e^y = N_x$. From $f_x = 3x^2y + e^y$ we obtain $f = x^3y + xe^y + h(y)$, $h'(y) = -2y$, and $h(y) = -y^2$. A solution is $x^3y + xe^y - y^2 = c$.

15. Let $M = x^2y^3 - 1/(1 + 9x^2)$ and $N = x^3y^2$ so that $M_y = 3x^2y^2 = N_x$. From $f_x = x^2y^3 - 1/(1 + 9x^2)$ we obtain $f = \frac{1}{3}x^3y^3 - \frac{1}{3}\arctan(3x) + h(y)$, $h'(y) = 0$, and $h(y) = 0$. A solution is $x^3y^3 - \arctan(3x) = c$.

18. Let $M = 2y\sin x\cos x - y + 2y^2e^{xy^2}$ and $N = -x + \sin^2 x + 4xye^{xy^2}$ so that

$$M_y = 2\sin x\cos x - 1 + 4xy^3e^{xy^2} + 4ye^{xy^2} = N_x.$$

From $f_x = 2y\sin x\cos x - y + 2y^2e^{xy^2}$ we obtain $f = y\sin^2 x - xy + 2e^{xy^2} + h(y)$, $h'(y) = 0$, and $h(y) = 0$. A solution is $y\sin^2 x - xy + 2e^{xy^2} = c$.

21. Let $M = x^2 + 2xy + y^2$ and $N = 2xy + x^2 - 1$ so that $M_y = 2(x + y) = N_x$. From $f_x = x^2 + 2xy + y^2$ we obtain $f = \frac{1}{3}x^3 + x^2y + xy^2 + h(y)$, $h'(y) = -1$, and $h(y) = -y$. The solution is $\frac{1}{3}x^3 + x^2y + xy^2 - y = c$. If $y(1) = 1$ then $c = 4/3$ and a solution of the initial-value problem is $\frac{1}{3}x^3 + x^2y + xy^2 - y = \frac{4}{3}$.

24. Let $M = t/2y^4$ and $N = (3y^2 - t^2)/y^5$ so that $M_y = -2t/y^5 = N_t$. From $f_t = t/2y^4$ we obtain $f = \dfrac{t^2}{4y^4} + h(y)$, $h'(y) = \dfrac{3}{y^3}$, and $h(y) = -\dfrac{3}{2y^2}$. The solution is $\dfrac{t^2}{4y^4} - \dfrac{3}{2y^2} = c$. If $y(1) = 1$ then $c = -5/4$ and a solution of the initial-value problem is $\dfrac{t^2}{4y^4} - \dfrac{3}{2y^2} = -\dfrac{5}{4}$.

27. Equating $M_y = 3y^2 + 4kxy^3$ and $N_x = 3y^2 + 40xy^3$ we obtain $k = 10$.

30. Let $M = \left(x^2 + 2xy - y^2\right)/\left(x^2 + 2xy + y^2\right)$ and $N = \left(y^2 + 2xy - x^2\right)/\left(y^2 + 2xy + x^2\right)$ so that $M_y = -4xy/(x+y)^3 = N_x$. From $f_x = \left(x^2 + 2xy + y^2 - 2y^2\right)/(x+y)^2$ we obtain $f = x + \dfrac{2y^2}{x+y} + h(y)$, $h'(y) = -1$, and $h(y) = -y$. A solution of the differential equation is $x^2 + y^2 = c(x+y)$.

33. We note that $(N_x - M_y)/M = 2/y$, so an integrating factor is $e^{\int 2\,dy/y} = y^2$. Let $M = 6xy^3$ and $N = 4y^3 + 9x^2y^2$ so that $M_y = 18xy^2 = N_x$. From $f_x = 6xy^3$ we obtain $f = 3x^2y^3 + h(y)$, $h'(y) = 4y^3$, and $h(y) = y^4$. A solution of the differential equation is $3x^2y^3 + y^4 = c$.

36. We note that $(N_x - M_y)/M = -3/y$, so an integrating factor is $e^{-3\int dy/y} = 1/y^3$. Let

$$M = (y^2 + xy^3)/y^3 = 1/y + x \quad \text{and} \quad N = (5y^2 - xy + y^3 \sin y)/y^3 = 5/y - x/y^2 + \sin y,$$

so that $M_y = -1/y^2 = N_x$. From $f_x = 1/y + x$ we obtain $f = x/y + \frac{1}{2}x^2 + h(y)$, $h'(y) = 5/y + \sin y$, and $h(y) = 5\ln|y| - \cos y$. A solution of the differential equation is $x/y + \frac{1}{2}x^2 + 5\ln|y| - \cos y = c$.

39. (a) Implicitly differentiating $x^3 + 2x^2y + y^2 = c$ and solving for dy/dx we obtain

$$3x^2 + 2x^2 \frac{dy}{dx} + 4xy + 2y\frac{dy}{dx} = 0 \quad \text{and} \quad \frac{dy}{dx} = -\frac{3x^2 + 4xy}{2x^2 + 2y}.$$

By writing the last equation in differential form we get $(4xy + 3x^2)dx + (2y + 2x^2)dy = 0$.

(b) Setting $x = 0$ and $y = -2$ in $x^3 + 2x^2y + y^2 = c$ we find $c = 4$, and setting $x = y = 1$ we also find $c = 4$. Thus, both initial conditions determine the same implicit solution.

(c) Solving $x^3 + 2x^2y + y^2 = 4$ for y we get

$$y_1(x) = -x^2 - \sqrt{4 - x^3 + x^4}$$

and

$$y_2(x) = -x^2 + \sqrt{4 - x^3 + x^4}.$$

Observe in the figure that $y_1(0) = -2$ and $y_2(1) = 1$.

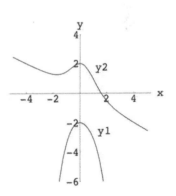

2.5 | Solutions by Substitutions

The terminology and concepts listed below provide an outline of the main ideas encountered in this section. These can be useful when preparing for a quiz or test.

Terminology and Concepts

- homogeneous DE

- Bernoulli's equation

- use of a substitution to reduce a DE of the form $dy/dx = f(Ax + By + C)$ to a separable DE

The basic skills listed below summarize the more mechanical types of problems encountered in the exercise set for this section.

Basic Skills

- recognize and solve a homogeneous DE using an appropriate substitution

- recognize a Bernoulli equation and solve it by reducing it to a linear first-order DE

- solve a DE of the form $dy/dx = f(Ax + By + C)$ using an appropriate substitution

Homogeneous DEs
As mentioned in the text, either substitution $y = ux$ or $x = vy$ will reduce a homogeneous DE $M(x, y)\, dx + N(x, y)\, dy = 0$ to one with separable variables. As a rule of thumb, try the substitution $y = ux$ whenever the coefficient N is simpler than M. On the other hand, try the substitution $x = vy$ whenever the coefficient M is simpler than N.

3. Letting $x = vy$ we have

$$vy(v\, dy + y\, dv) + (y - 2vy)\, dy = 0$$

$$vy^2\, dv + y\left(v^2 - 2v + 1\right) dy = 0$$

$$\frac{v\, dv}{(v - 1)^2} + \frac{dy}{y} = 0$$

$$\ln|v - 1| - \frac{1}{v - 1} + \ln|y| = c$$

$$\ln\left|\frac{x}{y} - 1\right| - \frac{1}{x/y - 1} + \ln y = c$$

$$(x - y)\ln|x - y| - y = c(x - y).$$

6. Letting $y = ux$ and using partial fractions, we have

$$\left(u^2 x^2 + ux^2\right) dx + x^2 (u\,dx + x\,du) = 0$$

$$x^2 \left(u^2 + 2u\right) dx + x^3\,du = 0$$

$$\frac{dx}{x} + \frac{du}{u(u+2)} = 0$$

$$\ln|x| + \frac{1}{2}\ln|u| - \frac{1}{2}\ln|u+2| = c$$

$$\frac{x^2 u}{u+2} = c_1$$

$$x^2 \frac{y}{x} = c_1 \left(\frac{y}{x} + 2\right)$$

$$x^2 y = c_1 (y + 2x).$$

9. Letting $y = ux$ we have

$$-ux\,dx + (x + \sqrt{u}\,x)(u\,dx + x\,du) = 0$$

$$(x^2 + x^2 \sqrt{u}\,)\,du + xu^{3/2}\,dx = 0$$

$$\left(u^{-3/2} + \frac{1}{u}\right) du + \frac{dx}{x} = 0$$

$$-2u^{-1/2} + \ln|u| + \ln|x| = c$$

$$\ln|y/x| + \ln|x| = 2\sqrt{x/y} + c$$

$$y(\ln|y| - c)^2 = 4x.$$

12. Letting $y = ux$ we have

$$(x^2 + 2u^2 x^2)dx - ux^2(u\,dx + x\,du) = 0$$

$$x^2(1 + u^2)dx - ux^3\,du = 0$$

$$\frac{dx}{x} - \frac{u\,du}{1 + u^2} = 0$$

$$\ln|x| - \frac{1}{2}\ln(1 + u^2) = c$$

$$\frac{x^2}{1 + u^2} = c_1$$

$$x^4 = c_1(x^2 + y^2).$$

Using $y(-1) = 1$ we find $c_1 = 1/2$. The solution of the initial-value problem is $2x^4 = y^2 + x^2$.

15. From $y' + \dfrac{1}{x}y = \dfrac{1}{x}y^{-2}$ and $w = y^3$ we obtain $\dfrac{dw}{dx} + \dfrac{3}{x}w = \dfrac{3}{x}$. An integrating factor is x^3 so that $x^3 w = x^3 + c$ or $y^3 = 1 + cx^{-3}$.

18. From $y' - \left(1 + \dfrac{1}{x}\right)y = y^2$ and $w = y^{-1}$ we obtain $\dfrac{dw}{dx} + \left(1 + \dfrac{1}{x}\right)w = -1$. An integrating factor is xe^x so that $xe^x w = -xe^x + e^x + c$ or $y^{-1} = -1 + \dfrac{1}{x} + \dfrac{c}{x}e^{-x}$.

21. From $y' - \dfrac{2}{x}y = \dfrac{3}{x^2}y^4$ and $w = y^{-3}$ we obtain $\dfrac{dw}{dx} + \dfrac{6}{x}w = -\dfrac{9}{x^2}$. An integrating factor is x^6 so that $x^6 w = -\frac{9}{5}x^5 + c$ or $y^{-3} = -\frac{9}{5}x^{-1} + cx^{-6}$. If $y(1) = \frac{1}{2}$ then $c = \frac{49}{5}$ and $y^{-3} = -\frac{9}{5}x^{-1} + \frac{49}{5}x^{-6}$.

24. Let $u = x + y$ so that $du/dx = 1 + dy/dx$. Then $\dfrac{du}{dx} - 1 = \dfrac{1 - u}{u}$ or $u\,du = dx$. Thus $\frac{1}{2}u^2 = x + c$ or $u^2 = 2x + c_1$, and $(x + y)^2 = 2x + c_1$.

27. Let $u = y - 2x + 3$ so that $du/dx = dy/dx - 2$. Then $\dfrac{du}{dx} + 2 = 2 + \sqrt{u}$ or $\dfrac{1}{\sqrt{u}}\,du = dx$. Thus $2\sqrt{u} = x + c$ and $2\sqrt{y - 2x + 3} = x + c$.

30. Let $u = 3x + 2y$ so that $du/dx = 3 + 2\,dy/dx$. Then $\dfrac{du}{dx} = 3 + \dfrac{2u}{u + 2} = \dfrac{5u + 6}{u + 2}$ and $\dfrac{u + 2}{5u + 6}\,du = dx$. Now by long division

$$\frac{u + 2}{5u + 6} = \frac{1}{5} + \frac{4}{25u + 30}$$

so we have

$$\int \left(\frac{1}{5} + \frac{4}{25u + 30}\right) du = dx$$

and $\frac{1}{5}u + \frac{4}{25}\ln|25u + 30| = x + c$. Thus

$$\frac{1}{5}(3x + 2y) + \frac{4}{25}\ln|75x + 50y + 30| = x + c.$$

Setting $x = -1$ and $y = -1$ we obtain $c = \frac{4}{25} \ln 95$. The solution is

$$\frac{1}{5}(3x + 2y) + \frac{4}{25} \ln |75x + 50y + 30| = x + \frac{4}{25} \ln 95$$

or $5y - 5x + 2\ln|75x + 50y + 30| = 2\ln 95$.

2.6 A Numerical Method

The terminology and concepts listed below provide an outline of the main ideas encountered in this section. These can be useful when preparing for a quiz or test.

Terminology and Concepts

- direction field
- linearization of a function at a point
- Euler's method
- absolute error
- percentage relative error
- numerical solver

The basic skills listed below summarize the more mechanical types of problems encountered in the exercise set for this section.

Basic Skills

- use Euler's method to approximate the solution of a first-order DE

Use of Computers Both *Mathematica* and *Maple* contain sophisticated programming languages. Below are examples of how each CAS can be used to implement Euler's method and display a table similar to the ones shown in Example 2 in this section of the text. Here we will consider the IVP $y' = f(x, y) = 0.2xy$, $y(1) = 1$, and use Euler's method to approximate $y(1.5)$ using the step size $h = 0.1$. To compute the absolute error at each step we will use the fact that the analytic solution is $y = g(x) = e^{0.1(x^2 - 1)}$.

```
Clear[f, g, u, v, t, err]                           (Mathematica)
f[x_, y_]:= 0.2x y        (* Input line *)
Print["y'= ", f[x, y]]
g[x_]:= Exp[0.1(x^2 - 1)]        (* Input line *)
Print["y(x) = ", g[x]]
h = 0.1; a = 1; b = 1.5; y0 = 1;        (* Input line *)
```

u[0] = a; v[0] = y0;

u[n_]:= u[n] = u[n - 1] + h

v[n_]:= v[n] = v[n - 1] + h f[u[n - 1], v[n - 1]]

t[n_]:= t[n] = g[u[n]]

err[n_]:= err[n] = Abs[t[n] - v[n]]

approxtab = Table[{u[n], v[n], t[n], err[n]}, {n, 0, Round[(b - a)/h]}];

tableheader = {"x", "y", "actual value", "abs. error"};

Prepend[approxtab, tableheader]//TableForm

The use of the form **u[n_]:= u[n] = u[n - 1] + h** in the *Mathematica* code above is a way to define a function that remembers a value once it has been computed. This greatly speeds up the running of a program that involves a recursively defined function. To make it easier to reuse the routine for subsequent problems, the lines containing inputs to the routine can be commented, as shown above, or color coded.

```
f:=(x,y)->0.2*x*y;          #Input line            (Maple)
g:=x->exp(0.1*(x^2-1));        #Input line
h:=0.1; a:=1; b:=1.5; y0:=1;      #Input line
u:=n->u(n-1)+h; u(0):=a;
v:=n->v(n-1)+h*f(u(n-1),v(n-1)); v(0):=y0;
t:=n->g(u(n));
err:=n->abs(t(n)-v(n));
u(-1):='x'; v(-1):='y'; t(-1):='actual_value';
err(-1):='abs._error';
with(linalg)
matrix(2+round((b-a)/h),4,(i,j)->piecewise(j=1,u(i-2),
    j=2,v(i-2),j=3,t(i-2),j=4,err(i-2)));
```

3. Separating variables and integrating, we have

$$\frac{dy}{y} = dx \qquad \text{and} \qquad \ln|y| = x + c.$$

Thus $y = c_1 e^x$ and, using $y(0) = 1$, we find $c = 1$, so $y = e^x$ is the solution of the initial-value problem.

$h=0.1$

x_n	y_n	Actual Value	Abs. Error	% Rel. Error
0.00	1.0000	1.0000	0.0000	0.00
0.10	1.1000	1.1052	0.0052	0.47
0.20	1.2100	1.2214	0.0114	0.93
0.30	1.3310	1.3499	0.0189	1.40
0.40	1.4641	1.4918	0.0277	1.86
0.50	1.6105	1.6487	0.0382	2.32
0.60	1.7716	1.8221	0.0506	2.77
0.70	1.9487	2.0138	0.0650	3.23
0.80	2.1436	2.2255	0.0820	3.68
0.90	2.3579	2.4596	0.1017	4.13
1.00	2.5937	2.7183	0.1245	4.58

$h=0.05$

x_n	y_n	Actual Value	Abs. Error	% Rel. Error
0.00	1.0000	1.0000	0.0000	0.00
0.05	1.0500	1.0513	0.0013	0.12
0.10	1.1025	1.1052	0.0027	0.24
0.15	1.1576	1.1618	0.0042	0.36
0.20	1.2155	1.2214	0.0059	0.48
0.25	1.2763	1.2840	0.0077	0.60
0.30	1.3401	1.3499	0.0098	0.72
0.35	1.4071	1.4191	0.0120	0.84
0.40	1.4775	1.4918	0.0144	0.96
0.45	1.5513	1.5683	0.0170	1.08
0.50	1.6289	1.6487	0.0198	1.20
0.55	1.7103	1.7333	0.0229	1.32
0.60	1.7959	1.8221	0.0263	1.44
0.65	1.8856	1.9155	0.0299	1.56
0.70	1.9799	2.0138	0.0338	1.68
0.75	2.0789	2.1170	0.0381	1.80
0.80	2.1829	2.2255	0.0427	1.92
0.85	2.2920	2.3396	0.0476	2.04
0.90	2.4066	2.4596	0.0530	2.15
0.95	2.5270	2.5857	0.0588	2.27
1.00	2.6533	2.7183	0.0650	2.39

6. $h=0.1$

x_n	y_n
0.00	1.0000
0.10	1.1000
0.20	1.2220
0.30	1.3753
0.40	1.5735
0.50	1.8371

$h=0.05$

x_n	y_n
0.00	1.0000
0.05	1.0500
0.10	1.1053
0.15	1.1668
0.20	1.2360
0.25	1.3144
0.30	1.4039
0.35	1.5070
0.40	1.6267
0.45	1.7670
0.50	1.9332

9. $h=0.1$

x_n	y_n
1.00	1.0000
1.10	1.0000
1.20	1.0191
1.30	1.0588
1.40	1.1231
1.50	1.2194

$h=0.05$

x_n	y_n
1.00	1.0000
1.05	1.0000
1.10	1.0049
1.15	1.0147
1.20	1.0298
1.25	1.0506
1.30	1.0775
1.35	1.1115
1.40	1.1538
1.45	1.2057
1.50	1.2696

12. Tables of values were computed using the Euler and RK4 methods. The resulting points were plotted and joined using **ListPlot** in *Mathematica*.

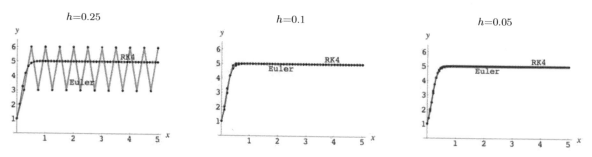

| **2.R** | Chapter 2 in Review |

We see that when $x = 0$, $y = 0$, so the initial-value problem has an infinite number of solutions for $k = 0$ and no solutions for $k \neq 0$.

3. True; $y = k_2/k_1$ is always a solution for $k_1 \neq 0$.

6. False, because $r\theta + r + \theta + 1 = (r+1)(\theta+1)$ and the differential equation can be written as

$$\frac{dr}{r+1} = (\theta+1)d\theta.$$

9. The differential equation is separable so

$$\frac{dy}{y} = e^x dx \qquad \text{implies} \qquad \ln|y| = e^{e^x} + c,$$

and thus $y = c_1 e^{e^x}$ is the general solution of the differential equation.

12. An example of an autonomous linear first-order differential equation with a single critical point at -3 is $\dfrac{dy}{dx} = y + 3$, whereas an autonomous nonlinear first-order differential equation with a single critical point -3 is $\dfrac{dy}{dx} = (y+3)^2$.

15. When n is odd, $x^n < 0$ for $x < 0$ and $x^n > 0$ for $x > 0$. In this case 0 is unstable. When n is even, $x^n > 0$ for $x < 0$ and for $x > 0$. In this case 0 is semi-stable. When n is odd, $-x^n > 0$ for $x < 0$ and $-x^n < 0$ for $x > 0$. In this case 0 is asymptotically stable. When n is even, $-x^n < 0$ for $x < 0$ and for $x > 0$. In this case 0 is semi-stable. Technically, 0^0 is an indeterminant form; however for all values of x except 0, $x^0 = 1$. Thus, we define 0^0 to be 1 in this case.

18. (a) linear in y, homogeneous, exact

(b) linear in x

(c) separable, exact, linear in x and y

(d) Bernoulli in x

(e) separable

(f) separable, linear in x, Bernoulli

(g) linear in x

(h) homogeneous

(i) Bernoulli

(j) homogeneous, exact, Bernoulli

(k) linear in x and y, exact, separable, homogeneous

(l) exact, linear in y

(m) homogeneous

(n) separable

21. The differential equation

$$\frac{dy}{dx} + \frac{2}{6x+1}y = -\frac{3x^2}{6x+1}y^{-2}$$

is Bernoulli. Using $w = y^3$, we obtain the linear equation

$$\frac{dw}{dx} + \frac{6}{6x+1}w = -\frac{9x^2}{6x+1}.$$

An integrating factor is $6x + 1$, so

$$\frac{d}{dx}[(6x+1)w] = -9x^2,$$

$$w = -\frac{3x^3}{6x+1} + \frac{c}{6x+1},$$

and

$$(6x+1)y^3 = -3x^3 + c.$$

Note: The differential equation is also exact.

24. Letting $u = 2x + y + 1$ we have

$$\frac{du}{dx} = 2 + \frac{dy}{dx},$$

and so the given differential equation is transformed into

$$u\left(\frac{du}{dx} - 2\right) = 1 \quad \text{or} \quad \frac{du}{dx} = \frac{2u+1}{u}.$$

Separating variables and integrating we get

$$\frac{u}{2u+1} \, du = dx$$

$$\left(\frac{1}{2} - \frac{1}{2}\frac{1}{2u+1}\right) du = dx$$

$$\frac{1}{2}u - \frac{1}{4}\ln|2u+1| = x + c$$

$$2u - \ln|2u+1| = 2x + c_1.$$

Resubstituting for u gives the solution

$$4x + 2y + 2 - \ln|4x + 2y + 3| = 2x + c_1$$

or

$$2x + 2y + 2 - \ln|4x + 2y + 3| = c_1.$$

27. The differential equation has the form $(d/dx)\,[(\sin x)y] = 0$. Integrating, we have $(\sin x)y = c$ or $y = c/\sin x$. The initial condition implies $c = -2\sin(7\pi/6) = 1$. Thus, $y = 1/\sin x = \csc x$, where the interval $\pi < x < 2\pi$ is chosen to include $x = 7\pi/6$.

30. (a) The differential equation is homogeneous and we let $y = ux$. Then

$$\left(x^2 - y^2\right) dx + xy \, dy = 0$$

$$\left(x^2 - u^2 x^2\right) dx + ux^2(u \, dx + x \, du) = 0$$

$$dx + ux \, du = 0$$

$$u \, du = -\frac{dx}{x}$$

$$\frac{1}{2}u^2 = -\ln|x| + c$$

$$\frac{y^2}{x^2} = -2\ln|x| + c_1.$$

The initial condition gives $c_1 = 2$, so an implicit solution is $y^2 = x^2(2 - 2\ln|x|)$.

(b) Solving for y in part (a) and being sure that the initial condition is still satisfied, we have $y = -\sqrt{2}\,|x|(1 - \ln|x|)^{1/2}$, where $-e \le x \le e$ so that $1 - \ln|x| \ge 0$. The graph of this function indicates that the derivative is not defined at $x = 0$ and $x = e$. Thus, the solution of the initial-value problem is $y = -\sqrt{2}\,x(1 - \ln x)^{1/2}$, for $0 < x < e$.

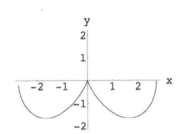

33. Since the differential equation is autonomous, all lineal elements on a given horizontal line have the same slope. The direction field is then as shown in the figure at the right. It appears from the figure that the differential equation has critical points at -2 (an attractor) and at 2 (a repeller). Thus, -2 is an asymptotically stable critical point and 2 is an unstable critical point.

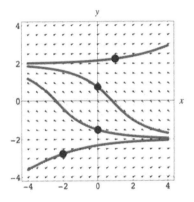

3 MODELING WITH FIRST-ORDER DIFFERENTIAL EQUATIONS

3.1 | Linear Models

The terminology and concepts listed below provide an outline of the main ideas encountered in this section. These can be useful when preparing for a quiz or test.

Terminology and Concepts

- growth and decay
- growth and decay constant
- half-life
- carbon dating
- Newton's law of cooling/warming
- mixtures
- series circuits
- steady-state part of a solution
- transient term in solution

The basic skills listed below summarize the more mechanical types of problems encountered in the exercise set for this section.

Basic Skills

- set up and solve DEs modeling populations that exhibit exponential growth or decay
- set up and solve DEs involving Newton's law of cooling/warming
- set up and solve DEs modeling mixtures of fluids
- set up and solve DEs modeling series circuits
- set up and solve DEs modeling falling objects both with and without air resistance

Half-life   To drive the point home that "the usual carbon-14 technique is limited to about nine half-lives of the isotope," we will estimate the amount of C-14 that would remain in a piece of wood from a tree about 50,000 years after the tree is cut down. We use the fact that the half-life of C-14 is approximately 5,600 years. Then, in $2(5,600) = 12,200$ years, $\frac{1}{4}$ of the original amount of C-14 remains; in $3(5,600) = 16,800$ years, $\frac{1}{8}$ of the original amount remains. Continuing in this fashion, in $9(5,600) = 50,400$ years, $\frac{1}{512}$ (or only about 0.2%) of the original amount of C-14 remains.

Growth and Decay

3. Let $P = P(t)$ be the population at time t. Then $dP/dt = kP$ and $P = ce^{kt}$. From $P(0) = c = 500$ we see that $P = 500e^{kt}$. Since 15% of 500 is 75, we have $P(10) = 500e^{10k} = 575$. Solving for k, we get $k = \frac{1}{10} \ln \frac{575}{500} = \frac{1}{10} \ln 1.15$. When $t = 30$,

$$P(30) = 500e^{(1/10)(\ln 1.15)30} = 500e^{3 \ln 1.15} = 760 \text{ years}$$

and

$$P'(30) = kP(30) = \frac{1}{10}(\ln 1.15)760 = 10.62 \text{ persons/year.}$$

6. Let $A = A(t)$ be the amount present at time t. From $dA/dt = kA$ and $A(0) = 100$ we obtain $A = 100e^{kt}$. Using $A(6) = 97$ we find $k = \frac{1}{6} \ln 0.97$. Then $A(24) = 100e^{(1/6)(\ln 0.97)24} = 100(0.97)^4 \approx 88.5$ mg.

9. Let $I = I(t)$ be the intensity, t the thickness, and $I(0) = I_0$. If $dI/dt = kI$ and $I(3) = 0.25I_0$, then $I = I_0 e^{kt}$, $k = \frac{1}{3} \ln 0.25$, and $I(15) = 0.00098I_0$.

Carbon Dating

12. From Example 3 in the text, the amount of carbon present at time t is $A(t) = A_0 e^{-0.00012378t}$. Letting $t = 660$ and solving for A_0 we have $A(660) = A_0 e^{-0.0001237(660)} = 0.921553A_0$. Thus, approximately 92% of the original amount of C-14 remained in the cloth as of 1988.

Newton's Law of Cooling/Warming

15. We use the fact that the boiling temperature for water is $100°$ C. Now assume that $dT/dt = k(T - 100)$ so that $T = 100 + ce^{kt}$. If $T(0) = 20°$ and $T(1) = 22°$, then $c = -80$ and $k = \ln(39/40) \approx -0.0253$. Then $T(t) = 100 - 80e^{-0.0253t}$, and when $T = 90$, $t = 82.1$ seconds. If $T(t) = 98°$ then $t = 145.7$ seconds.

18. (a) The initial temperature of the bath is $T_m(0) = 60°$, so in the short term the temperature of the chemical, which starts at $80°$, should decrease or cool. Over time, the temperature of the bath will increase toward $100°$ since $e^{-0.1t}$ decreases from 1 toward 0 as t increases from 0. Thus, in the long term, the temperature of the chemical should increase or warm toward $100°$.

(b) Adapting the model for Newton's law of cooling, we have

$$\frac{dT}{dt} = -0.1(T - 100 + 40e^{-0.1t}), \quad T(0) = 80.$$

Writing the differential equation in the form

$$\frac{dT}{dt} + 0.1T = 10 - 4e^{-0.1t}$$

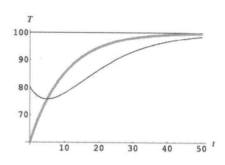

we see that it is linear with integrating factor $e^{\int 0.1\,dt} = e^{0.1t}$. Thus

$$\frac{d}{dt}[e^{0.1t}T] = 10e^{0.1t} - 4$$

$$e^{0.1t}T = 100e^{0.1t} - 4t + c$$

and

$$T(t) = 100 - 4te^{-0.1t} + ce^{-0.1t}.$$

Now $T(0) = 80$ so $100 + c = 80$, $c = -20$ and

$$T(t) = 100 - 4te^{-0.1t} - 20e^{-0.1t} = 100 - (4t + 20)e^{-0.1t}.$$

The thinner curve verifies the prediction of cooling followed by warming toward $100°$. The wider curve shows the temperature T_m of the liquid bath.

Mixtures

21. From $dA/dt = 4 - A/50$ we obtain $A = 200 + ce^{-t/50}$. If $A(0) = 30$ then $c = -170$ and $A = 200 - 170e^{-t/50}$.

24. From Problem 23 the number of pounds of salt in the tank at time t is $A(t) = 1000 - 1000e^{-t/100}$. The concentration at time t is $c(t) = A(t)/500 = 2 - 2e^{-t/100}$. Therefore $c(5) = 2 - 2e^{-1/20} = 0.0975\,\text{lb/gal}$ and $\lim_{t\to\infty} c(t) = 2$. Solving $c(t) = 1 = 2 - 2e^{-t/100}$ for t we obtain $t = 100\ln 2 \approx 69.3\,\text{min}$.

27. From

$$\frac{dA}{dt} = 3 - \frac{4A}{100 + (6-4)t} = 3 - \frac{2A}{50 + t}$$

we obtain $A = 50 + t + c(50 + t)^{-2}$. If $A(0) = 10$ then $c = -100{,}000$ and $A(30) = 64.38$ pounds.

Series Circuits

30. Assume $L\, di/dt + Ri = E(t)$, $E(t) = E_0 \sin \omega t$, and $i(0) = i_0$ so that

$$i = \frac{E_0 R}{L^2 \omega^2 + R^2} \sin \omega t - \frac{E_0 L \omega}{L^2 \omega^2 + R^2} \cos \omega t + ce^{-Rt/L}.$$

Since $i(0) = i_0$ we obtain $c = i_0 + \dfrac{E_0 L \omega}{L^2 \omega^2 + R^2}$.

33. For $0 \le t \le 20$ the differential equation is $20\, di/dt + 2i = 120$. An integrating factor is $e^{t/10}$, so $(d/dt)[e^{t/10} i] = 6e^{t/10}$ and $i = 60 + c_1 e^{-t/10}$. If $i(0) = 0$ then $c_1 = -60$ and $i = 60 - 60e^{-t/10}$. For $t > 20$ the differential equation is $20\, di/dt + 2i = 0$ and $i = c_2 e^{-t/10}$. At $t = 20$ we want $c_2 e^{-2} = 60 - 60e^{-2}$ so that $c_2 = 60\left(e^2 - 1\right)$. Thus

$$\begin{cases} 60 - 60e^{-t/10}, & 0 \le t \le 20 \\ 60\left(e^2 - 1\right)e^{-t/10}, & t > 20. \end{cases}$$

Additional Linear Models

36. (a) Integrating $d^2 s/dt^2 = -g$ we get $v(t) = ds/dt = -gt + c$. From $v(0) = 300$ we find $c = 300$, and we are given $g = 32$, so the velocity is $v(t) = -32t + 300$.

(b) Integrating again and using $s(0) = 0$ we get $s(t) = -16t^2 + 300t$. The maximum height is attained when $v = 0$, that is, at $t_a = 9.375$. The maximum height will be $s(9.375) = 1406.25\,\text{ft}$.

39. (a) The differential equation is first-order and linear. Letting $b = k/\rho$, the integrating factor is $e^{\int 3b\, dt/(bt + r_0)} = (r_0 + bt)^3$. Then

$$\frac{d}{dt}[(r_0 + bt)^3 v] = g(r_0 + bt)^3 \quad \text{and} \quad (r_0 + bt)^3 v = \frac{g}{4b}(r_0 + bt)^4 + c.$$

The solution of the differential equation is $v(t) = (g/4b)(r_0 + bt) + c(r_0 + bt)^{-3}$. Using $v(0) = 0$ we find $c = -gr_0^4/4b$, so that

$$v(t) = \frac{g}{4b}(r_0 + bt) - \frac{gr_0^4}{4b(r_0 + bt)^3} = \frac{g\rho}{4k}\left(r_0 + \frac{k}{\rho}t\right) - \frac{g\rho r_0^4}{4k(r_0 + kt/\rho)^3}.$$

(b) Integrating $dr/dt = k/\rho$ we get $r = kt/\rho + c$. Using $r(0) = r_0$ we have $c = r_0$, so $r(t) = kt/\rho + r_0$.

(c) If $r = 0.007\,\text{ft}$ when $t = 10\,\text{s}$, then solving $r(10) = 0.007$ for k/ρ, we obtain $k/\rho = -0.0003$ and $r(t) = 0.01 - 0.0003t$. Solving $r(t) = 0$ we get $t = 33.3$, so the raindrop will have evaporated completely at 33.3 seconds.

42. (a) The solution of the differential equation is $P(t) = c_1 e^{kt} + h/k$. If we let the initial population of fish be P_0 then $P(0) = P_0$ which implies that

$$c_1 = P_0 - \frac{h}{k} \quad \text{and} \quad P(t) = \left(P_0 - \frac{h}{k}\right) e^{kt} + \frac{h}{k}.$$

(b) For $P_0 > h/k$ all terms in the solution are positive. In this case $P(t)$ increases as time t increases. That is, $P(t) \to \infty$ as $t \to \infty$. For $P_0 = h/k$ the population remains constant for all time t:

$$P(t) = \left(\frac{h}{k} - \frac{h}{k}\right) e^{kt} + \frac{h}{k} = \frac{h}{k}.$$

For $0 < P_0 < h/k$ the coefficient of the exponential function is negative and so the function decreases as time t increases.

(c) Since the function decreases and is concave down, the graph of $P(t)$ crosses the t-axis. That is, there exists a time $T > 0$ such that $P(T) = 0$. Solving

$$\left(P_0 - \frac{h}{k}\right) e^{kT} + \frac{h}{k} = 0$$

for T shows that the time of extinction is

$$T = \frac{1}{k} \ln\left(\frac{h}{h - kP_0}\right).$$

45. (a) For $0 \le t < 4$, $6 \le t < 10$ and $12 \le t < 16$, no voltage is applied to the heart and $E(t) = 0$. At the other times, the differential equation is $dE/dt = -E/RC$. Separating variables, integrating, and solving for e, we get $E = ke^{-t/RC}$, subject to $E(4) = E(10) = E(16) = 12$. These intitial conditions yield, respectively, $k = 12e^{4/RC}$, $k = 12e^{10/RC}$, $k = 12e^{16/RC}$, and $k = 12e^{22/RC}$. Thus

$$\begin{cases} 0, & 0 \le t < 4,\ 6 \le t < 10,\ 12 \le t < 16 \\ 12e^{(4-t)/RC}, & 4 \le t < 6 \\ 12e^{(10-t)/RC}, & 10 \le t < 12 \\ 12e^{(16-t)/RC}, & 16 \le t < 18 \\ 12e^{(22-t)/RC}, & 22 \le t < 24. \end{cases}$$

(b) E

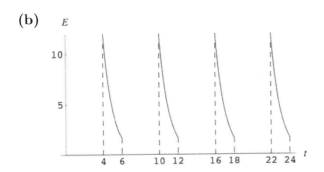

48. **(a)** We saw in part **(b)** of Problem 36 that the ascent time is $t_a = 9.375$. To find when the cannonball hits the ground we solve $s(t) = -16t^2 + 300t = 0$, getting a total time in flight of $t = 18.75\,\text{s}$. Thus, the time of descent is $t_d = 18.75 - 9.375 = 9.375$. The impact velocity is $v_i = v(18.75) = -300$, which has the same magnitude as the initial velocity.

(b) We saw in Problem 37 that the ascent time in the case of air resistance is $t_a = 9.162$. Solving $s(t) = 1{,}340{,}000 - 6{,}400t - 1{,}340{,}000e^{-0.005t} = 0$ we see that the total time of flight is $18.466\,\text{s}$. Thus, the descent time is $t_d = 18.466 - 9.162 = 9.304$. The impact velocity is $v_i = v(18.466) = -290.91$, compared to an initial velocity of $v_0 = 300$.

3.2 Nonlinear Models

The terminology and concepts listed below provide an outline of the main ideas encountered in this section. These can be useful when preparing for a quiz or test.

Terminology and Concepts

- logistic DE
- logistic growth
- harvesting
- Gompertz DE
- second-order chemical reaction

The basic skills listed below summarize the more mechanical types of problems encountered in the exercise set for this section.

Basic Skills

- set up and solve DEs involving the logistic model

- set up and solve DEs involving chemical reactions

- set up and solve DEs involving falling bodies subject to air resistance that is proportional to the square of the instantaneous velocity

Use of Computers

A scatter plot in two dimensions is a set of points in the plane. Both *Mathematica* and *Maple* have commands for plotting a set of points. In the examples below we plot the points $(-1, -2.7)$, $(-0.5, -2.5)$, $(0, -1.8)$, $(0.5, -1.6)$, $(1, -2.3)$, $(1.5, 0)$, $(2, 1.3)$, $(2.5, 1.9)$, $(3, 3.1)$.

$$\textbf{data} = \{\{\textbf{-1, -2.7}\}, \{\textbf{-0.5, -2.5}\}, \{\textbf{0, -1.8}\}, \{\textbf{0.5, -1.6}\}, \qquad (\textit{Mathematica})$$
$$\{\textbf{1, -2.3}\}, \{\textbf{1.5, 0}\}, \{\textbf{2, 1.3}\}, \{\textbf{2.5, 1.9}\}, \{\textbf{3, 3.1}\}\}$$
$$\textbf{grdata} = \textbf{ListPlot[data]}$$

The most commonly used options for **ListPlot** cause the dots to be enlarged or the dots to be connected with straight line segments. Although these two commands cannot be combined in a single call to **PlotStyle**, the same effect can be achieved as shown below.

$$\textbf{gr0} = \textbf{ListPlot[data, PlotStyle->PointSize[0.02]]} \qquad (\textit{Mathematica})$$
$$\textbf{gr1} = \textbf{ListPlot[data, PlotJoined->True]}$$
$$\textbf{Show[gr0, gr1]}$$

In **Maple** a scatter plot is obtained using the **plots** package.

```
data:=[[-1,-2.7],[-0.5,-2.5],[0,-1.8],          (Maple)
    [0.5,-1.6],[1,-2.3],[1.5,0],[2,1.3],
    [2.5,1.9],[3,3.1]];
with(plots);
pointplot(data);
```

In **Maple** it is possible to show the points in a variety of symbols and symbol sizes. You can also connect the points with straight line segments. You cannot, however, do both in a single plot. The routine below shows how to accomplish this. It is assumed that the input data is as above and that the **plots** package has been activated.

```
gr0:=pointplot(data,style=point,            (Maple)
    symbol=circle);
gr1:=pointplot(data,style=line);
display([gr0,gr1]);
```

Algorithms that fit relatively simple functions to a set of data points are not difficult to develop but they tend to involve a large number of routine computations. This means that a CAS is particularly well suited to implementing a curve-fitting algorithm. The following routines use a least squares method to fit a line to the data set given above over the interval $[-1, 3]$. The resulting line is then plotted over the data points.

linfit = fit[data, {1, x}, x]   (*Mathematica*)
gr2 = Plot[linfit, {x, -1, 3}]
Show[gr0, gr2]

The format of **data** in the routines above is as a set of points with two coordinates. In **Maple**, to apply a least squares fit of a line to the set of points, we need to reformat the data as a set of x-coordinates, which we will call **xdata**, and a set of y-coordinates, which we will call **ydata**. To accomplish this we first use the **nops** command to count the number of data points. Then we activate the **stats** pacakage which contains commands for fitting a line to a set of points in the plane.

```
n:=nops(data);                              (Maple)
xdata:=[seq(data[i,1],i=1..n)];
ydata:=[seq(data[i,2],i=1..n)];
with(stats);
linfit:=fit[leastsquare[[x,y],y=a*x+b,{a,b}]]([xdata,ydata]);
gr3:=implicitplot(linfit,x=-1..3,y=-5..5);
display([gr0,gr3]);
```

In the above *Maple* routine, `implicitplot`, which is part of the **plots** package, is used because the result of the **fit** command is an equation rather than an expression or function. The range of the equation, in this case **y=-5..5**, is specified to be large enough to include all of the data points.

Logistic Equation

3. From $dP/dt = P\left(10^{-1} - 10^{-7}P\right)$ and $P(0) = 5000$ we obtain $P = 500/(0.0005 + 0.0995e^{-0.1t})$ so that $P \to 1{,}000{,}000$ as $t \to \infty$. If $P(t) = 500{,}000$ then $t = 52.9$ months.

Modifications of the Logistic Model

6. Solving $P(5 - P) - \frac{25}{4} = 0$ for P we obtain the equilibrium solution $P = \frac{5}{2}$. For $P \neq \frac{5}{2}$, $dP/dt < 0$. Thus, if $P_0 < \frac{5}{2}$, the population becomes extinct (otherwise there would be another equilibrium solution.) Using separation of variables to solve the initial-value problem, we get

$$P(t) = [4P_0 + (10P_0 - 25)t]/[4 + (4P_0 - 10)t].$$

To find when the population becomes extinct for $P_0 < \frac{5}{2}$ we solve $P(t) = 0$ for t. We see that the time of extinction is $t = 4P_0/5(5 - 2P_0)$.

Chemical Reactions

9. Let $X = X(t)$ be the amount of C at time t and $dX/dt = k(120 - 2X)(150 - X)$. If $X(0) = 0$ and $X(5) = 10$, then

$$X(t) = \frac{150 - 150e^{180kt}}{1 - 2.5e^{180kt}},$$

where $k = .0001259$ and $X(20) = 29.3$ grams. Now by L'Hôpital's rule, $X \to 60$ as $t \to \infty$, so that the amount of $A \to 0$ and the amount of $B \to 30$ as $t \to \infty$.

Additional nonlinear Models

12. To obtain the solution of this differential equation we use $h(t)$ from Problem 13 in Exercises 1.3. Then $h(t) = (A_w\sqrt{H} - 4cA_h t)^2/A_w^2$. Solving $h(t) = 0$ with $c = 0.6$ and the values from Problem 11 we see that the tank empties in 3035.79 seconds or 50.6 minutes.

15. (a) After separating variables we obtain

$$\frac{m\,dv}{mg - kv^2} = dt$$

$$\frac{1}{g}\frac{dv}{1 - (\sqrt{k}\,v/\sqrt{mg}\,)^2} = dt$$

$$\frac{\sqrt{mg}}{\sqrt{k}\,g}\frac{\sqrt{k/mg}\,dv}{1 - (\sqrt{k}\,v/\sqrt{mg}\,)^2} = dt$$

$$\sqrt{\frac{m}{kg}}\tanh^{-1}\frac{\sqrt{k}\,v}{\sqrt{mg}} = t + c$$

$$\tanh^{-1}\frac{\sqrt{k}\,v}{\sqrt{mg}} = \sqrt{\frac{kg}{m}}\,t + c_1.$$

Thus the velocity at time t is

$$v(t) = \sqrt{\frac{mg}{k}}\tanh\left(\sqrt{\frac{kg}{m}}t + c_1\right).$$

Setting $t = 0$ and $v = v_0$ we find $c_1 = \tanh^{-1}(\sqrt{k}\,v_0/\sqrt{mg}\,)$.

(b) Since $\tanh t \to 1$ as $t \to \infty$, we have $v \to \sqrt{mg/k}$ as $t \to \infty$.

(c) Integrating the expression for $v(t)$ in part **(a)** we obtain an integral of the form $\int du/u$:

$$s(t) = \sqrt{\frac{mg}{k}}\int\tanh\left(\sqrt{\frac{kg}{m}}t + c_1\right)dt = \frac{m}{k}\ln\left[\cosh\left(\sqrt{\frac{kg}{m}}t + c_1\right)\right] + c_2.$$

Setting $t = 0$ and $s = 0$ we find $c_2 = -(m/k)\ln(\cosh c_1)$, where c_1 is given in part **(a)**.

18. (a) Writing the equation in the form $(x - \sqrt{x^2 + y^2}\,)dx + y\,dy = 0$ we identify $M = x - \sqrt{x^2 + y^2}$ and $N = y$. Since M and N are both homogeneous functions of degree 1 we use the substitution $y = ux$. It follows that

$$\left(x - \sqrt{x^2 + u^2 x^2}\,\right)dx + ux(u\,dx + x\,du) = 0$$

$$x\left[1 - \sqrt{1 + u^2} + u^2\right]dx + x^2 u\,du = 0$$

$$-\frac{u\,du}{1 + u^2 - \sqrt{1 + u^2}} = \frac{dx}{x}$$

$$\frac{u\,du}{\sqrt{1 + u^2}\,(1 - \sqrt{1 + u^2}\,)} = \frac{dx}{x}.$$

Letting $w = 1 - \sqrt{1 + u^2}$ we have $dw = -u\, du/\sqrt{1 + u^2}$ so that

$$-\ln\left|1 - \sqrt{1 + u^2}\right| = \ln|x| + c$$

$$\frac{1}{1 - \sqrt{1 + u^2}} = c_1 x$$

$$1 - \sqrt{1 + u^2} = -\frac{c_2}{x} \qquad (-c_2 = 1/c_1)$$

$$1 + \frac{c_2}{x} = \sqrt{1 + \frac{y^2}{x^2}}$$

$$1 + \frac{2c_2}{x} + \frac{c_2^2}{x^2} = 1 + \frac{y^2}{x^2}\,.$$

Solving for y^2 we have

$$y^2 = 2c_2 x + c_2^2 = 4\left(\frac{c_2}{2}\right)\left(x + \frac{c_2}{2}\right)$$

which is a family of parabolas symmetric with respect to the x-axis with vertex at $(-c_2/2, 0)$ and focus at the origin.

(b) Let $u = x^2 + y^2$ so that

$$\frac{du}{dx} = 2x + 2y\,\frac{dy}{dx}\,.$$

Then

$$y\,\frac{dy}{dx} = \frac{1}{2}\,\frac{du}{dx} - x$$

and the differential equation can be written in the form

$$\frac{1}{2}\,\frac{du}{dx} - x = -x + \sqrt{u} \quad \text{or} \quad \frac{1}{2}\,\frac{du}{dx} = \sqrt{u}\,.$$

Separating variables and integrating gives

$$\frac{du}{2\sqrt{u}} = dx$$

$$\sqrt{u} = x + c$$

$$u = x^2 + 2cx + c^2$$

$$x^2 + y^2 = x^2 + 2cx + c^2$$

$$y^2 = 2cx + c^2\,.$$

21. (a) With $c = 0.01$ the differential equation is $dP/dt = kP^{1.01}$. Separating variables and integrating we obtain

$$P^{-1.01}dP = k\,dt$$

$$\frac{P^{-0.01}}{-0.01} = kt + c_1$$

$$P^{-0.01} = -0.01kt + c_2$$

$$P(t) = (-0.01kt + c_2)^{-100}$$

$$P(0) = c_2^{-100} = 10$$

$$c_2 = 10^{-0.01}.$$

Then

$$P(t) = \frac{1}{(-0.01kt + 10^{-0.01})^{100}}$$

and, since P doubles in 5 months from 10 to 20,

$$P(5) = \frac{1}{(-0.01k(5) + 10^{-0.01})^{100}} = 20,$$

so

$$\left(-0.01k(5) + 10^{-0.01}\right)^{100} = \frac{1}{20}$$

$$-0.01k = \frac{\left[\left(\frac{1}{20}\right)^{1/100} - \left(\frac{1}{10}\right)^{1/100}\right]}{5}$$

$$= -0,001350.$$

Thus $P(t) = 1/\left(-0.001350t + 10^{-0.01}\right)^{100}$.

(b) Define $T = \left(\frac{1}{10}\right)^{1/100}/0.001350 \approx 724$ months $= 60$ years. As $t \to 724$ (from the left), $P \to \infty$.

(c) $P(50) = 1/\left[-0.001350(50) + 10^{-0.01}\right]^{100} \approx 12,839$ and

$P(100) = 1/\left[-0.001350(100) + 10^{-0.01}\right]^{100} \approx 28,630,966$

3.3 | Modeling with Systems of First-Order Differential Equations

The terminology and concepts listed below provide an outline of the main ideas encountered in this section. These can be useful when preparing for a quiz or test.

Terminology and Concepts

- systems of first-order DEs – linear and nonlinear
- radioactive decay series
- mixture models involving a system of DEs
- Lotka-Volterra predator-prey model
- competition models
- electrical networks involving a system of first-order DEs

The basic skills listed below summarize the more mechanical types of problems encountered in the exercise set for this section.

Basic Skills

- solve a system of DEs where one of the equations involves only a single unknown
- construct a mathematical model for mixtures involving multiple tanks
- use a computer program to numerically analyze solutions of a system of DEs

Use of Computers

There are a number of computer programs that can solve a DE or system of DEs numerically and graph the resulting numerical solution curves. We show below how *Mathematica* and *Maple* can produce graphs of numerical solutions.

In *Mathematica* **NDSolve** is used to obtain a numerical solution of an IVP which can then be graphed.

```
Clear[y]                                          (Mathematica)
sol=Flatten[NDSolve[{y'[x]==Exp[x/2]Sin[x^2], y[0]==0}, y, {x, -5, 5}]];
y[x_]:=Evaluate[y[x]/.sol]
y[x]
Plot[y[x], {x, -5, 5}]
```

Note that the interval $[-5, 5]$ specified in the **Plot** command can be no larger than the interval used in the **NDSolve** command. Following is an example of how *Mathematica* can be used to obtain a numerical solution of a system of DEs with initial conditions.

Clear[y]

sol=Flatten[NDSolve[{x'[t]== -0.3x[t] + 0.1x[t]y[t],

y'[t]==0.2y[t] - 0.1x[t]y[t], x[0]==6, y[0]==4}, {x, y}, {t, -2, 2}]];

x[t_]:=Evaluate[x[t]/.sol]

y[t_]:=Evaluate[y[t]/.sol]

x[t]

y[t]

Plot[{x[t], y[t]}, {t, -2, 2}]

In *Maple* there are two ways to plot the numerical solution curve of an IVP. The first way shown below returns a procedure that can be used with the **plots** package to graph the solution. It is also possible to define a function that can evaluate the solution at any point.

```
sol:=dsolve({diff(y(x),x)=exp(x/t)*sin(x^2),    (Maple)

y(0)=0},{y(x)}, type=numeric);

with(plots);

odeplots(sol,[x,y(x)],-5..5);

y1:=x1->eval(y(x),sol(x1));

y1(2.67);
```

The second way to plot a numerical solution of the DE uses the **DEtools** package.

```
with(DEtools);                                  (Maple)

DEplot(diff(y(x),x)=exp(x/2)*sin(x^2),y(x),-5..5,

    [[y(0)=0]],arrows=NONE,stepsize=0.05);
```

The **arrows** option suppresses a direction field in the background of the graph and the **stepsize** option smooths the graph out.

Maple can also graph the numerical solutions of a system of DEs with initial conditions. In this case we first specify the DEs and initial conditions before calling the **DEtools** package.

```
eq1:=diff(x(t),t)=-0.3*x(t)+0.1*x(t)*y(t);      (Maple)

eq2:=diff(y(t),t)=0.2*y(t)-0.1*x(t)*y(t);

with(DEtools);

gr1:=DEplot({eq1,eq2},[x(t),y(t)],-5..5,[[x(0)=6,y(0)=4]],

    scene=[t,x]):

gr2:=DEplot({eq1,eq2},[x(t),y(t)],-5..5,[[x(0)=6,y(0)=4]],

    scene=[t,y]):

display([gr1,gr2]);
```

Note the use of colons, rather than the usual `semicolons`, in the lines above defining `gr1` and `gr2`. This prevents the display of a very long list of uninteresting expressions.

Radioactive Series

3. The amounts x and y are the same at about $t = 5$ days. The amounts x and z are the same at about $t = 20$ days. The amounts y and z are the same at about $t = 147$ days. The time when y and z are the same makes sense because most of A and half of B are gone, so half of C should have been formed.

Mixtures

6. Let x_1, x_2, and x_3 be the amounts of salt in tanks A, B, and C, respectively, so that

$$x_1' = \frac{1}{100}x_2 \cdot 2 - \frac{1}{100}x_1 \cdot 6 = \frac{1}{50}x_2 - \frac{3}{50}x_1$$

$$x_2' = \frac{1}{100}x_1 \cdot 6 + \frac{1}{100}x_3 - \frac{1}{100}x_2 \cdot 2 - \frac{1}{100}x_2 \cdot 5 = \frac{3}{50}x_1 - \frac{7}{100}x_2 + \frac{1}{100}x_3$$

$$x_3' = \frac{1}{100}x_2 \cdot 5 - \frac{1}{100}x_3 - \frac{1}{100}x_3 \cdot 4 = \frac{1}{20}x_2 - \frac{1}{20}x_3.$$

Predator-Prey Models

9. Zooming in on the graph it can be seen that the populations are first equal at about $t = 5.6$. The approximate periods of x and y are both 45.

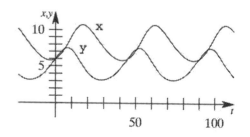

Networks

12. By Kirchhoff's first law we have $i_1 = i_2 + i_3$. By Kirchhoff's second law, on each loop we have $E(t) = Li_1' + R_1 i_2$ and $E(t) = Li_1' + R_2 i_3 + q/C$ so that $q = CR_1 i_2 - CR_2 i_3$. Then $i_3 = q' = CR_1 i_2' - CR_2 i_3$ so that the system is

$$Li_2' + Li_3' + R_1 i_2 = E(t)$$

$$-R_1 i_2' + R_2 i_3' + \frac{1}{C}i_3 = 0.$$

Additional Nonlinear Models

15. We first note that $s(t) + i(t) + r(t) = n$. Now the rate of change of the number of susceptible persons, $s(t)$, is proportional to the number of contacts between the number of people infected and the number who are susceptible; that is, $ds/dt = -k_1 si$. We use $-k_1 < 0$ because $s(t)$ is decreasing. Next, the rate of change of the number of persons who have recovered is proportional to the number infected; that is, $dr/dt = k_2 i$ where $k_2 > 0$ since r is increasing. Finally, to obtain di/dt we use

$$\frac{d}{dt}(s + i + r) = \frac{d}{dt}n = 0.$$

This gives

$$\frac{di}{dt} = -\frac{dr}{dt} - \frac{ds}{dt} = -k_2 i + k_1 si.$$

The system of differential equations is then

$$\frac{ds}{dt} = -k_1 si$$

$$\frac{di}{dt} = -k_2 i + k_1 si$$

$$\frac{dr}{dt} = k_2 i.$$

A reasonable set of initial conditions is $i(0) = i_0$, the number of infected people at time 0, $s(0) = n - i_0$, and $r(0) = 0$.

3.R Chapter 3 in Review

3. From $\dfrac{dP}{dt} = 0.018P$ and $P(0) = 4$ billion we obtain $P = 4e^{0.018t}$ so that $P(45) = 8.99$ billion.

6. From $V\, dC/dt = kA(C_s - C)$ and $C(0) = C_0$ we obtain $C = C_s + (C_0 - C_s)e^{-kAt/V}$.

9. We first solve $(1 - t/10)di/dt + 0.2i = 4$. Separating variables we obtain $di/(40 - 2i) = dt/(10 - t)$. Then

$$-\frac{1}{2}\ln|40 - 2i| = -\ln|10 - t| + c \quad \text{or} \quad \sqrt{40 - 2i} = c_1(10 - t).$$

Since $i(0) = 0$ we must have $c_1 = 2/\sqrt{10}$. Solving for i we get $i(t) = 4t - \frac{1}{5}t^2$, $0 \le t < 10$. For $t \ge 10$ the equation for the current becomes

$$i(t) = \begin{cases} 4t - \frac{1}{5}t^2, & 0 \le t < 10 \\ 20, & t \ge 10. \end{cases}$$

The graph of $i(t)$ is given in the figure.

12. In tank A the salt input is

$$\left(7\,\frac{\text{gal}}{\text{min}}\right)\left(2\,\frac{\text{lb}}{\text{gal}}\right) + \left(1\,\frac{\text{gal}}{\text{min}}\right)\left(\frac{x_2}{100}\,\frac{\text{lb}}{\text{gal}}\right) = \left(14 + \frac{1}{100}x_2\right)\frac{\text{lb}}{\text{min}}.$$

The salt output is

$$\left(3\,\frac{\text{gal}}{\text{min}}\right)\left(\frac{x_1}{100}\,\frac{\text{lb}}{\text{gal}}\right) + \left(5\,\frac{\text{gal}}{\text{min}}\right)\left(\frac{x_1}{100}\,\frac{\text{lb}}{\text{gal}}\right) = \frac{2}{25}x_1\,\frac{\text{lb}}{\text{min}}.$$

In tank B the salt input is

$$\left(5\,\frac{\text{gal}}{\text{min}}\right)\left(\frac{x_1}{100}\,\frac{\text{lb}}{\text{gal}}\right) = \frac{1}{20}x_1\,\frac{\text{lb}}{\text{min}}.$$

The salt output is

$$\left(1\,\frac{\text{gal}}{\text{min}}\right)\left(\frac{x_2}{100}\,\frac{\text{lb}}{\text{gal}}\right) + \left(4\,\frac{\text{gal}}{\text{min}}\right)\left(\frac{x_2}{100}\,\frac{\text{lb}}{\text{gal}}\right) = \frac{1}{20}x_2\,\frac{\text{lb}}{\text{min}}.$$

The system of differential equations is then

$$\frac{dx_1}{dt} = 14 + \frac{1}{100}x_2 - \frac{2}{25}x_1$$

$$\frac{dx_2}{dt} = \frac{1}{20}x_1 - \frac{1}{20}x_2.$$

15. (a) From the third differential equation in the statement of the problem in the text we have

$$\frac{dK}{dt} = -(\lambda_1 + \lambda_2)K \qquad \text{so} \qquad K(t) = c_1 e^{-(\lambda_1 + \lambda_2)t}.$$

Since $K(0) = K_0$ we have $K(t) = K_0 e^{-(\lambda_1 + \lambda_2)t}$. Next,

$$\frac{dC}{dt} = \lambda_1 K = \lambda_1 K_0 e^{-(\lambda_1 + \lambda_2)t} \qquad \text{so} \qquad C(t) = -\frac{\lambda_1}{\lambda_1 + \lambda_2} + c_2.$$

Since $C(0) = 0$ we have

$$c_2 = \frac{\lambda_1}{\lambda_1 + \lambda_2}K_0 \qquad \text{and} \qquad C(t) = \frac{\lambda_1}{\lambda_1 + \lambda_2}K_0\left[1 - e^{-(\lambda_1 + \lambda_2)t}\right].$$

Finally

$$\frac{dA}{dt} = \lambda_2 K = \lambda_2 K_0 e^{-(\lambda_1 + \lambda_2)t} \qquad \text{so} \qquad A(t) = -\frac{\lambda_1}{\lambda_1 + \lambda_2}K_0 e^{-(\lambda_1 + \lambda_2)t} + c_3.$$

Since $A(0) = 0$

$$c_3 = \frac{\lambda_2}{\lambda_1 + \lambda_2}K_0 \qquad \text{and} \qquad A(t) = \frac{\lambda_2}{\lambda_1 + \lambda_2}K_0\left[1 - e^{-(\lambda_1 + \lambda_2)t}\right].$$

(b) Since $\lambda_1 + \lambda_2 = 5.34 \times 10^{-10}$ we have

$$K(t) = K_0 e^{-0.000000000534t} = \frac{1}{2} K_0 \quad \text{and} \quad t = \frac{\ln \frac{1}{2}}{-0.000000000534} \approx 1.3 \times 10^9 \text{ years.}$$

(c) Since

$$\lim_{t \to \infty} C(t) = \frac{\lambda_1}{\lambda_1 + \lambda_2} \lim_{t \to \infty} K_0 \left[1 - e^{-(\lambda_1 + \lambda_2)t} \right] = \frac{\lambda_1}{\lambda_1 + \lambda_2} K_0$$

and

$$\lim_{t \to \infty} A(t) = \frac{\lambda_1}{\lambda_1 + \lambda_2} \lim_{t \to \infty} K_0 \left[1 - e^{-(\lambda_2 + \lambda_2)t} \right] = \frac{\lambda_2}{\lambda_1 + \lambda_2} K_0,$$

we have

$$\lim_{t \to \infty} C(t) = \frac{4.7526 \times 10^{-10}}{5.34 \times 10^{-10}} K_0 = 0.89 K_0 \quad \text{or} \quad 89\%$$

and

$$\lim_{t \to \infty} A(t) = \frac{0.5874 \times 10^{-10}}{5.34 \times 10^{-10}} K_0 = 0.11 K_0 \quad \text{or} \quad 11\%.$$

4 Higher-Order Differential Equations

4.1 | Preliminary Theory – Linear Equations

The terminology and concepts listed below provide an outline of the main ideas encountered in this section. These can be useful when preparing for a quiz or test.

Terminology and Concepts

- linear nth-order initial-value problem

- existence-uniqueness theorem for linear nth-order IVPs

- boundary-value problems for linear nth-order DEs

- no existence-uniqueness theorem for linear nth-order boundary-value problems

- homogeneous linear nth-order DEs

- representation of a DE using differential operators

- superposition principles

- linear dependence and linear independence of a set of functions

- Wronskian

- fundamental set of solutions

- general solution of a linear nth-order DE

- particular solution of a nonhomogeneous linear nth-order DE

- complementary function for a nonhomogeneous linear nth-order DE

The basic skills listed below summarize the more mechanical types of problems encountered in the exercise set for this section.

Basic Skills

- given the general solution of a linear nth-order DE, find a solution satisfying n initial conditions
- given a two-parameter family of solutions of a linear second-order DE, determine a member or members of the family satisfying specified boundary conditions (or show that no solution exists)
- determine whether a given set of functions is linearly independent on an interval
- show that a given set of functions forms a fundamental set of solutions on an interval

74

4.1.1 Initial-Value and Boundary-Value Problems

3. From $y = c_1 x + c_2 x \ln x$ we find $y' = c_1 + c_2(1 + \ln x)$. Then $y(1) = c_1 = 3$, $y'(1) = c_1 + c_2 = -1$ so that $c_1 = 3$ and $c_2 = -4$. The solution is $y = 3x - 4x \ln x$.

6. In this case we have $y(0) = c_1 = 0$, $y'(0) = 2c_2 \cdot 0 = 0$ so $c_1 = 0$ and c_2 is arbitrary. Two solutions are $y = x^2$ and $y = 2x^2$.

9. Since $a_2(x) = x - 2$ and $x_0 = 0$ the problem has a unique solution for $-\infty < x < 2$.

12. In this case we have $y(0) = c_1 = 1$, $y'(1) = 2c_2 = 6$ so that $c_1 = 1$ and $c_2 = 3$. The solution is $y = 1 + 3x^2$.

4.1.2 Homogeneous Equations

15. Since $(-4)x + (3)x^2 + (1)(4x - 3x^2) = 0$ the set of functions is linearly dependent.

18. Since $(1)\cos 2x + (1)1 + (-2)\cos^2 x = 0$ the set of functions is linearly dependent.

21. Suppose $c_1(1 + x) + c_2 x + c_3 x^2 = 0$. Then $c_1 + (c_1 + c_2)x + c_3 x^2 = 0$ and so $c_1 = 0$, $c_1 + c_2 = 0$, and $c_3 = 0$. Since $c_1 = 0$ we also have $c_2 = 0$. Thus, the set of functions is linearly independent.

24. The functions satisfy the differential equation and are linearly independent since

$$W(\cosh 2x,\ \sinh 2x) = 2$$

for $-\infty < x < \infty$. The general solution is

$$y = c_1 \cosh 2x + c_2 \sinh 2x.$$

27. The functions satisfy the differential equation and are linearly independent since

$$W\left(x^3,\ x^4\right) = x^6 \neq 0$$

for $0 < x < \infty$. The general solution on this interval is

$$y = c_1 x^3 + c_2 x^4.$$

30. The functions satisfy the differential equation and are linearly independent since

$$W(1,\ x,\ \cos x,\ \sin x) = 1$$

for $-\infty < x < \infty$. The general solution on this interval is

$$y = c_1 + c_2 x + c_3 \cos x + c_4 \sin x.$$

4.1.3 Nonhomogeneous Equations

33. The functions $y_1 = e^{2x}$ and $y_2 = xe^{2x}$ form a fundamental set of solutions of the associated homogeneous equation, and $y_p = x^2 e^{2x} + x - 2$ is a particular solution of the nonhomogeneous equation.

36. (a) $y_{p_1} = 5$ **(b)** $y_{p_2} = -2x$ **(c)** $y_p = y_{p_1} + y_{p_2} = 5 - 2x$ **(d)** $y_p = \frac{1}{2}y_{p_1} - 2y_{p_2} = \frac{5}{2} + 4x$

4.2 │ Reduction of Order

The terminology and concepts listed below provide an outline of the main ideas encountered in this section. These can be useful when preparing for a quiz or test.

Terminology and Concepts

- reduction of order

The basic skills listed below summarize the more mechanical types of problems encountered in the exercise set for this section.

Basic Skills

- given a solution y_1 of a linear second-order homogeneous DE, find a second solution y_2 using reduction of order

- given a solution y_1 of a linear second-order homogeneous DE, find a second solution y_2 using formula (5) in the text

In Problems 3 and 6 we use reduction of order to find a second solution. In Problems 9-18 we use formula (5) from the text.

3. Define $y = u(x)\cos 4x$ so

$$y' = -4u\sin 4x + u'\cos 4x, \quad y'' = u''\cos 4x - 8u'\sin 4x - 16u\cos 4x$$

and

$$y'' + 16y = (\cos 4x)u'' - 8(\sin 4x)u' = 0 \quad \text{or} \quad u'' - 8(\tan 4x)u' = 0.$$

If $w = u'$ we obtain the linear first-order equation $w' - 8(\tan 4x)w = 0$ which has the integrating factor $e^{-8\int \tan 4x\, dx} = \cos^2 4x$. Now

$$\frac{d}{dx}\left[(\cos^2 4x)w\right] = 0 \quad \text{gives} \quad (\cos^2 4x)w = c.$$

Therefore $w = u' = c \sec^2 4x$ and $u = c_1 \tan 4x$. A second solution is $y_2 = \tan 4x \cos 4x = \sin 4x$.

6. Define $y = u(x)e^{5x}$ so

$$y' = 5e^{5x}u + e^{5x}u', \quad y'' = e^{5x}u'' + 10e^{5x}u' + 25e^{5x}u$$

and

$$y'' - 25y = e^{5x}(u'' + 10u') = 0 \quad \text{or} \quad u'' + 10u' = 0.$$

If $w = u'$ we obtain the linear first-order equation $w' + 10w = 0$ which has the integrating factor $e^{10\int dx} = e^{10x}$. Now

$$\frac{d}{dx}\left[e^{10x}w\right] = 0 \quad \text{gives} \quad e^{10x}w = c.$$

Therefore $w = u' = ce^{-10x}$ and $u = c_1 e^{-10x}$. A second solution is $y_2 = e^{-10x}e^{5x} = e^{-5x}$.

9. Identifying $P(x) = -7/x$ we have

$$y_2 = x^4 \int \frac{e^{-\int(-7/x)\, dx}}{x^8}\, dx = x^4 \int \frac{1}{x}\, dx = x^4 \ln|x|.$$

A second solution is $y_2 = x^4 \ln|x|$.

12. Identifying $P(x) = 0$ we have

$$y_2 = x^{1/2} \ln x \int \frac{e^{-\int 0\, dx}}{x(\ln x)^2}\, dx = x^{1/2} \ln x \left(-\frac{1}{\ln x}\right) = -x^{1/2}.$$

A second solution is $y_2 = x^{1/2}$.

15. Identifying $P(x) = 2(1+x)/\left(1 - 2x - x^2\right)$ we have

$$y_2 = (x+1) \int \frac{e^{-\int 2(1+x)dx/\left(1-2x-x^2\right)}}{(x+1)^2}\, dx = (x+1) \int \frac{e^{\ln\left(1-2x-x^2\right)}}{(x+1)^2}\, dx$$

$$= (x+1) \int \frac{1 - 2x - x^2}{(x+1)^2}\, dx = (x+1) \int \left[\frac{2}{(x+1)^2} - 1\right] dx$$

$$= (x+1)\left[-\frac{2}{x+1} - x\right] = -2 - x^2 - x.$$

A second solution is $y_2 = x^2 + x + 2$.

18. Define $y = u(x) \cdot 1$ so

$$y' = u', \quad y'' = u'' \quad \text{and} \quad y'' + y' = u'' + u' = 1.$$

If $w = u'$ we obtain the linear first-order equation $w' + w = 1$ which has the integrating factor $e^{\int dx} = e^x$. Now

$$\frac{d}{dx}[e^x w] = e^x \quad \text{gives} \quad e^x w = e^x + c.$$

Therefore $w = u' = 1 + ce^{-x}$ and $u = x + c_1 e^{-x} + c_2$. The general solution is

$$y = u = x + c_1 e^{-x} + c_2.$$

4.3 | Homogeneous Linear Equations with Constant Coefficients

The terminology and concepts listed below provide an outline of the main ideas encountered in this section. These can be useful when preparing for a quiz or test.

Terminology and Concepts

- auxiliary equation

The basic skills listed below summarize the more mechanical types of problems encountered in the exercise set for this section.

Basic Skills

- find the general solution of a homogeneous linear DE with constant coefficients when the roots of the auxiliary equation are real and distinct, real and repeated, conjugate complex, and repeated conjugate complex

- find the solution of an nth-order homogeneous linear DE with constant coefficients and n initial conditions

Factorization of Polynomials   In Section 4.3 in the text the solution method for homogeneous linear DEs with constant coefficents relies on your ability to solve polynomial equations. For example, you know that a quadratic polynomial equation

$$am^2 + bm + c = 0, \quad a \neq 0, \ a, b, c \text{ real constants},$$

can always be solved using the quadratic formula. The two roots m_1 and m_2 are

$$m_1 = \frac{-b + \sqrt{b^2 - 4ac}}{2a} \quad \text{and} \quad m_2 = \frac{-b - \sqrt{b^2 - 4ac}}{2a}.$$

The roots m_1 and m_2 are also called the **zeros** of the polynomial $am^2 + bm + c$. Note that m_1 and m_2 are real and distinct ($m_1 \neq m_2$) if $b^2 - 4ac > 0$, real and equal ($m_1 = m_2$) if $b^2 - 4ac = 0$, and finally, m_1 and m_2 are complex conjugates ($m_1 = \alpha + i\beta$, $m_1 = \alpha - i\beta$, α and β are positive real numbers) if $b^2 - 4ac < 0$. The Factor Theorem of Algebra states that finding zeros of a polynomial and factoring the polynomial are equivalent problems. Thus, in the case of a quadratic polynomial $am^2 + bm + c$ we can write

$$am^2 + bm + c = a(m - m_1)(m - m_2).$$

A polynomial of degree n, $a_n m^n + a_{n-1} m^{n-1} + \cdots + a_1 m + a_0$ with real coefficients has n zeros (counting multiplicities) and hence can be factored as

$$a_n m^n + a_{n-1} m^{n-1} + \cdots + a_1 m + a_0 = a_n(m - m_1)(m - m_2) \cdots (m - m_n).$$

Finding the zeros requires that we find roots of a polynomial equation of degree > 2, and *that* is the fundamental problem here. Analogous to quadratic equations, cubic and quartic polynomial equations can be solved by (complicated) formulas involving radicals, but there exist no such formulas for solving polynomial equations of degree $n > 4$. But if the polynomial has *integer* coefficients and *if* p/q is a rational zero (p and q are integers and $q \neq 0$), then it can be proved that

$$p \text{ must be an integer factor of } a_0,$$

and

$$q \text{ must be an integer factor of } a_n.$$

By forming all possible quotients of each factor of a_0 to each factor of a_n, we can construct a list of all *possible* rational zeros. For example, consider the fourth-degree polynomial equation $3m^4 - 10m^3 - 3m^2 + 8m - 2 = 0$. We identify $a_0 = -2$ and $a_n = 3$ and list all integer factors of a_0 and a_n, respectively:

$$p : \pm 1, \pm 2 \qquad \text{and} \qquad q : \pm 1, \pm 3.$$

The eight possible rational roots of the equation are then

$$\frac{p}{q} : -1, 1, -\frac{1}{3}, \frac{1}{3}, -2, 2, -\frac{2}{3}, \frac{2}{3}.$$

To determine which, if any, of these numbers are roots, we could use brute force substitution of a number p/q into the polynomial to see whether it satisfies the equation. We could also use long division. If the term $m - p/q$ divides the polynomial evenly, that is, the remainder r is zero, then $m - p/q$ is a factor and hence p/q is a root. A short-hand version of long division, called **synthetic division**, is an efficient way of proceeding. We begin by testing -1 by synthetically dividing by $m - (-1) = m + 1$:

$$
\begin{array}{r|rrrrr}
-1 & 3 & -10 & -3 & 8 & -2 \\
 & & -3 & 13 & -10 & 2 \\
\hline
 & 3 & -13 & 10 & -2 & \boxed{\,0 = r} \\
\end{array}
$$

Since the remainder upon division is $r = 0$, we can conclude $m + 1$ is a factor of the polynomial and so -1 is a root. Now the division shows that

$$3m^4 - 10m^3 - 3m^2 + 8m - 2 = (m+1)(3m^3 - 13m^2 + 10m - 2).$$

Observe that if the original polynomial has additional rational zeros they must be zeros of the second, or cubic, factor $3m^3 - 13m^2 + 10m - 2$. Thus we continue the testing by dividing this factor by $m - 1$:

$$
\begin{array}{r|rrrr}
1 & 3 & -13 & 10 & -2 \\
 & & 3 & -10 & 0 \\
\hline
 & 3 & -10 & 0 & \boxed{-2 = r} \\
\end{array}
$$

The foregoing division shows that $m - 1$ is not a factor since the remainder r is not zero, and so 1 is not a rational root. Checking $\frac{1}{3}$ we have

$$
\begin{array}{r|rrrr}
\frac{1}{3} & 3 & -13 & 10 & -2 \\
 & & 1 & -4 & 2 \\
\hline
 & 3 & -12 & 6 & \boxed{\,0 = r} \\
\end{array}
$$

Thus, $\frac{1}{3}$ is a root, and we now have the factorization

$$
\begin{aligned}
3m^4 - 10m^3 - 3m^2 + 8m - 2 &= (m+1)\left(m - \frac{1}{3}\right)(3m^2 - 12m + 6) \\
&= (m+1)\left(m - \frac{1}{3}\right)(3)(m^2 - 4m + 2) \\
&= (m+1)(3m - 1)(m^2 - 4m + 2).
\end{aligned}
$$

Since the factor $m^2 - 4m + 2$ discovered by the division is quadratic, the remaining roots, $2 + \sqrt{2}$ and $2 - \sqrt{2}$, can be determined from the quadratic formula. Therefore, the original polynomial equation has four real roots: two rational and two irrational.

3. From $m^2 - m - 6 = 0$ we obtain $m = 3$ and $m = -2$ so that $y = c_1 e^{3x} + c_2 e^{-2x}$.

6. From $m^2 - 10m + 25 = 0$ we obtain $m = 5$ and $m = 5$ so that $y = c_1 e^{5x} + c_2 x e^{5x}$.

9. From $m^2 + 9 = 0$ we obtain $m = 3i$ and $m = -3i$ so that $y = c_1 \cos 3x + c_2 \sin 3x$.

12. From $2m^2 + 2m + 1 = 0$ we obtain $m = -1/2 \pm i/2$ so that

$$y = e^{-x/2}[c_1 \cos(x/2) + c_2 \sin(x/2)].$$

15. From $m^3 - 4m^2 - 5m = 0$ we obtain $m = 0$, $m = 5$, and $m = -1$ so that

$$y = c_1 + c_2 e^{5x} + c_3 e^{-x}.$$

18. From $m^3 + 3m^2 - 4m - 12 = 0$ we obtain $m = -2$, $m = 2$, and $m = -3$ so that

$$y = c_1 e^{-2x} + c_2 e^{2x} + c_3 e^{-3x}.$$

21. From $m^3 + 3m^2 + 3m + 1 = 0$ we obtain $m = -1$, $m = -1$, and $m = -1$ so that

$$y = c_1 e^{-x} + c_2 x e^{-x} + c_3 x^2 e^{-x}.$$

24. From $m^4 - 2m^2 + 1 = 0$ we obtain $m = 1$, $m = 1$, $m = -1$, and $m = -1$ so that

$$y = c_1 e^x + c_2 x e^x + c_3 e^{-x} + c_4 x e^{-x}.$$

27. From $m^5 + 5m^4 - 2m^3 - 10m^2 + m + 5 = 0$ we obtain $m = -1$, $m = -1$, $m = 1$, and $m = 1$, and $m = -5$ so that

$$u = c_1 e^{-r} + c_2 r e^{-r} + c_3 e^r + c_4 r e^r + c_5 e^{-5r}.$$

30. From $m^2 + 1 = 0$ we obtain $m = \pm i$ so that $y = c_1 \cos\theta + c_2 \sin\theta$. If $y(\pi/3) = 0$ and $y'(\pi/3) = 2$ then

$$\frac{1}{2}c_1 + \frac{\sqrt{3}}{2}c_2 = 0$$

$$-\frac{\sqrt{3}}{2}c_1 + \frac{1}{2}c_2 = 2,$$

so $c_1 = -\sqrt{3}$, $c_2 = 1$, and $y = -\sqrt{3}\cos\theta + \sin\theta$.

33. From $m^2 + m + 2 = 0$ we obtain $m = -1/2 \pm \sqrt{7}\,i/2$ so that

$$y = e^{-x/2}[c_1 \cos(\sqrt{7}\,x/2) + c_2 \sin(\sqrt{7}\,x/2)].$$

If $y(0) = 0$ and $y'(0) = 0$ then $c_1 = 0$ and $c_2 = 0$ so that $y = 0$.

36. From $m^3 + 2m^2 - 5m - 6 = 0$ we obtain $m = -1$, $m = 2$, and $m = -3$ so that

$$y = c_1 e^{-x} + c_2 e^{2x} + c_3 e^{-3x}.$$

If $y(0) = 0$, $y'(0) = 0$, and $y''(0) = 1$ then

$$c_1 + c_2 + c_3 = 0, \quad -c_1 + 2c_2 - 3c_3 = 0, \quad c_1 + 4c_2 + 9c_3 = 1,$$

so $c_1 = -1/6$, $c_2 = 1/15$, $c_3 = 1/10$, and

$$y = -\frac{1}{6} e^{-x} + \frac{1}{15} e^{2x} + \frac{1}{10} e^{-3x}.$$

39. From $m^2 + 1 = 0$ we obtain $m = \pm i$ so that $y = c_1 \cos x + c_2 \sin x$ and $y' = -c_1 \sin x + c_2 \cos x$. From $y'(0) = c_1(0) + c_2(1) = c_2 = 0$ and $y'(\pi/2) = -c_1(1) = 0$ we find $c_1 = c_2 = 0$. A solution of the boundary-value problem is $y = 0$.

42. The auxiliary equation is $m^2 - 1 = 0$ which has roots -1 and 1. By (10) the general solution is $y = c_1 e^x + c_2 e^{-x}$. By (11) the general solution is $y = c_1 \cosh x + c_2 \sinh x$. For $y = c_1 e^x + c_2 e^{-x}$ the boundary conditions imply $c_1 + c_2 = 1$, $c_1 e - c_2 e^{-1} = 0$. Solving for c_1 and c_2 we find $c_1 = 1/(1+e^2)$ and $c_2 = e^2/(1+e^2)$ so $y = e^x/(1+e^2) + e^2 e^{-x}/(1+e^2)$. For $y = c_1 \cosh x + c_2 \sinh x$ the boundary conditions imply $c_1 = 1$, $c_2 = -\tanh 1$, so $y = \cosh x - (\tanh 1) \sinh x$.

45. The auxiliary equation should have a pair of complex roots $\alpha \pm \beta i$ where $\alpha < 0$, so that the solution has the form $e^{\alpha x}(c_1 \cos \beta x + c_2 \sin \beta x)$. Thus, the differential equation is **(e)**.

48. The differential equation should have the form $y'' + k^2 y = 0$ where $k = 2$ so that the period of the solution is π. Thus, the differential equation is **(b)**.

51. We have $m(m - 2) = m^2 - 2m$, so the differential equation is $y'' - 2y' = 0$.

54. We have $(m - 7)(m + 7) = m^2 - 49$, so the differential equation is $y'' - 49y = 0$.

57. We have $m^2(m - 8) = m3 - 8m^2$, so the differential equation is $y''' - 8y'' = 0$.

4.4 | Undetermined Coefficients—Superposition Approach

The terminology and concepts listed below provide an outline of the main ideas encountered in this section. These can be useful when preparing for a quiz or test.

Terminology and Concepts

- undetermined coefficients
- superposition principle
- multiplication rule

The basic skills listed below summarize the more mechanical types of problems encountered in the exercise set for this section.

Basic Skills

- use undetermined coefficients with the superposition approach to find a particular solution of a nonhomogeneous linear DE with constant coefficients when (1) no function in the assumed particular solution is a solution of the associated homogeneous DE, and (2) a function in the assumed particular solution is also a solution of the associated homogeneous DE

- recognize when the input function (right-hand side of the DE) is appropriate for the method of undetermined coefficients.

3. From $m^2 - 10m + 25 = 0$ we find $m_1 = m_2 = 5$. Then $y_c = c_1 e^{5x} + c_2 x e^{5x}$ and we assume that $y_p = Ax + B$. Substituting into the differential equation we obtain $25A = 30$ and $-10A + 25B = 3$. Then $A = \frac{6}{5}$, $B = \frac{3}{5}$, $y_p = \frac{6}{5}x + \frac{3}{5}$, and

$$y = c_1 e^{5x} + c_2 x e^{5x} + \frac{6}{5}x + \frac{3}{5}.$$

6. From $m^2 - 8m + 20 = 0$ we find $m_1 = 4 + 2i$ and $m_2 = 4 - 2i$. Then $y_c = e^{4x}(c_1 \cos 2x + c_2 \sin 2x)$ and we assume $y_p = Ax^2 + Bx + C + (Dx + E)e^x$. Substituting into the differential equation we obtain

$$2A - 8B + 20C = 0$$

$$-6D + 13E = 0$$

$$-16A + 20B = 0$$

$$13D = -26$$

$$20A = 100.$$

Then $A = 5$, $B = 4$, $C = \frac{11}{10}$, $D = -2$, $E = -\frac{12}{13}$, $y_p = 5x^2 + 4x + \frac{11}{10} + \left(-2x - \frac{12}{13}\right)e^x$ and

$$y = e^{4x}(c_1 \cos 2x + c_2 \sin 2x) + 5x^2 + 4x + \frac{11}{10} + \left(-2x - \frac{12}{13}\right)e^x.$$

9. From $m^2 - m = 0$ we find $m_1 = 1$ and $m_2 = 0$. Then $y_c = c_1 e^x + c_2$ and we assume $y_p = Ax$. Substituting into the differential equation we obtain $-A = -3$. Then $A = 3$, $y_p = 3x$ and $y = c_1 e^x + c_2 + 3x$.

12. From $m^2 - 16 = 0$ we find $m_1 = 4$ and $m_2 = -4$. Then $y_c = c_1 e^{4x} + c_2 e^{-4x}$ and we assume $y_p = Axe^{4x}$. Substituting into the differential equation we obtain $8A = 2$. Then $A = \frac{1}{4}$, $y_p = \frac{1}{4}xe^{4x}$ and

$$y = c_1 e^{4x} + c_2 e^{-4x} + \frac{1}{4}xe^{4x}.$$

15. From $m^2 + 1 = 0$ we find $m_1 = i$ and $m_2 = -i$. Then $y_c = c_1 \cos x + c_2 \sin x$ and we assume $y_p = (Ax^2 + Bx)\cos x + (Cx^2 + Dx)\sin x$. Substituting into the differential equation we obtain $4C = 0$, $2A + 2D = 0$, $-4A = 2$, and $-2B + 2C = 0$. Then $A = -\frac{1}{2}$, $B = 0$, $C = 0$, $D = \frac{1}{2}$, $y_p = -\frac{1}{2}x^2 \cos x + \frac{1}{2}x \sin x$, and

$$y = c_1 \cos x + c_2 \sin x - \frac{1}{2}x^2 \cos x + \frac{1}{2}x \sin x.$$

18. From $m^2 - 2m + 2 = 0$ we find $m_1 = 1 + i$ and $m_2 = 1 - i$. Then $y_c = e^x(c_1 \cos x + c_2 \sin x)$ and we assume $y_p = Ae^{2x}\cos x + Be^{2x}\sin x$. Substituting into the differential equation we obtain $A + 2B = 1$ and $-2A + B = -3$. Then $A = \frac{7}{5}$, $B = -\frac{1}{5}$, $y_p = \frac{7}{5}e^{2x}\cos x - \frac{1}{5}e^{2x}\sin x$ and

$$y = e^x(c_1 \cos x + c_2 \sin x) + \frac{7}{5}e^{2x}\cos x - \frac{1}{5}e^{2x}\sin x.$$

21. From $m^3 - 6m^2 = 0$ we find $m_1 = m_2 = 0$ and $m_3 = 6$. Then $y_c = c_1 + c_2 x + c_3 e^{6x}$ and we assume $y_p = Ax^2 + B\cos x + C\sin x$. Substituting into the differential equation we obtain $-12A = 3$, $6B - C = -1$, and $B + 6C = 0$. Then $A = -\frac{1}{4}$, $B = -\frac{6}{37}$, $C = \frac{1}{37}$, $y_p = -\frac{1}{4}x^2 - \frac{6}{37}\cos x + \frac{1}{37}\sin x$, and

$$y = c_1 + c_2 x + c_3 e^{6x} - \frac{1}{4}x^2 - \frac{6}{37}\cos x + \frac{1}{37}\sin x.$$

24. From $m^3 - m^2 - 4m + 4 = 0$ we find $m_1 = 1$, $m_2 = 2$, and $m_3 = -2$. Then

$$y_c = c_1 e^x + c_2 e^{2x} + c_3 e^{-2x}$$

and we assume $y_p = A + Bxe^x + Cxe^{2x}$. Substituting into the differential equation we obtain $4A = 5$, $-3B = -1$, and $4C = 1$. Then $A = \frac{5}{4}$, $B = \frac{1}{3}$, $C = \frac{1}{4}$, $y_p = \frac{5}{4} + \frac{1}{3}xe^x + \frac{1}{4}xe^{2x}$, and

$$y = c_1 e^x + c_2 e^{2x} + c_3 e^{-2x} + \frac{5}{4} + \frac{1}{3}xe^x + \frac{1}{4}xe^{2x}.$$

27. We have $y_c = c_1 \cos 2x + c_2 \sin 2x$ and we assume $y_p = A$. Substituting into the differential equation we find $A = -\frac{1}{2}$. Thus $y = c_1 \cos 2x + c_2 \sin 2x - \frac{1}{2}$. From the initial conditions we obtain $c_1 = 0$ and $c_2 = \sqrt{2}$, so $y = \sqrt{2} \sin 2x - \frac{1}{2}$.

30. We have $y_c = c_1 e^{-2x} + c_2 x e^{-2x}$ and we assume $y_p = (Ax^3 + Bx^2)e^{-2x}$. Substituting into the differential equation we find $A = \frac{1}{6}$ and $B = \frac{3}{2}$. Thus $y = c_1 e^{-2x} + c_2 x e^{-2x} + \left(\frac{1}{6}x^3 + \frac{3}{2}x^2\right)e^{-2x}$. From the initial conditions we obtain $c_1 = 2$ and $c_2 = 9$, so

$$y = 2e^{-2x} + 9xe^{-2x} + \left(\frac{1}{6}x^3 + \frac{3}{2}x^2\right)e^{-2x}.$$

33. We have $x_c = c_1 \cos \omega t + c_2 \sin \omega t$ and we assume $x_p = At \cos \omega t + Bt \sin \omega t$. Substituting into the differential equation we find $A = -F_0/2\omega$ and $B = 0$. Thus

$$x = c_1 \cos \omega t + c_2 \sin \omega t - (F_0/2\omega)t \cos \omega t.$$

36. We have $y_c = c_1 e^{-2x} + e^x(c_2 \cos \sqrt{3}\,x + c_3 \sin \sqrt{3}\,x)$ and we assume $y_p = Ax + B + Cxe^{-2x}$. Substituting into the differential equation we find $A = \frac{1}{4}$, $B = -\frac{5}{8}$, and $C = \frac{2}{3}$. Thus

$$y = c_1 e^{-2x} + e^x(c_2 \cos \sqrt{3}\,x + c_3 \sin \sqrt{3}\,x) + \frac{1}{4}x - \frac{5}{8} + \frac{2}{3}xe^{-2x}.$$

From the initial conditions we obtain $c_1 = -\frac{23}{12}$, $c_2 = -\frac{59}{24}$, and $c_3 = \frac{17}{72}\sqrt{3}$, so

$$y = -\frac{23}{12}e^{-2x} + e^x\left(-\frac{59}{24}\cos \sqrt{3}\,x + \frac{17}{72}\sqrt{3}\sin \sqrt{3}\,x\right) + \frac{1}{4}x - \frac{5}{8} + \frac{2}{3}xe^{-2x}.$$

39. The general solution of the differential equation $y'' + 3y = 6x$ is $y = c_1 \cos \sqrt{3}x + c_2 \sin \sqrt{3}x + 2x$. The condition $y(0) = 0$ implies $c_1 = 0$ and so $y = c_2 \sin \sqrt{3}x + 2x$. The condition $y(1) + y'(1) = 0$ implies $c_2 \sin \sqrt{3} + 2 + c_2\sqrt{3}\cos \sqrt{3} + 2 = 0$ so $c_2 = -4/(\sin \sqrt{3} + \sqrt{3}\cos \sqrt{3})$. The solution is

$$y = \frac{-4\sin \sqrt{3}x}{\sin \sqrt{3} + \sqrt{3}\cos \sqrt{3}} + 2x.$$

42. We have $y_c = e^x(c_1 \cos 3x + c_2 \sin 3x)$ and we assume $y_p = A$ on $[0, \pi]$. Substituting into the differential equation we find $A = 2$. Thus, $y = e^x(c_1 \cos 3x + c_2 \sin 3x) + 2$ on $[0, \pi]$. On (π, ∞) we have $y = e^x(c_3 \cos 3x + c_4 \sin 3x)$. From $y(0) = 0$ and $y'(0) = 0$ we obtain

$$c_1 = -2, \qquad c_1 + 3c_2 = 0.$$

Solving this system, we find $c_1 = -2$ and $c_2 = \frac{2}{3}$. Thus $y = e^x(-2\cos 3x + \frac{2}{3}\sin 3x) + 2$ on $[0, \pi]$. Now, continuity of y at $x = \pi$ implies

$$e^\pi\left(-2\cos 3\pi + \frac{2}{3}\sin 3\pi\right) + 2 = e^\pi(c_3 \cos 3\pi + c_4 \sin 3\pi)$$

or $2 + 2e^\pi = -c_3 e^\pi$ or $c_3 = -2e^{-\pi}(1 + e^\pi)$. Continuity of y' at π implies

$$\frac{20}{3} e^\pi \sin 3\pi = e^\pi [(c_3 + 3c_4) \cos 3\pi + (-3c_3 + c_4) \sin 3\pi]$$

or $-c_3 e^\pi - 3c_4 e^\pi = 0$. Since $c_3 = -2e^{-\pi}(1 + e^\pi)$ we have $c_4 = \frac{2}{3} e^{-\pi}(1 + e^\pi)$. The solution of the initial-value problem is

$$y(x) = \begin{cases} e^x(-2 \cos 3x + \frac{2}{3} \sin 3x) + 2, & 0 \le x \le \pi \\ (1 + e^\pi)e^{x-\pi}(-2 \cos 3x + \frac{2}{3} \sin 3x), & x > \pi. \end{cases}$$

4.5 | Undetermined Coefficients—Annihilator Approach

The terminology and concepts listed below provide an outline of the main ideas encountered in this section. These can be useful when preparing for a quiz or test.

Terminology and Concepts

- annihilator operator

- method of undetermined coefficients

The basic skills listed below summarize the more mechanical types of problems encountered in the exercise set for this section.

Basic Skills

- write a DE using differential operator notation

- find a differential operator that annihilates a given function

- find linearly independent functions that are annihilated by a given differential operator

- solve a nonhomogeneous linear DE with constant coefficients using annihilators and undetermined coefficients

3. $(D^2 - 4D - 12)y = (D - 6)(D + 2)y = x - 6$

6. $(D^3 + 4D)y = D(D^2 + 4)y = e^x \cos 2x$

9. $(D^4 + 8D)y = D(D + 2)(D^2 - 2D + 4)y = 4$

12. $(2D - 1)y = (2D - 1)4e^{x/2} = 8De^{x/2} - 4e^{x/2} = 4e^{x/2} - 4e^{x/2} = 0$

15. D^4 because of x^3

18. $D^2(D - 6)^2$ because of x and xe^{6x}

21. $D^3(D^2 + 16)$ because of x^2 and $\sin 4x$

24. $D(D-1)(D-2)$ because of 1, e^x, and e^{2x}

27. 1, x, x^2, x^3, x^4

30. $D^2 - 9D - 36 = (D-12)(D+3)$; e^{12x}, e^{-3x}

33. $D^3 - 10D^2 + 25D = D(D-5)^2$; 1, e^{5x}, xe^{5x}

36. Applying D to the differential equation we obtain

$$D(2D^2 - 7D + 5)y = 0.$$

Then

$$y = \underbrace{c_1 e^{5x/2} + c_2 e^x}_{y_c} + c_3$$

and $y_p = A$. Substituting y_p into the differential equation yields $5A = -29$ or $A = -29/5$. The general solution is

$$y = c_1 e^{5x/2} + c_2 e^x - \frac{29}{5}.$$

39. Applying D^2 to the differential equation we obtain

$$D^2(D^2 + 4D + 4)y = D^2(D+2)^2 y = 0.$$

Then

$$y = \underbrace{c_1 e^{-2x} + c_2 x e^{-2x}}_{y_c} + c_3 + c_4 x$$

and $y_p = Ax + B$. Substituting y_p into the differential equation yields $4Ax + (4A+4B) = 2x+6$. Equating coefficients gives

$$4A = 2$$

$$4A + 4B = 6.$$

Then $A = 1/2$, $B = 1$, and the general solution is

$$y = c_1 e^{-2x} + c_2 x e^{-2x} + \frac{1}{2}x + 1.$$

42. Applying D^4 to the differential equation we obtain

$$D^4(D^2 - 2D + 1)y = D^4(D-1)^2 y = 0.$$

Then

$$y = \underbrace{c_1 e^x + c_2 x e^x}_{y_c} + c_3 x^3 + c_4 x^2 + c_5 x + c_6$$

and $y_p = Ax^3 + Bx^2 + Cx + E$. Substituting y_p into the differential equation yields

$$Ax + (B - 6A)x + (6A - 4B + C)x + (2B - 2C + E) = x + 4x.$$

Equating coefficients gives

$$A = 1$$

$$B - 6A = 0$$

$$6A - 4B + C = 4$$

$$2B - 2C + E = 0.$$

Then $A = 1$, $B = 6$, $C = 22$, $E = 32$, and the general solution is

$$y = c_1 e^x + c_2 x e^x + x^3 + 6x^2 + 22x + 32.$$

45. Applying $D(D-1)$ to the differential equation we obtain

$$D(D-1)(D^2 - 2D - 3)y = D(D-1)(D+1)(D-3)y = 0.$$

Then

$$y = \underbrace{c_1 e^{3x} + c_2 e^{-x}}_{y_c} + c_3 e^x + c_4$$

and $y_p = A e^x + B$. Substituting y_p into the differential equation yields $-4A e^x - 3B = 4e^x - 9$. Equating coefficients gives $A = -1$ and $B = 3$. The general solution is

$$y = c_1 e^{3x} + c_2 e^{-x} - e^x + 3.$$

48. Applying $D(D^2 + 1)$ to the differential equation we obtain

$$D(D^2 + 1)(D^2 + 4)y = 0.$$

Then

$$y = \underbrace{c_1 \cos 2x + c_2 \sin 2x}_{y_c} + c_3 \cos x + c_4 \sin x + c_5$$

and $y_p = A \cos x + B \sin x + C$. Substituting y_p into the differential equation yields

$$3A \cos x + 3B \sin x + 4C = 4 \cos x + 3 \sin x - 8.$$

Equating coefficients gives $A = 4/3$, $B = 1$, and $C = -2$. The general solution is

$$y = c_1 \cos 2x + c_2 \sin 2x + \frac{4}{3} \cos x + \sin x - 2.$$

51. Applying $D(D-1)^3$ to the differential equation we obtain

$$D(D-1)^3(D^2 - 1)y = D(D-1)^4(D+1)y = 0.$$

Then

$$y = \underbrace{c_1 ex + c_2 e-x}_{y_c} + c_3 x3ex + c_4 x2ex + c_5 xex + c_6$$

and $y_p = A x^3 e^x + B x^2 e^x + C x e^x + E$. Substituting y_p into the differential equation yields

$$6Ax^2 e^x \qquad A + 4B \; xe^x \qquad B + 2C \; e^x \qquad E = x^2 e^x + 5$$

Equating coefficients gives

$$6A = 1$$

$$6A + 4B = 0$$

$$2B + 2C = 0$$

$$-E = 5.$$

Then $A = 1/6$, $B = -1/4$, $C = 1/4$, $E = -5$, and the general solution is

$$y = c_1 e^x + c_2 e^{-x} + \frac{1}{6}x^3 e^x - \frac{1}{4}x^2 e^x + \frac{1}{4}xe^x - 5.$$

54. Applying $D^2 - 2D + 10$ to the differential equation we obtain

$$(D^2 - 2D + 10)\left(D^2 + D + \frac{1}{4}\right)y = (D^2 - 2D + 10)\left(D + \frac{1}{2}\right)^2 y = 0.$$

Then

$$y = \underbrace{c_1 e^{-x/2} + c_2 xe^{-x/2}}_{y_c} + c_3 e^x \cos 3x + c_4 e^x \sin 3x$$

and $y_p = Ae^x \cos 3x + Be^x \sin 3x$. Substituting y_p into the differential equation yields

$$(9B - 27A/4)e^x \cos 3x - (9A + 27B/4)e^x \sin 3x = -e^x \cos 3x + e^x \sin 3x.$$

Equating coefficients gives

$$-\frac{27}{4}A + 9B = -1$$

$$-9A - \frac{27}{4}B = 1.$$

Then $A = -4/225$, $B = -28/225$, and the general solution is

$$y = c_1 e^{-x/2} + c_2 xe^{-x/2} - \frac{4}{225}e^x \cos 3x - \frac{28}{225}e^x \sin 3x.$$

57. Applying $(D^2 + 1)^2$ to the differential equation we obtain

$$(D^2 + 1)^2(D^2 + D + 1) = 0.$$

Then

$$y = e^{-x/2}\underbrace{\left[c_1 \cos \frac{\sqrt{3}}{2}x + c_2 \sin \frac{\sqrt{3}}{2}x\right] + c_3 \cos x + c_4 \sin x + c_5 x \cos x + c_6 x \sin x}_{y_c}$$

and $y_p = A\cos x + B\sin x + Cx\cos x + Ex\sin x$. Substituting y_p into the differential equation

yields

$$(B + C + 2E)\cos x + Ex\cos x + (-A - 2C + E)\sin x - Cx\sin x = x\sin x.$$

Equating coefficients gives

$$B + C + 2E = 0$$

$$E = 0$$

$$-A - 2C + E = 0$$

$$-C = 1.$$

Then $A = 2$, $B = 1$, $C = -1$, and $E = 0$, and the general solution is

$$y = e^{-x/2}\left[c_1 \cos \frac{\sqrt{3}}{2}x + c_2 \sin \frac{\sqrt{3}}{2}x\right] + 2\cos x + \sin x - x\cos x.$$

60. Applying $D(D-1)^2(D+1)$ to the differential equation we obtain

$$D(D-1)^2(D+1)(D^3 - D^2 + D - 1) = D(D-1)^3(D+1)(D^2+1) = 0.$$

Then

$$y = \underbrace{c_1 e^x + c_2 \cos x + c_3 \sin x}_{y_c} + c_4 + c_5 e^{-x} + c_6 x e^x + c_7 x^2 e^x$$

and $y_p = A + Be^{-x} + Cxe^x + Ex^2 e^x$. Substituting y_p into the differential equation yields

$$4Exe^x + (2C + 4E)e^x - 4Be^{-x} - A = xe^x - e^{-x} + 7.$$

Equating coefficients gives

$$4E = 1$$

$$2C + 4E = 0$$

$$-4B = -1$$

$$-A = 7.$$

Then $A = -7$, $B = 1/4$, $C = -1/2$, and $E = 1/4$, and the general solution is

$$y = c_1 e^x + c_2 \cos x + c_3 \sin x - 7 + \frac{1}{4}e^{-x} - \frac{1}{2}xe^x + \frac{1}{4}x^2 e^x.$$

63. Applying $D(D-1)$ to the differential equation we obtain

$$D(D-1)(D^4 - 2D^3 + D^2) = D^3(D-1)^3 = 0.$$

Then

$$y = \underbrace{c_1 + c_2 x + c_3 e^x + c_4 x e^x}_{y_c} + c_5 x^2 + c_6 x^2 e^x$$

and $y_p = Ax^2 + Bx^2 e^x$. Substituting y_p into the differential equation yields $2A + 2Be^x = 1 + e^x$.
Equating coefficients gives $A = 1/2$ and $B = 1/2$. The general solution is

$$y = c_1 + c_2 x + c_3 e^x + c_4 x e^x + \frac{1}{2}x^2 + \frac{1}{2}x^2 e^x.$$

66. The complementary function is $y_c = c_1 + c_2 e^{-x}$. Using D^2 to annihilate x we find $y_p = Ax + Bx^2$ Substituting y_p into the differential equation we obtain $(A + 2B) + 2Bx = x$. Thus $A = -1$ and $B = 1/2$, and

$$y = c_1 + c_2 e^{-x} - x + \frac{1}{2}x^2$$
$$y' = -c_2 e^{-x} - 1 + x.$$

The initial conditions imply

$$c_1 + c_2 = 1$$
$$-c_2 = 1.$$

Thus $c_1 = 2$ and $c_2 = -1$, and

$$y = 2 - e^{-x} - x + \frac{1}{2}x^2.$$

69. The complementary function is $y_c = c_1 \cos x + c_2 \sin x$. Using $(D^2 + 1)(D^2 + 4)$ to annihilate $8 \cos 2x - 4 \sin x$ we find $y_p = Ax \cos x + Bx \sin x + C \cos 2x + E \sin 2x$. Substituting y_p into the differential equation we obtain $2B \cos x - 3C \cos 2x - 2A \sin x - 3E \sin 2x = 8 \cos 2x - 4 \sin x$. Thus $A = 2$, $B = 0$, $C = -8/3$, and $E = 0$, and

$$y = c_1 \cos x + c_2 \sin x + 2x \cos x - \frac{8}{3} \cos 2x$$
$$y' = -c_1 \sin x + c_2 \cos x + 2 \cos x - 2x \sin x + \frac{16}{3} \sin 2x.$$

The initial conditions imply

$$c_2 + \frac{8}{3} = -1$$
$$-c_1 - \pi = 0.$$

Thus $c_1 = -\pi$ and $c_2 = -11/3$, and

$$y = -\pi \cos x - \frac{11}{3} \sin x + 2x \cos x - \frac{8}{3} \cos 2x.$$

72. The complementary function is $y_c = c_1 + c_2 x + c_3 x^2 + c_4 e^x$. Using $D^2(D - 1)$ to annihilate $x + e^x$ we find $y_p = Ax^3 + Bx^4 + Cxe^x$. Substituting y_p into the differential equation we obtain $(-6A + 24B) - 24Bx + Ce^x = x + e^x$. Thus $A = -1/6$, $B = -1/24$, and $C = 1$, and

$$y = c_1 + c_2 x + c_3 x^2 + c_4 e^x - \frac{1}{6}x^3 - \frac{1}{24}x^4 + xe^x$$

$$y' = c_2 + 2c_3 x + c_4 e^x - \frac{1}{2}x^2 - \frac{1}{6}x^3 + e^x + xe^x$$

$$y'' = 2c_3 + c_4 e^x - x - \frac{1}{2}x^2 + 2e^x + xe^x.$$

$$y''' = c_4 e^x - 1 - x + 3e^x + xe^x.$$

The initial conditions imply

$$c_1 + c_4 = 0$$

$$c_2 + c_4 + 1 = 0$$

$$2c_3 + c_4 + 2 = 0$$

$$2 + c_4 = 0.$$

Thus $c_1 = 2$, $c_2 = 1$, $c_3 = 0$, and $c_4 = -2$, and

$$y = 2 + x - 2e^x - \frac{1}{6}x^3 - \frac{1}{24}x^4 + xe^x.$$

4.6 | Variation of Parameters

The terminology and concepts listed below provide an outline of the main ideas encountered in this section. These can be useful when preparing for a quiz or test.

Terminology and Concepts

- variation of parameters

The basic skills listed below summarize the more mechanical types of problems encountered in the exercise set for this section.

Basic Skills

- solve a nonhomogeneous linear second-order DE with constant coefficients using variation of parameters

Integrals of Trig Functions In Example 2 in the text we had to integrate

$$\int \cot 3x \, dx = \int \frac{\cos 3x}{\sin 3x} \, dx.$$

If we identify $u = \sin 3x$, then $du = 3\cos 3x \, dx$. The integral is recognized as one of the form $\int du/u = \ln|u| + c$:

$$\frac{1}{3} \int \frac{\overbrace{3\cos 3x \, dx}^{du}}{\underbrace{\sin 3x}_{u}} = \frac{1}{3} \ln|\sin 3x| + c.$$

In general, we can integrate $\tan u$ and $\cot u$ in terms of logarithms:

$$\int \tan u \, du = -\ln|\cos u| + c \qquad (1)$$

$$\int \cot u \, du = \ln|\sin u| + c. \qquad (2)$$

The integrals in (1) and (2), along with

$$\int \sec u \, du = \ln|\sec u + \tan u| + c \qquad (3)$$

$$\int \csc u \, du = \ln|\csc u - \cot u| + c \qquad (4)$$

occur frequently in the solution of differential equations by variation of parameters. Watch for them in Exercises 4.6.

The particular solution, $y_p = u_1 y_1 + u_2 y_2$, in the following problems can take on a variety of forms, especially where trigonometric functions are involved. The validity of a particular form can best be checked by substituting it back into the differential equation.

In the remaining problems in this section the value of the Wronskian is given, but the determinant which is used to evaluate the Wronskian is not shown.

3. The auxiliary equation is $m^2 + 1 = 0$, so $y_c = c_1 \cos x + c_2 \sin x$ and $W = 1$. Identifying $f(x) = \sin x$ we obtain

$$u_1' = -\sin^2 x$$
$$u_2' = \cos x \sin x.$$

Then

$$u_1 = \frac{1}{4}\sin 2x - \frac{1}{2}x = \frac{1}{2}\sin x \cos x - \frac{1}{2}x$$

$$u_2 = -\frac{1}{2}\cos^2 x.$$

and

$$y = c_1 \cos x + c_2 \sin x + \frac{1}{2}\sin x \cos^2 x - \frac{1}{2}x \cos x - \frac{1}{2}\cos^2 x \sin x$$

$$= c_1 \cos x + c_2 \sin x - \frac{1}{2}x \cos x.$$

6. The auxiliary equation is $m^2 + 1 = 0$, so $y_c = c_1 \cos x + c_2 \sin x$ and $W = 1$. Identifying $f(x) = \sec^2 x$ we obtain

$$u_1' = -\frac{\sin x}{\cos^2 x}$$

$$u_2' = \sec x.$$

Then

$$u_1 = -\frac{1}{\cos x} = -\sec x$$

$$u_2 = \ln|\sec x + \tan x|$$

and

$$y = c_1 \cos x + c_2 \sin x - \cos x \sec x + \sin x \ln|\sec x + \tan x|$$
$$= c_1 \cos x + c_2 \sin x - 1 + \sin x \ln|\sec x + \tan x|.$$

9. The auxiliary equation is $m^2 - 4 = 0$, so $y_c = c_1 e^{2x} + c_2 e^{-2x}$ and $W = -4$. Identifying $f(x) = e^{2x}/x$ we obtain $u_1' = 1/4x$ and $u_2' = -e^{4x}/4x$. Then

$$u_1 = \frac{1}{4}\ln|x|,$$

$$u_2 = -\frac{1}{4}\int_{x_0}^{x} \frac{e^{4t}}{t}\, dt$$

and

$$y = c_1 e^{2x} + c_2 e^{-2x} + \frac{1}{4}\left(e^{2x}\ln|x| - e^{-2x}\int_{x_0}^{x} \frac{e^{4t}}{t}\, dt\right), \qquad x_0 > 0.$$

12. The auxiliary equation is $m^2 - 2m + 1 = (m-1)^2 = 0$, so $y_c = c_1 e^x + c_2 x e^x$ and $W = e^{2x}$. Identifying $f(x) = e^x/(1 + x^2)$ we obtain

$$u_1' = -\frac{x e^x e^x}{e^{2x}(1 + x^2)} = -\frac{x}{1 + x^2}$$

$$u_2' = \frac{e^x e^x}{e^{2x}(1 + x^2)} = \frac{1}{1 + x^2}.$$

Then $u_1 = -\frac{1}{2}\ln\left(1 + x^2\right)$, $u_2 = \tan^{-1} x$, and

$$y = c_1 e^x + c_2 x e^x - \frac{1}{2} e^x \ln\left(1 + x^2\right) + x e^x \tan^{-1} x.$$

15. The auxiliary equation is $m^2 + 2m + 1 = (m+1)^2 = 0$, so $y_c = c_1 e^{-t} + c_2 t e^{-t}$ and $W = e^{-2t}$. Identifying $f(t) = e^{-t} \ln t$ we obtain

$$u_1' = -\frac{t e^{-t} e^{-t} \ln t}{e^{-2t}} = -t \ln t$$

$$u_2' = \frac{e^{-t} e^{-t} \ln t}{e^{-2t}} = \ln t.$$

Then

$$u_1 = -\frac{1}{2} t^2 \ln t + \frac{1}{4} t^2$$

$$u_2 = t \ln t - t$$

and

$$y = c_1 e^{-t} + c_2 t e^{-t} - \frac{1}{2} t^2 e^{-t} \ln t + \frac{1}{4} t^2 e^{-t} + t^2 e^{-t} \ln t - t^2 e^{-t}$$

$$= c_1 e^{-t} + c_2 t e^{-t} + \frac{1}{2} t^2 e^{-t} \ln t - \frac{3}{4} t^2 e^{-t}.$$

18. The auxiliary equation is $4m^2 - 4m + 1 = (2m-1)^2 = 0$, so $y_c = c_1 e^{x/2} + c_2 x e^{x/2}$ and $W = e^x$. Identifying $f(x) = \frac{1}{4} e^{x/2} \sqrt{1-x^2}$ we obtain

$$u_1' = -\frac{x e^{x/2} e^{x/2} \sqrt{1-x^2}}{4 e^x} = -\frac{1}{4} x \sqrt{1-x^2}$$

$$u_2' = \frac{e^{x/2} e^{x/2} \sqrt{1-x^2}}{4 e^x} = \frac{1}{4} \sqrt{1-x^2}.$$

To find u_1 and u_2 we use the substitution $v = 1 - x^2$ and the trig substitution $x = \sin\theta$, respectively:

$$u_1 = \frac{1}{12} \left(1 - x^2\right)^{3/2}$$

$$u_2 = \frac{x}{8} \sqrt{1-x^2} + \frac{1}{8} \sin^{-1} x.$$

Thus

$$y = c_1 e^{x/2} + c_2 x e^{x/2} + \frac{1}{12} e^{x/2} \left(1 - x^2\right)^{3/2} + \frac{1}{8} x^2 e^{x/2} \sqrt{1-x^2} + \frac{1}{8} x e^{x/2} \sin^{-1} x.$$

21. The auxiliary equation is $m^2 + 2m - 8 = (m - 2)(m + 4) = 0$, so $y_c = c_1 e^{2x} + c_2 e^{-4x}$ and $W = -6e^{-2x}$. Identifying $f(x) = 2e^{-2x} - e^{-x}$ we obtain

$$u_1' = \frac{1}{3}e^{-4x} - \frac{1}{6}e^{-3x}$$

$$u_2' = \frac{1}{6}e^{3x} - \frac{1}{3}e^{2x}.$$

Then

$$u_1 = -\frac{1}{12}e^{-4x} + \frac{1}{18}e^{-3x}$$

$$u_2 = \frac{1}{18}e^{3x} - \frac{1}{6}e^{2x}.$$

Thus

$$y = c_1 e^{2x} + c_2 e^{-4x} - \frac{1}{12}e^{-2x} + \frac{1}{18}e^{-x} + \frac{1}{18}e^{-x} - \frac{1}{6}e^{-2x}$$

$$= c_1 e^{2x} + c_2 e^{-4x} - \frac{1}{4}e^{-2x} + \frac{1}{9}e^{-x}$$

and

$$y' = 2c_1 e^{2x} - 4c_2 e^{-4x} + \frac{1}{2}e^{-2x} - \frac{1}{9}e^{-x}.$$

The initial conditions imply

$$c_1 + c_2 - \frac{5}{36} = 1$$

$$2c_1 - 4c_2 + \frac{7}{18} = 0.$$

Thus $c_1 = 25/36$ and $c_2 = 4/9$, and

$$y = \frac{25}{36}e^{2x} + \frac{4}{9}e^{-4x} - \frac{1}{4}e^{-2x} + \frac{1}{9}e^{-x}.$$

24. Write the equation in the form

$$y'' + \frac{1}{x}y' + \frac{1}{x^2}y = \frac{\sec(\ln x)}{x^2}$$

and identify $f(x) = \sec(\ln x)/x^2$. From $y_1 = \cos(\ln x)$ and $y_2 = \sin(\ln x)$ we compute $W = 1/x$.

Now

$$u_1' = -\frac{\tan(\ln x)}{x} \quad \text{so} \quad u_1 = \ln|\cos(\ln x)|$$

and

$$u_2' = \frac{1}{x} \quad \text{so} \quad u_2 = \ln x.$$

Thus, a particular solution is

$$y_p = \cos(\ln x) \ln|\cos(\ln x)| + (\ln x)\sin(\ln x),$$

and the general solution is

$$y = c_1 \cos(\ln x) + c_2 \sin(\ln x) + \cos(\ln x) \ln|\cos(\ln x)| + (\ln x)\sin(\ln x).$$

27. The auxiliary equation is $m^3 - 2m^2 - m + 2 = 0$ so $y_c = c_1 e^x + c_2 e^{-x} + c_3 e^{2x}$ and $W = -6e^{2x}$. Identifying $f(x) = e^{4x}$ we find $W_1 = 3e^{5x}$, $W_2 = -e^{7x}$, and $W_3 = -2e^{4x}$. Then

$$u_1' = \frac{W_1}{W} = \frac{3e^{5x}}{-6e^{2x}} = -\frac{1}{2}e^{3x}$$

$$u_2' = \frac{W_2}{W} = \frac{-e^{7x}}{-6e^{2x}} = \frac{1}{6}e^{5x}$$

$$u_3' = \frac{W_3}{W} = \frac{-2e^{4x}}{-6e^{2x}} = \frac{1}{3}e^{2x},$$

and integrating we find that

$$u_1 = -\frac{1}{6}e^{3x}$$

$$u_2 = \frac{1}{30}e^{5x}$$

$$u_3 = \frac{1}{6}e^{2x}.$$

Thus $y_p = -\frac{1}{6}e^{3x} \cdot e^x + \frac{1}{30}e^{5x} \cdot e^{-x} + \frac{1}{6}e^{2x} \cdot e^{2x} = \frac{1}{30}e^{4x}$, and the general solution is

$$y = c_1 e^x + c_2 e^{-x} + c_3 e^{2x} + \frac{1}{30}e^{4x}.$$

4.7 | Cauchy-Euler Equation

The terminology and concepts listed below provide an outline of the main ideas encountered in this section. These can be useful when preparing for a quiz or test.

Terminology and Concepts

- Cauchy-Euler equation

- auxiliary equation

- variation of parameters

The basic skills listed below summarize the more mechanical types of problems encountered in the exercise set for this section.

Basic Skills

- find the general solution of a homogeneous Cauchy-Euler equation

- use variation of parameters to find a particular solution of a nonhomogeneous Cauchy-Euler equation

- use the substitution $x = e^t$ to transform a Cauchy-Euler equation into a linear DE with constant coefficients

3. The auxiliary equation is $m^2 = 0$ so that $y = c_1 + c_2 \ln x$.

6. The auxiliary equation is $m^2 + 4m + 3 = (m + 1)(m + 3) = 0$ so that $y = c_1 x^{-1} + c_2 x^{-3}$.

9. The auxiliary equation is $25m^2 + 1 = 0$ so that $y = c_1 \cos\left(\frac{1}{5}\ln x\right) + c_2 \sin\left(\frac{1}{5}\ln x\right)$.

12. The auxiliary equation is $m^2 + 7m + 6 = (m + 1)(m + 6) = 0$ so that $y = c_1 x^{-1} + c_2 x^{-6}$.

15. Assuming that $y = x^m$ and substituting into the differential equation we obtain

$$m(m - 1)(m - 2) - 6 = m^3 - 3m^2 + 2m - 6 = (m - 3)(m^2 + 2) = 0.$$

Thus

$$y = c_1 x^3 + c_2 \cos\left(\sqrt{2}\ln x\right) + c_3 \sin\left(\sqrt{2}\ln x\right).$$

18. Assuming that $y = x^m$ and substituting into the differential equation we obtain

$$m(m - 1)(m - 2)(m - 3) + 6m(m - 1)(m - 2) + 9m(m - 1) + 3m + 1 = m^4 + 2m^2 + 1$$
$$= (m^2 + 1)^2 = 0.$$

Thus

$$y = c_1 \cos(\ln x) + c_2 \sin(\ln x) + c_3(\ln x)\cos(\ln x) + c_4(\ln x)\sin(\ln x).$$

21. The auxiliary equation is $m^2 - 2m + 1 = (m-1)^2 = 0$ so that $y_c = c_1 x + c_2 x \ln x$ and the Wronskian is $W(x, x \ln x) = x$. Identifying $f(x) = 2/x$ we obtain $u_1' = -2 \ln x / x$ and $u_2' = 2/x$. Then $u_1 = -(\ln x)^2$, $u_2 = 2 \ln x$, and

$$y = c_1 x + c_2 x \ln x - x(\ln x)^2 + 2x(\ln x)^2$$
$$= c_1 x + c_2 x \ln x + x(\ln x)^2, \qquad x > 0.$$

24. The auxiliary equation $m(m-1) + m - 1 = m^2 - 1 = 0$ has roots $m_1 = -1$, $m_2 = 1$, so $y_c = c_1 x^{-1} + c_2 x$. With $y_1 = x^{-1}$, $y_2 = x$, and the identification $f(x) = 1/x^2(x+1)$, we get

$$W = 2x^{-1}, \qquad W_1 = -1/x(x+1), \qquad \text{and} \qquad W_2 = 1/x^3(x+1).$$

Then $u_1' = W_1/W = -1/2(x+1)$, $u_2' = W_2/W = 1/2x^2(x+1)$, and integration by partial fractions for u_2' gives

$$u_1 = -\frac{1}{2} \ln(x+1)$$
$$u_2 = -\frac{1}{2}x^{-1} - \frac{1}{2}\ln x + \frac{1}{2}\ln(x+1),$$

so

$$y_p = u_1 y_1 + u_2 y_2 = \left[-\frac{1}{2}\ln(x+1) \right] x^{-1} + \left[-\frac{1}{2}x^{-1} - \frac{1}{2}\ln x + \frac{1}{2}\ln(x+1) \right] x$$
$$= -\frac{1}{2} - \frac{1}{2}x \ln x + \frac{1}{2}x \ln(x+1) - \frac{\ln(x+1)}{2x} = -\frac{1}{2} + \frac{1}{2}x \ln\left(1 + \frac{1}{x}\right) - \frac{\ln(x+1)}{2x}$$

and

$$y = y_c + y_p = c_1 x^{-1} + c_2 x - \frac{1}{2} + \frac{1}{2}x \ln\left(1 + \frac{1}{x}\right) - \frac{\ln(x+1)}{2x}, \qquad x > 0.$$

27. The auxiliary equation is $m^2 + 1 = 0$, so that

$$y = c_1 \cos(\ln x) + c_2 \sin(\ln x)$$

and

$$y' = -c_1 \frac{1}{x} \sin(\ln x) + c_2 \frac{1}{x} \cos(\ln x).$$

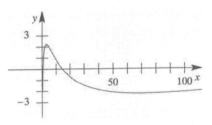

The initial conditions imply $c_1 = 1$ and $c_2 = 2$. Thus $y = \cos(\ln x) + 2\sin(\ln x)$. The graph is given to the right.

30. The auxiliary equation is

$$m^2 - 6m + 8 = (m-2)(m-4) = 0,$$

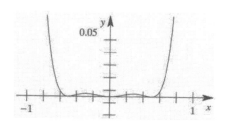

so that $y_c = c_1 x^2 + c_2 x^4$ and the Wronskian is $W = 2x^5$.
Identifying $f(x) = 8x^4$ we obtain $u_1' = -4x^3$ and $u_2' = 4x$. Then $u_1 = -x^4$, $u_2 = 2x^2$, and $y = c_1 x^2 + c_2 x^4 + x^6$.
The initial conditions imply

$$\frac{1}{4}c_1 + \frac{1}{16}c_2 = -\frac{1}{64}$$
$$c_1 + \frac{1}{2}c_2 = -\frac{3}{16}.$$

Thus $c_1 = \frac{1}{16}$, $c_2 = -\frac{1}{2}$, and $y = \frac{1}{16}x^2 - \frac{1}{2}x^4 + x^6$. The graph is given above.

33. Substituting $x = e^t$ into the differential equation we obtain

$$\frac{d^2 y}{dt^2} + 9\frac{dy}{dt} + 8y = e^{2t}.$$

The auxiliary equation is $m^2 + 9m + 8 = (m+1)(m+8) = 0$ so that $y_c = c_1 e^{-t} + c_2 e^{-8t}$. Using undetermined coefficients we try $y_p = Ae^{2t}$. This leads to $30Ae^{2t} = e^{2t}$, so that $A = 1/30$ and

$$y = c_1 e^{-t} + c_2 e^{-8t} + \frac{1}{30}e^{2t} = c_1 x^{-1} + c_2 x^{-8} + \frac{1}{30}x^2.$$

36. From

$$\frac{d^2 y}{dx^2} = \frac{1}{x^2}\left(\frac{d^2 y}{dt^2} - \frac{dy}{dt}\right)$$

it follows that

$$\frac{d^3 y}{dx^3} = \frac{1}{x^2}\frac{d}{dx}\left(\frac{d^2 y}{dt^2} - \frac{dy}{dt}\right) - \frac{2}{x^3}\left(\frac{d^2 y}{dt^2} - \frac{dy}{dt}\right)$$

$$= \frac{1}{x^2}\frac{d}{dx}\left(\frac{d^2 y}{dt^2}\right) - \frac{1}{x^2}\frac{d}{dx}\left(\frac{dy}{dt}\right) - \frac{2}{x^3}\frac{d^2 y}{dt^2} + \frac{2}{x^3}\frac{dy}{dt}$$

$$= \frac{1}{x^2}\frac{d^3 y}{dt^3}\left(\frac{1}{x}\right) - \frac{1}{x^2}\frac{d^2 y}{dt^2}\left(\frac{1}{x}\right) - \frac{2}{x^3}\frac{d^2 y}{dt^2} + \frac{2}{x^3}\frac{dy}{dt}$$

$$= \frac{1}{x^3}\left(\frac{d^3 y}{dt^3} - 3\frac{d^2 y}{dt^2} + 2\frac{dy}{dt}\right).$$

Substituting into the differential equation we obtain

$$\frac{d^3y}{dt^3} - 3\frac{d^2y}{dt^2} + 2\frac{dy}{dt} - 3\left(\frac{d^2y}{dt^2} - \frac{dy}{dt}\right) + 6\frac{dy}{dt} - 6y = 3 + 3t$$

or

$$\frac{d^3y}{dt^3} - 6\frac{d^2y}{dt^2} + 11\frac{dy}{dt} - 6y = 3 + 3t.$$

The auxiliary equation is $m^3 - 6m^2 + 11m - 6 = (m-1)(m-2)(m-3) = 0$ so that $y_c = c_1e^t + c_2e^{2t} + c_3e^{3t}$. Using undetermined coefficients we try $y_p = A + Bt$. This leads to $(11B - 6A) - 6Bt = 3 + 3t$, so that $A = -17/12$, $B = -1/2$, and

$$y = c_1e^t + c_2e^{2t} + c_3e^{3t} - \frac{17}{12} - \frac{1}{2}t = c_1x + c_2x^2 + c_3x^3 - \frac{17}{12} - \frac{1}{2}\ln x.$$

39. Letting $y = (x+3)^m$ we have $y' = m(x+3)^{m-1}$, and $y'' = m(m-1)(x+3)^{m-2}$. Substituting into the differential equation we find

$$(x+3)^2 m(m-1)(x+3)^{m-2} - 8(x+3)m(x+3)^{m-1} + 14(x+3)^m$$
$$= m(m-1)(x+3)^m - 8m(x+3)^m + 14(x+3)^m = 0,$$

so $m(m-1) - 8m + 14 = m^2 - 9m + 14 = (m-2)(m-7) = 0$. Thus, $m_1 = 2$ and $m_2 = 7$, and the general solution is

$$y = c_1(x+3)^2 + c_2(x+3)^7.$$

42. Letting $t = x - 4$ we obtain $dy/dx = dy/dt$ and, using the Chain Rule,

$$\frac{d^2y}{dx^2} = \frac{d}{dx}\left(\frac{dy}{dt}\right) = \frac{d^2y}{dt^2}\frac{dt}{dx} = \frac{d^2y}{dt^2}(1) = \frac{d^2y}{dt^2}.$$

Substituting into the differential equation we obtain

$$t^2\frac{d^2y}{dt^2} - 5t\frac{dy}{dt} + 9y = 0.$$

The auxiliary equation is $m(m-1) - 5m + 9 = m^2 - 6m + 9 = (m-3)^2 = 0$ so that $m_1 = m_2 = 3$. Then

$$y = c_1t^3 + c_2t^3 \ln t = c_1(x-4)^3 + c_2(x-4)^3 \ln(x-4).$$

4.8 | Green's Functions

The terminology and concepts listed below provide an outline of the main ideas encountered in this section. These can be useful when preparing for a quiz or test.

Terminology and Concepts

- rest solution

- Green's function (or Green's function for the second-order differential operator)

- response of system due to initial conditions

- response of system due to the forcing function f

- Green's function for a boundary-value problem

Basic Skills

- use Green's functions to solve initial-value problems

- use Green's functions to solve initial-value problems with piecewise-defined forcing functions

- use Green's functions to solve boundary-value problems

4.8.1 Initial-Value Problems

3. From the homogeneous differential equation $y'' + 2y' + y = 0$, we find $y_1 = e^{-x}$ and $y_2 = xe^{-x}$ so

$$W\left(e^{-x},\, xe^{-x}\right) = \begin{vmatrix} e^{-x} & xe^{-x} \\ -e^{-x} & -xe^{-x} + e^{-x} \end{vmatrix} = e^{-2x}$$

Then

$$G(x,\, t) = \frac{e^{-t}xe^{-x} - e^{-x}te^{-t}}{e^{-2t}} = (x - t)e^{-(x-t)}$$

and

$$y_p(x) = \int_{x_0}^{x} (x - t)e^{-(x-t)} f(t)\, dt.$$

6. From the homogeneous differential equation $y'' - 2y' + 2y = 0$, we find $y_1 = e^x \cos x$ and $y_2 = e^x \sin x$ so

$$W(e^x \cos x,\, e^x \sin x) = \begin{vmatrix} e^x \cos x & e^x \sin x \\ -e^x \sin x + e^x \cos x & e^x \cos x + e^x \sin x \end{vmatrix} = e^{2x}$$

Then

$$G(x,\, t) = \frac{e^t \cos t\, e^x \sin x - e^x \cos x\, e^t \sin t}{e^{2t}} = e^{x-t} \sin(x - t)$$

and

$$y_p(x) = \int_{x_0}^{x} e^{x-t} \sin(x - t) f(t)\, dt.$$

9. From $y'' + 2y' + y = e^{-x}$ we have

$$y = y_c + y_p = c_1 e^{-x} + c_2 x e^{-x} + y_p,$$

so

$$y = c_1 e^{-x} + c_2 x e^{-x} + \int_{x_0}^{x} (x - t) e^{-(x-t)} e^{-t}\, dt.$$

12. From $y'' - 2y' + 2y = \cos^2 x$ we have

$$y = y_c + y_p = c_1 e^x \cos x + c_2 e^x \sin x + y_p,$$

so

$$y = c_1 e^x \cos x + c_2 e^x \sin x = \int_{x_0}^{x} e^{(x-t)} \sin(x - t) \cos^2 t\, dt.$$

15. The initial-value problem is $y'' - 10y' + 25y = e^{5x}, \quad y(0) = 0,\ y'(0) = 0$. Then we find that

$$y_1 = e^{5x},\ y_2 = x e^{5x} \qquad \text{so} \qquad W(e^{5x},\, x e^{5x}) = \begin{vmatrix} e^{5x} & x e^{5x} \\ 5 e^{5x} & 5x e^{5x} + e^{5x} \end{vmatrix} = e^{10x}.$$

Then $G(x,\, t) = \dfrac{e^{5t} x e^{5x} - e^{5x} t e^{5t}}{e^{10t}} = (x - t) e^{5(x-t)}$ and the solution of the initial-value problem is

$$y_p(x) = \int_{0}^{x} (x - t) e^{5(x-t)} e^{5t}\, dt = x e^{5x} \int_{0}^{x} dt - e^{5x} \int_{0}^{x} t\, dt = x^2 e^{5x} - \frac{1}{2} x^2 e^{5x} = \frac{1}{2} x^2 e^{5x}.$$

18. The initial-value problem is $y'' + y = \sec^2 x,\quad y(\pi) = 0,\ y'(\pi) = 0.$ Then we find that

$$y_1 = \cos x,\ y_2 = \sin x \qquad \text{so} \qquad W(\cos x,\ \sin x) = \begin{vmatrix} \cos x & \sin x \\ -\sin x & \cos x \end{vmatrix} = 1.$$

Then $G(x, t) = \cos t \sin x - \cos x \sin t = \sin(x - t)$ and the solution of the initial-value problem is

$$y_p(x) = \int_\pi^x (\cos t \sin x - \cos x \sin t) \sec^2 t\, dt$$

$$= \sin x \int_\pi^x \sec t\, dt - \cos x \int_\pi^x \sec t \tan t\, dt$$

$$= -\cos x - 1 + \sin x \ln|\sec x + \tan x|.$$

21. The initial-value problem is $y'' - 10y' + 25y = e^{5x},\quad y(0) = -1,\ y'(0) = 1,$ so $y(x) = c_1 e^{5x} + c_2 x e^{5x}.$ The initial conditions give

$$c_1 \qquad = -1$$
$$5c_1 + c_2 = \quad 1,$$

so $c_1 = -1$ and $c_2 = 6$, which implies that $y_h = -e^{5x} + 6x e^{5x}.$ Now, y_p found in the solution of Problem 15 in this section gives

$$y = y_h + y_p = -e^{5x} + 6x e^{5x} + \frac{1}{2} x^2 e^{5x}.$$

24. The initial-value problem is $y'' + y = \sec^2 x,\quad y(\pi) = \frac{1}{2},\ y'(\pi) = -1,$ so $y(x) = c_1 \cos x + c_2 \sin x.$ The initial conditions give

$$-c_1 \quad = \quad \frac{1}{2}$$
$$-c_2 = -1,$$

so $c_1 = -\frac{1}{2}$ and $c_2 = 1$, which implies that $y_h = -\frac{1}{2}\cos x + \sin x.$ Now, y_p found in the solution of Problem 18 in this section gives

$$y = y_h + y_p = -\frac{1}{2}\cos x + \sin x + (-\cos x - 1 + \sin x \ln|\sec x + \tan x|)$$

$$= -\frac{3}{2}\cos x + \sin x - 1 + \sin x \ln|\sec x + \tan x|.$$

27. The Cauchy-Euler initial-value problem $x^2 y'' - 2xy' + 2y = x$, $y(1) = 2$, $y'(1) = -1$, has auxiliary equation $m(m-1) - 2m + 2 = m^2 - 3m + 2 = (m-1)(m-2) = 0$ so $m_1 = 1$, $m_2 = 2$, $y(x) = c_1 x + c_2 x^2$, and $y' = c_1 + 2c_2 x$. The initial conditions give

$$
\begin{aligned}
c_1 + \ c_2 &= \ 2 \\
c_1 + 2c_2 &= -1,
\end{aligned}
$$

so $c_1 = 5$ and $c_2 = -3$, which implies that $y_h = 5x - 3x^2$. The Wronskian is

$$
W\left(x, x^2\right) = \begin{vmatrix} x & x^2 \\ 1 & 2x \end{vmatrix} = x^2.
$$

Then $G(x, t) = \dfrac{tx^2 - xt^2}{t^2} = \dfrac{x(x-t)}{t}$. From the standard form of the differential equation we identify the forcing function $f(t) = \dfrac{1}{t}$. Then, for $x > 1$,

$$
\begin{aligned}
y_p(x) &= \int_1^x \frac{x(x-t)}{t} \frac{1}{t}\, dt = x^2 \int_1^x \frac{1}{t^2}\, dt - x \int_1^x \frac{1}{t}\, dt \\
&= x^2 \left(-\frac{1}{x} + 1\right) - x(\ln x - \ln 1) = x^2 - x - x \ln x.
\end{aligned}
$$

The solution of the initial-value problem is

$$
y = y_h + y_p = (5x - 3x^2) + (x^2 - x - x \ln x) = 4x - 2x^2 - x \ln x.
$$

30. The Cauchy-Euler initial-value problem $x^2 y'' - xy' + y = x^2$, $y(1) = 4$, $y'(1) = 3$, has auxiliary equation $m(m-1) - m + 1 = m^2 - 2m + 1 = (m-1)^2 = 0$ so $m_1 = m_2 = 1$, $y(x) = c_1 x + c_2 x \ln x$, and $y' = c_1 + c_2(1 + \ln x)$. The initial conditions give

$$
\begin{aligned}
c_1 \qquad &= 4 \\
c_1 + c_2 &= 3,
\end{aligned}
$$

so $c_1 = 4$ and $c_2 = -1$, which implies that $y_h = 4x - x \ln x$. The Wronskian is

$$
W(x, x \ln x) = \begin{vmatrix} x & x \ln x \\ 1 & 1 + \ln x \end{vmatrix} = x.
$$

Then $G(x, t) = \dfrac{tx \ln x - xt \ln t}{t} = x \ln x - x \ln t$. From the standard form of the differential equation we identify the forcing function $f(t) = 1$. Then, for $x > 1$,

$$y_p(x) = \int_1^x (x \ln x - x \ln t)\, dt = x \ln x \int_1^x dt - x \int_1^x \ln t\, dt$$

$$= x \ln x \,(x - 1) - x(x \ln x - x + 1) = x^2 - x - x \ln x.$$

The solution of the initial-value problem is

$$y = y_h + y_p = 4x - x \ln x + \left(x^2 - x - x \ln x\right) = x^2 + 3x - 2x \ln x.$$

33. The initial-value problem is $y'' + y = f(x)$, $y(0) = 1$, $y'(0) = -1$, where

$$f(x) = \begin{cases} 0, & x < 0 \\ 10, & 0 \le x \le 3\pi \\ 0, & x > 3\pi. \end{cases}$$

We first find

$$y_h(x) = \cos x - \sin x \qquad \text{and} \qquad y_p(x) = \int_0^x \sin(x - t)\, f(t)\, dt.$$

Then for $x < 0$

$$y_p(x) = \int_0^x \sin(x - t)\, 0\, dt = 0,$$

for $0 \le x \le 3\pi$

$$y_p(x) = 10 \int_0^x \sin(x - t)\, dt = 10 - 10 \cos x,$$

and for $x > 3\pi$

$$y_p(x) = 10 \int_0^{3\pi} \sin(x - t)\, dt + \int_{3\pi}^x \sin(x - t)\, 0\, dt = -20 \cos x.$$

The solution is

$$y(x) = y_h(x) + y_p(x) = \cos x - \sin x + y_p(x),$$

where

$$y_p(x) = \begin{cases} 0, & x < 0 \\ 10 - 10 \cos x, & 0 \le x \le 3\pi \\ -20 \cos x, & x > 3\pi. \end{cases}$$

4.8.2 Boundary-Value Problems

36. The boundary-value problem is $y'' = f(x)$, $y(0) = 0$, $y(1) + y'(1) = 0$. The solution of the associated homogeneous equation is $y = c_1 + c_2 x$.

(a) To satisfy $y(0) = 0$ we take $y_1(x) = x$ and to satisfy $y(1) + y'(1) = 0$ we note that $y(1) + y'(1) = (c_1 + c_2) + c_2 = 0$ which implies that $c_1 = -2c_2$. Taking $c_2 = 1$, we find that $c_1 = -2$, so we have $y_2(x) = -2 + x$. The Wronskian of y_1 and y_2 is

$$W(y_1, y_2) = \begin{vmatrix} x & -2+x \\ 1 & 1 \end{vmatrix} = 2,$$

so

$$G(x, t) = \begin{cases} \frac{1}{2} t(x-2), & 0 \le t \le x \\ \frac{1}{2} x(t-2), & x \le t \le 1. \end{cases}$$

Therefore

$$y_p(x) = \int_0^1 G(x, t)\, f(t)\, dt = \frac{1}{2}(x-2) \int_0^x t f(t)\, dt + \frac{1}{2} x \int_x^1 (t-2) f(t)\, dt.$$

(b) By the Product Rule and the Fundamental Theorem of Calculus, the first two derivative of $y_p(x)$ are

$$y_p'(x) = \frac{1}{2}(x-2)\, x f(x) + \frac{1}{2} \int_0^x t f(t)\, dt + \frac{1}{2} x [-(x-2) f(x)] + \frac{1}{2} \int_x^1 (t-2) f(t)\, dt$$

$$y_p''(x) = \frac{1}{2}(x-2)[x f'(x) + f(x)] + \frac{1}{2} x f(x) + \frac{1}{2} x f(x)$$

$$- \frac{1}{2}(x^2 - 2x) f'(x) - \frac{1}{2}(2x - 2) f(x) - \frac{1}{2}(x-2) f(x)$$

$$= \frac{1}{2} \left[x^2 f'(x) + x f(x) - 2x f'(x) - 2f(x) + x f(x) + x f(x) - x^2 f'(x) \right.$$

$$\left. + 2x f'(x) - 2x f(x) + 2f(x) - x f(x) + 2f(x) \right] = f(x)$$

Thus, $y_p(x)$ satisfies the differential equation. To see that the boundary conditions are satisfied we first compute

$$y_p(0) = \frac{1}{2}(0 - 2) \int_0^0 t f(t)\, dt + \frac{1}{2}(0) \int_0^1 (t-2) f(t)\, dt = 0.$$

Next we use $y_p'(x)$ found at the beginning of this part of the solution to compute

$$y_p(1) + y_p'(1) = \frac{1}{2}(1-2)\int_0^1 tf(t)\,dt + \frac{1}{2}(1)\int_1^1 (t-2)f(t)\,dt + \frac{1}{2}(1-2)f(x)$$

$$+ \frac{1}{2}\int_0^1 tf(t)\,dt + \frac{1}{2}(1)[-(1-2)f(x)] + \frac{1}{2}(1)\int_1^1 (1-2)f(t)\,dt$$

$$= \frac{1}{2}(-1)\int_0^1 tf(t)\,dt + \frac{1}{2}(-1)f(1) + \frac{1}{2}\int_0^1 tf(t)\,dt + \frac{1}{2}f(1) = 0.$$

39. The boundary-value problem is $y'' + y = 1$, $y(0) = 0$, $y(1) = 0$. The solution of the associated homogeneous equation is $y = c_1\cos x + c_2\sin x$. Since $y(0) = c_1\cos 0 + c_2\sin 0 = c_1 = 0$, we take $y_1(x) = \sin x$. To satisfy $y(1) = 0$ we note that $y(1) = c_1\cos 1 + c_2\sin 1 = 0$ which implies that $c_1 = -c_2\sin 1/\cos 1$ so

$$y(x) = -c_2\frac{\sin 1}{\cos 1}\cos x + c_2\sin x = -\frac{c_2}{\cos 1}\left(\sin x\cos 1 - \cos x\sin 1\right).$$

Taking $c_2 = -\cos 1$, we have

$$y_2(x) = \sin x\cos 1 - \cos x\sin 1 = \sin(x-1).$$

The Wronskian of y_1 and y_2 is

$$W(y_1, y_2) = \begin{vmatrix} \sin x & \sin(x-1) \\ \cos x & \cos(x-1) \end{vmatrix} = \sin x\cos(x-1) - \cos x\sin(x-1) = \sin[x-(x-1)] = \sin 1,$$

so

$$G(x, t) = \begin{cases} \dfrac{\sin t\sin(x-1)}{\sin 1}, & 0 \le t \le x \\[2mm] \dfrac{\sin x\sin(t-1)}{\sin 1}, & x \le t \le 1. \end{cases}$$

Therefore, taking $f(t) = 1$,

$$y_p(x) = \int_0^1 G(x, t)\,dt = \frac{\sin(x-1)}{\sin 1}\int_0^x \sin t\,dt + \frac{\sin x}{\sin 1}\int_x^1 \sin(t-1)\,dt$$

$$= \frac{\sin(x-1)}{\sin 1}(-\cos x + 1) + \frac{\sin x}{\sin 1}[-1 + \cos(x-1)] = \frac{\sin(x-1)}{\sin 1} - \frac{\sin x}{\sin 1} + 1.$$

42. The boundary-value problem is $y'' - y' = e^{2x}$, $y(0) = 0$, $y(1) = 0$. The solution of the associated homogeneous equation is $y = c_1 + c_2 e^x$. Since $y(0) = c_1 + c_2 = 0$, we see that $c_1 = -c_2$ and we take $y_1(x) = 1 - e^x$. To satisfy $y(1) = 0$ we note that $y(1) = c_1 + c_2 e = 0$ which implies that $c_1 = -c_2 e$ so

$$y(x) = -c_2 e + c_2 e^x = -c_2 e (1 - e^{x-1}).$$

Taking $c_2 = -1/e$, we have $y_2(x) = 1 - e^{x-1}$. The Wronskian of y_1 and y_2 is

$$W(y_1, y_2) = \begin{vmatrix} 1 - e^x & 1 - e^{x-1} \\ -e^x & -e^{x-1} \end{vmatrix} = -e^{x-1} + e^{2x-1} + e^x - e^{2x-1} = e^x - e^{x-1} = e^x(1 - e^{-1}),$$

so

$$G(x, t) = \begin{cases} \dfrac{(1 - e^t)(1 - e^{x-1})}{e^t(1 - e^{-1})}, & 0 \le t \le x \\[3mm] \dfrac{(1 - e^x)(1 - e^{t-1})}{e^t(1 - e^{-1})}, & x \le t \le 1. \end{cases}$$

Therefore, taking $f(t) = e^{2t}$,

$$y_p(x) = \int_0^1 G(x, t)\, e^{2t} dt = \frac{1 - e^{x-1}}{1 - e^{-1}} \int_0^x \left(e^t - e^{2t}\right) dt + \frac{1 - e^x}{1 - e^{-1}} \int_x^1 \left(e^t - e^{2t-1}\right) dt$$

$$= \frac{1 - e^{x-1}}{1 - e^{-1}} \left(e^x - \frac{1}{2} e^{2x} - \frac{1}{2}\right) + \frac{1 - e^x}{1 - e^{-1}} \left(\frac{1}{2} e - e^x + \frac{1}{2} e^{2x-1}\right)$$

$$= \frac{1}{2} e^{2x} - \frac{1}{2} e^x = \frac{1}{2} e^{x+1} + \frac{1}{2} e.$$

4.9 Solving Systems of Linear DEs by Elimination

The terminology and concepts listed below provide an outline of the main ideas encountered in this section. These can be useful when preparing for a quiz or test.

Terminology and Concepts

- system of linear DEs
- systematic elimination to find the solution of a system of linear DEs

The basic skills listed below summarize the more mechanical types of problems encountered in the exercise set for this section.

Basic Skills

- use systematic elimnation to solve a system of linear DEs

3. From $Dx = -y + t$ and $Dy = x - t$ we obtain $y = t - Dx$, $Dy = 1 - D^2x$, and $(D^2 + 1)x = 1 + t$. The solution is

$$x = c_1 \cos t + c_2 \sin t + 1 + t$$
$$y = c_1 \sin t - c_2 \cos t + t - 1.$$

6. From $(D + 1)x + (D - 1)y = 2$ and $3x + (D + 2)y = -1$ we obtain $x = -\frac{1}{3} - \frac{1}{3}(D + 2)y$, $Dx = -\frac{1}{3}(D^2 + 2D)y$, and $(D^2 + 5)y = -7$. The solution is

$$y = c_1 \cos \sqrt{5}\,t + c_2 \sin \sqrt{5}\,t - \frac{7}{5}$$

$$x = \left(-\frac{2}{3}c_1 - \frac{\sqrt{5}}{3}c_2 \right) \cos \sqrt{5}\,t + \left(\frac{\sqrt{5}}{3}c_1 - \frac{2}{3}c_2 \right) \sin \sqrt{5}\,t + \frac{3}{5}.$$

9. From $Dx + D^2y = e^{3t}$ and $(D + 1)x + (D - 1)y = 4e^{3t}$ we obtain $D(D^2 + 1)x = 34e^{3t}$ and $D(D^2 + 1)y = -8e^{3t}$. The solution is

$$y = c_1 + c_2 \sin t + c_3 \cos t - \frac{4}{15}e^{3t}$$

$$x = c_4 + c_5 \sin t + c_6 \cos t + \frac{17}{15}e^{3t}.$$

Substituting into $(D + 1)x + (D - 1)y = 4e^{3t}$ gives

$$(c_4 - c_1) + (c_5 - c_6 - c_3 - c_2) \sin t + (c_6 + c_5 + c_2 - c_3) \cos t = 0$$

so that $c_4 = c_1$, $c_5 = c_3$, $c_6 = -c_2$, and

$$x = c_1 - c_2 \cos t + c_3 \sin t + \frac{17}{15}e^{3t}.$$

12. From $(2D^2 - D - 1)x - (2D + 1)y = 1$ and $(D - 1)x + Dy = -1$; $(2D + 1)(D - 1)(D + 1)x = -1$ and $(2D + 1)(D + 1)y = -2$. The solution is

$$x = c_1 e^{-t/2} + c_2 e^{-t} + c_3 e^t + 1$$
$$y = c_4 e^{-t/2} + c_5 e^{-t} - 2.$$

Substituting into $(D-1)x + Dy = -1$ gives

$$\left(-\frac{3}{2}c_1 - \frac{1}{2}c_4\right)e^{-t/2} + (-2c_2 - c_5)e^{-t} = 0$$

so that $c_4 = -3c_1$, $c_5 = -2c_2$, and

$$y = -3c_1 e^{-t/2} - 2c_2 e^{-t} - 2.$$

15. Multiplying the first equation by $D+1$ and the second equation by D^2+1 and subtracting we obtain $(D^4 - D^2)x = 1$. Then

$$x = c_1 + c_2 t + c_3 e^t + c_4 e^{-t} - \frac{1}{2}t^2.$$

Multiplying the first equation by $D+1$ and subtracting we obtain $D^2(D+1)y = 1$. Then

$$y = c_5 + c_6 t + c_7 e^{-t} - \frac{1}{2}t^2.$$

Substituting into $(D-1)x + (D^2+1)y = 1$ gives

$$(-c_1 + c_2 + c_5 - 1) + (-2c_4 + 2c_7)e^{-t} + (-1 - c_2 + c_6)t = 1$$

so that $c_5 = c_1 - c_2 + 2$, $c_6 = c_2 + 1$, and $c_7 = c_4$. The solution of the system is

$$x = c_1 + c_2 t + c_3 e^t + c_4 e^{-t} - \frac{1}{2}t^2$$

$$y = (c_1 - c_2 + 2) + (c_2 + 1)t + c_4 e^{-t} - \frac{1}{2}t^2.$$

18. From $Dx + z = e^t$, $(D-1)x + Dy + Dz = 0$, and $x + 2y + Dz = e^t$ we obtain $z = -Dx + e^t$, $Dz = -D^2 x + e^t$, and the system $(-D^2 + D - 1)x + Dy = -e^t$ and $(-D^2 + 1)x + 2y = 0$. Then $y = \frac{1}{2}(D^2 - 1)x$, $Dy = \frac{1}{2}D(D^2 - 1)x$, and $(D-2)(D^2 + 1)x = -2e^t$ so that the solution is

$$x = c_1 e^{2t} + c_2 \cos t + c_3 \sin t + e^t$$

$$y = \frac{3}{2}c_1 e^{2t} - c_2 \cos t - c_3 \sin t$$

$$z = -2c_1 e^{2t} - c_3 \cos t + c_2 \sin t.$$

21. From $(D+5)x + y = 0$ and $4x - (D+1)y = 0$ we obtain $y = -(D+5)x$ so that $Dy = -(D^2 + 5D)x$. Then $4x + (D^2 + 5D)x + (D+5)x = 0$ and $(D+3)^2 x = 0$. Thus

$$x = c_1 e^{-3t} + c_2 t e^{-3t}$$

$$y = -(2c_1 + c_2)e^{-3t} - 2c_2 t e^{-3t}.$$

Using $x(1) = 0$ and $y(1) = 1$ we obtain

$$c_1 e^{-3} + c_2 e^{-3} = 0$$

$$-(2c_1 + c_2)e^{-3} - 2c_2 e^{-3} = 1$$

or

$$c_1 + c_2 = 0$$

$$2c_1 + 3c_2 = -e^3.$$

Thus $c_1 = e^3$ and $c_2 = -e^3$. The solution of the initial value problem is

$$x = e^{-3t+3} - te^{-3t+3}$$

$$y = -e^{-3t+3} + 2te^{-3t+3}.$$

Mathematical Models

24. From Newton's second law in the x-direction we have

$$m\frac{d^2 x}{dt^2} = -k\cos\theta = -k\frac{1}{v}\frac{dx}{dt} = -|c|\frac{dx}{dt}\,.$$

In the y-direction we have

$$m\frac{d^2 y}{dt^2} = -mg - k\sin\theta = -mg - k\frac{1}{v}\frac{dy}{dt} = -mg - |c|\frac{dy}{dt}\,.$$

From $mD^2 x + |c|Dx = 0$ we have $D(mD + |c|)x = 0$. Then $(mD + |c|)x = c_1$ or $(D + |c|/m)x = c_2$. This is a linear first-order differential equation. An integrating factor is $e^{\int |c| dt/m} = e^{|c|t/m}$ so that

$$\frac{d}{dt}[e^{|c|t/m}x] = c_2 e^{|c|t/m}$$

and $e^{|c|t/m}x = (c_2 m/|c|)e^{|c|t/m} + c_3$. The general solution of this equation is

$$x(t) = c_4 + c_3 e^{-|c|t/m}.$$

From $(mD^2 + |c|D)y = -mg$ we have $D(mD + |c|)y = -mg$ so that $(mD + |c|)y = -mgt + c_1$ or $(D + |c|/m)y = -gt + c_2$. This is a linear first-order differential equation with integrating factor $e^{\int |c| dt/m} = e^{|c|t/m}$. Thus

$$\frac{d}{dt}[e^{|c|t/m}y] = (-gt + c_2)e^{|c|t/m}$$

$$e^{|c|t/m}y = -\frac{mg}{|c|}te^{|c|t/m} + \frac{m^2g}{c^2}e^{|c|t/m} + c_3e^{|c|t/m} + c_4$$

and

$$y(t) = -\frac{mg}{|c|}t + \frac{m^2g}{c^2} + c_3 + c_4e^{-|c|t/m}.$$

4.10 | Nonlinear Differential Equations

The terminology and concepts listed below provide an outline of the main ideas encountered in this section. These can be useful when preparing for a quiz or test.

Terminology and Concepts

- superposition of solutions does *not* apply to nonlinear DEs
- nonlinear DEs can possess singular solutions
- using a substitution, express a nonlinear second-order DE, $F(x, y', y'') = 0$, as a first-order DE
- using a substitution, express a nonlinear second-order DE, $F(y, y', y'') = 0$, as a first-order DE
- Taylor series solution of an IVP

The basic skills listed below summarize the more mechanical types of problems encountered in the exercise set for this section.

Basic Skills

- solve a nonlinear second-order DE using reduction of order when appropriate
- use a substitution to find a Taylor series solution of a nonlinear second-order IVP

3. Let $u = y'$ so that $u' = y''$. The equation becomes $u' = -u^2 - 1$ which is separable. Thus

$$\frac{du}{u^2 + 1} = -dx \quad \Longrightarrow \quad \tan^{-1} u = -x + c_1 \quad \Longrightarrow \quad y' = \tan(c_1 - x)$$

$$\Longrightarrow \quad y = \ln|\cos(c_1 - x)| + c_2.$$

6. Let $u = y'$ so that $y'' = u\, du/dy$. The equation becomes $(y+1)u\, du/dy = u^2$. Separating variables we obtain

$$\frac{du}{u} = \frac{dy}{y+1} \quad \Longrightarrow \quad \ln|u| = \ln|y+1| + \ln c_1 \quad \Longrightarrow \quad u = c_1(y+1)$$

$$\Longrightarrow \quad \frac{dy}{dx} = c_1(y+1) \quad \Longrightarrow \quad \frac{dy}{y+1} = c_1\, dx$$

$$\Longrightarrow \quad \ln|y+1| = c_1 x + c_2 \quad \Longrightarrow \quad y + 1 = c_3 e^{c_1 x}.$$

9. Letting $u = y'$ we have

$$y'' = \frac{du}{dx} = \frac{du}{dy}\frac{dy}{dx} = u\frac{du}{dy} \qquad \text{so} \qquad 2y'y'' = 1 \qquad \text{becomes} \qquad 2u^2\frac{du}{dy} = 1.$$

Then, separating variables, integrating, and simplifying, we have

$$2u^2\, du = dy$$

$$\frac{2}{3}u^3 = y + c_1$$

$$u = \left(\frac{3}{2}y + c_2\right)^{1/3} = \frac{dy}{dx}$$

$$\left(\frac{3}{2}y + c_2\right)^{-1/3} dy = dx$$

$$\left(\frac{3}{2}y + c_2\right)^{2/3} = x + c_3$$

$$\frac{3}{2}y + c_2 = \left(x + c_3\right)^{3/2}.$$

Now

$$y(0) = 2 \quad \text{implies} \quad 3 + c_2 = c_3^{3/2} \qquad \text{and} \qquad y'(0) = 1 \quad \text{implies} \quad c_3 = 1.$$

Thus $c_2 = -2$ and $\frac{3}{2}y - 2 = (x+1)^{3/2}$. The solution of the initial-value problem is

$$y = \frac{2}{3}(x+1)^{3/2} + \frac{4}{3}.$$

12. Let $u = y'$ so that $u' = y''$. The equation becomes $(u')^2 + u^2 = 1$ which results in $u' = \pm\sqrt{1 - u^2}$. To solve $u' = \sqrt{1 - u^2}$ we separate variables:

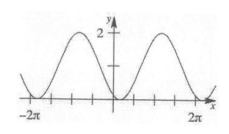

$$\frac{du}{\sqrt{1 - u^2}} = dx \quad \Longrightarrow \quad \sin^{-1} u = x + c_1 \quad \Longrightarrow \quad u = \sin(x + c_1)$$

$$\Longrightarrow \quad y' = \sin(x + c_1).$$

When $x = \pi/2$, $y' = \sqrt{3}/2$, so $\sqrt{3}/2 = \sin(\pi/2 + c_1)$ and $c_1 = -\pi/6$. Thus

$$y' = \sin\left(x - \frac{\pi}{6}\right) \quad \Longrightarrow \quad y = -\cos\left(x - \frac{\pi}{6}\right) + c_2.$$

When $x = \pi/2$, $y = 1/2$, so $1/2 = -\cos(\pi/2 - \pi/6) + c_2 = -1/2 + c_2$ and $c_2 = 1$. The solution of the initial-value problem is $y = 1 - \cos(x - \pi/6)$.

To solve $u' = -\sqrt{1 - u^2}$ we separate variables:

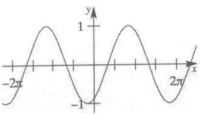

$$\frac{du}{\sqrt{1 - u^2}} = -dx \quad \Longrightarrow \quad \cos^{-1} u = x + c_1$$

$$\Longrightarrow \quad u = \cos(x + c_1) \quad \Longrightarrow \quad y' = \cos(x + c_1).$$

When $x = \pi/2$, $y' = \sqrt{3}/2$, so $\sqrt{3}/2 = \cos(\pi/2 + c_1)$ and $c_1 = -\pi/3$. Thus

$$y' = \cos\left(x - \frac{\pi}{3}\right) \quad \Longrightarrow \quad y = \sin\left(x - \frac{\pi}{3}\right) + c_2.$$

When $x = \pi/2$, $y = 1/2$, so $1/2 = \sin(\pi/2 - \pi/3) + c_2 = 1/2 + c_2$ and $c_2 = 0$. The solution of the initial-value problem is $y = \sin(x - \pi/3)$.

In Problems 15 and 18 the thinner curve is obtained using a numerical solver, while the thicker curve is the graph of the Taylor polynomial.

15. We look for a solution of the form

$$y(x) = y(0) + y'(0)x + \frac{1}{2!}y''(0)x^2 + \frac{1}{3!}y'''(0)x^3 +$$

$$\frac{1}{4!}y^{(4)}(0)x^4 + \frac{1}{5!}y^{(5)}(0)x^5.$$

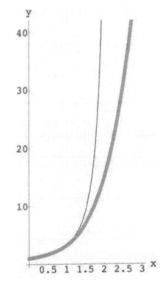

From $y''(x) = x + y^2$ we compute

$$y'''(x) = 1 + 2yy'$$

$$y^{(4)}(x) = 2yy'' + 2(y')^2$$

$$y^{(5)}(x) = 2yy''' + 6y'y''.$$

Using $y(0) = 1$ and $y'(0) = 1$ we find

$$y''(0) = 1, \quad y'''(0) = 3, \quad y^{(4)}(0) = 4, \quad y^{(5)}(0) = 12.$$

An approximate solution is

$$y(x) = 1 + x + \frac{1}{2}x^2 + \frac{1}{2}x^3 + \frac{1}{6}x^4 + \frac{1}{10}x^5.$$

18. We look for a solution of the form

$$y(x) = y(0) + y'(0)x + \frac{1}{2!}y''(0)x^2 + \frac{1}{3!}y'''(0)x^3 +$$

$$\frac{1}{4!}y^{(4)}(0)x^4 + \frac{1}{5!}y^{(5)}(0)x^5 + \frac{1}{6!}y^{(6)}(0)x^6.$$

From $y''(x) = e^y$ we compute

$$y'''(x) = e^y y'$$
$$y^{(4)}(x) = e^y(y')^2 + e^y y''$$
$$y^{(5)}(x) = e^y(y')^3 + 3e^y y' y'' + e^y y'''$$
$$y^{(6)}(x) = e^y(y')^4 + 6e^y(y')^2 y'' + 3e^y(y'')^2 + 4e^y y' y''' + e^y y^{(4)}.$$

Using $y(0) = 0$ and $y'(0) = -1$ we find

$$y''(0) = 1, \quad y'''(0) = -1, \quad y^{(4)}(0) = 2, \quad y^{(5)}(0) = -5, \quad y^{(6)}(0) = 16.$$

An approximate solution is

$$y(x) = -x + \frac{1}{2}x^2 - \frac{1}{6}x^3 + \frac{1}{12}x^4 + \frac{1}{24}x^5 + \frac{1}{45}x^6.$$

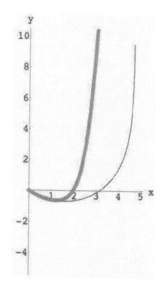

4.R | Chapter 4 in Review

3. False; it is not true unless the differential equation is homogeneous. For example, $y_1 = x$ is a solution of $y'' + y = x$, but $y_2 = 5x$ is not.

6. The roots are $m_1 = m_2 = m_3 = 0$ and $m_4 = 1$. To see this, note that the general solution of a homogeneous linear fourth-order differential equation with auxiliary equation $m^3(m - 1) = 0$ is $y = c_1 + c_2 x + c_3 x^2 + c_4 e^x$.

9. By the superposition principle for nonhomogeneous equations a particular solution is

$$y_p = y_{p_1} + y_{p_2} = x + x^2 - 2 = x^2 + x - 2.$$

12. (a) Since $f_2(x) = 2\ln x = 2f_1(x)$, the set of functions is linearly dependent.

(b) Since x^{n+1} is not a constant multiple of x^n, the set of functions is linearly independent.

(c) Since $x + 1$ is not a constant multiple of x, the set of functions is linearly independent.

(d) Since $f_1(x) = \cos x \cos(\pi/2) - \sin x \sin(\pi/2) = -\sin x = -f_2(x)$, the set of functions is linearly dependent.

(e) Since $f_1(x) = 0 \cdot f_2(x)$, the set of functions is linearly dependent.

(f) Since $2x$ is not a constant multiple of 2, the set of functions is linearly independent.

(g) Since $3(x^2) + 2(1 - x^2) - (2 + x^2) = 0$, the set of functions is linearly dependent.

(h) Since $xe^{x+1} + 0(4x - 5)e^x - exe^x = 0$, the set of functions is linearly dependent.

15. From $m^2 - 2m - 2 = 0$ we obtain $m = 1 \pm \sqrt{3}$ so that

$$y = c_1 e^{(1+\sqrt{3})x} + c_2 e^{(1-\sqrt{3})x}.$$

18. From $2m^3 + 9m^2 + 12m + 5 = 0$ we obtain $m = -1$, $m = -1$, and $m = -5/2$ so that

$$y = c_1 e^{-5x/2} + c_2 e^{-x} + c_3 x e^{-x}.$$

21. Applying D^4 to the differential equation we obtain $D^4(D^2 - 3D + 5) = 0$. Then

$$y = \underbrace{e^{3x/2}\left(c_1 \cos\frac{\sqrt{11}}{2}x + c_2 \sin\frac{\sqrt{11}}{2}x\right)}_{y_c} + c_3 + c_4 x + c_5 x^2 + c_6 x^3$$

and $y_p = A + Bx + Cx^2 + Dx^3$. Substituting y_p into the differential equation yields

$$(5A - 3B + 2C) + (5B - 6C + 6D)x + (5C - 9D)x^2 + 5Dx^3 = -2x + 4x^3.$$

Equating coefficients gives $A = -222/625$, $B = 46/125$, $C = 36/25$, and $D = 4/5$. The general solution is

$$y = e^{3x/2}\left(c_1 \cos\frac{\sqrt{11}}{2}x + c_2 \sin\frac{\sqrt{11}}{2}x\right) - \frac{222}{625} + \frac{46}{125}x + \frac{36}{25}x^2 + \frac{4}{5}x^3.$$

24. Applying D to the differential equation we obtain $D(D^3 - D^2) = D^3(D - 1) = 0$. Then

$$y = \underbrace{c_1 + c_2 x + c_3 e^x}_{y_c} + c_4 x^2$$

and $y_p = Ax^2$. Substituting y_p into the differential equation yields $-2A = 6$. Equating coefficients gives $A = -3$. The general solution is

$$y = c_1 + c_2 x + c_3 e^x - 3x^2.$$

27. The auxiliary equation is $6m^2 - m - 1 = 0$ so that

$$y = c_1 x^{1/2} + c_2 x^{-1/3}.$$

30. The auxiliary equation is $m^2 - 2m + 1 = (m-1)^2 = 0$ and a particular solution is $y_p = \frac{1}{4}x^3$ so that

$$y = c_1 x + c_2 x \ln x + \frac{1}{4}x^3.$$

33. (a) The auxiliary equation is $m^4 - 2m^2 + 1 = (m^2 - 1)^2 = 0$, so the general solution of the differential equation is

$$y = c_1 \sinh x + c_2 \cosh x + c_3 x \sinh x + c_4 x \cosh x.$$

(b) Since both $\sinh x$ and $x \sinh x$ are solutions of the associated homogeneous differential equation, a particular solution of $y^{(4)} - 2y'' + y = \sinh x$ has the form

$$y_p = Ax^2 \sinh x + Bx^2 \cosh x.$$

36. The auxiliary equation is $m^2 + 2m + 1 = (m+1)^2 = 0$, so that $y = c_1 e^{-x} + c_2 x e^{-x}$. Setting $y(-1) = 0$ and $y'(0) = 0$ we get $c_1 e - c_2 e = 0$ and $-c_1 + c_2 = 0$. Thus $c_1 = c_2$ and $y = c_1(e^{-x} + x e^{-x})$ is a solution of the boundary-value problem for any real number c_1.

39. Let $u = y'$ so that $u' = y''$. The equation becomes $u\, du/dx = 4x$. Separating variables we obtain

$$u\, du = 4x\, dx \quad \Longrightarrow \quad \frac{1}{2}u^2 = 2x^2 + c_1 \quad \Longrightarrow \quad u^2 = 4x^2 + c_2.$$

When $x = 1$, $y' = u = 2$, so $4 = 4 + c_2$ and $c_2 = 0$. Then

$$u^2 = 4x^2 \quad \Longrightarrow \quad \frac{dy}{dx} = 2x \quad \text{or} \quad \frac{dy}{dx} = -2x$$

$$\Longrightarrow \quad y = x^2 + c_3 \quad \text{or} \quad y = -x^2 + c_4.$$

When $x = 1$, $y = 5$, so $5 = 1 + c_3$ and $5 = -1 + c_4$. Thus $c_3 = 4$ and $c_4 = 6$. We have $y = x^2 + 4$ and $y = -x^2 + 6$. Note however that when $y = -x^2 + 6$, $y' = -2x$ and $y'(1) = -2 \neq 2$. Thus, the solution of the initial-value problem is $y = x^2 + 4$.

42. Consider $xy'' + y' = 0$ and look for a solution of the form $y = x^m$. Substituting into the differential equation we have

$$xy'' + y' = m(m-1)x^{m-1} + mx^{m-1} = m^2 x^{m-1}.$$

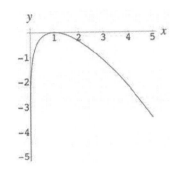

Thus, the general solution of $xy'' + y' = 0$ is $y_c = c_1 + c_2 \ln x$. To find a particular solution of $xy'' + y' = -\sqrt{x}$ we use variation of parameters. The Wronskian is $W = 1/x$. Identifying $f(x) = -x^{-1/2}$ we obtain

$$u_1' = \frac{x^{-1/2} \ln x}{1/x} = \sqrt{x} \ln x \quad \text{and} \quad u_2' = \frac{-x^{-1/2}}{1/x} = -\sqrt{x} ,$$

so that

$$u_1 = x^{3/2}\left(\frac{2}{3} \ln x - \frac{4}{9}\right) \quad \text{and} \quad u_2 = -\frac{2}{3}x^{3/2}.$$

Then

$$y_p = x^{3/2}\left(\frac{2}{3} \ln x - \frac{4}{9}\right) - \frac{2}{3}x^{3/2} \ln x = -\frac{4}{9}x^{3/2}$$

and the general solution of the differential equation is

$$y = c_1 + c_2 \ln x - \frac{4}{9}x^{3/2}.$$

The initial conditions are $y(1) = 0$ and $y'(1) = 0$. These imply that $c_1 = \frac{4}{9}$ and $c_2 = \frac{2}{3}$. The solution of the initial-value problem is

$$y = \frac{4}{9} + \frac{2}{3} \ln x - \frac{4}{9}x^{3/2}.$$

The graph is shown above.

45. From $(D-2)x - y = -e^t$ and $-3x + (D-4)y = -7e^t$ we obtain $(D-1)(D-5)x = -4e^t$ so that

$$x = c_1 e^t + c_2 e^{5t} + te^t.$$

Then

$$y = (D-2)x + e^t = -c_1 e^t + 3c_2 e^{5t} - te^t + 2e^t.$$

Modeling with Higher-Order Differential Equations

5.1 | Linear Models: Initial-Value Problems

The terminology and concepts listed below provide an outline of the main ideas encountered in this section. These can be useful when preparing for a quiz or test.

Terminology and Concepts

- Hooke's law

- spring constant (k)

- Newton's second law of motion

- mass (m) versus weight $(W = mg)$

- equilibrium position

- DE of free undamped motion: $m\, d^2x/dt^2 + kx = 0$ or $d^2x/dt^2 + \omega^2 x = 0$, where $\omega^2 = k/m$

- equation of free undamped (simple harmonic) motion: $x(t) = c_1 \cos \omega t + c_2 \sin \omega t$

- period of motion: $T = 2\pi/\omega$

- cycle

- frequency of motion: $f = \omega/2\pi$

- amplitude of free vibrations (A)

- phase angle (ϕ)

- damping constant (β)

- DE of free damped motion: $m\, d^2x/dt^2 + \beta\, dx/dt + kx = 0$ or $d^2x/dt^2 + 2\lambda\, dx/dt + \omega^2 x = 0$ where $2\lambda = \beta/m$ and $\omega^2 = k/m$

- equation of motion of an overdamped system: $x(t) = e^{-\lambda t}(c_1 e^{\sqrt{\lambda^2 - \omega^2}\, t} + c_2 e^{-\sqrt{\lambda^2 - \omega^2}\, t})$

- equation of motion of a critically damped system: $x(t) = e^{-\lambda t}(c_1 + c_2 t)$

- equation of motion of an underdamped system:

$$x(t) = e^{-\lambda t}\left(c_1 \cos \sqrt{\omega^2 - \lambda^2}\, t + c_2 \sin \sqrt{\omega^2 - \lambda^2}\, t\right)$$

- damped amplitude: $Ae^{-\lambda t}$

- quasi period: $2\pi/\sqrt{\omega^2 - \lambda^2}$

- DE of driven (forced) motion with damping: $m\, d^2x/dt^2 + \beta\, dx/dt + kx = f(t)$

- transient term (solution)

- steady-state term (solution)

- DE of driven motion without damping: $m\, d^2x/dt^2 + kx = f(t)$

- pure resonance

- aging spring

- Airy's differential equation

- LRC series electrical circuit

- free electrical vibrations

- reactance

- impedance

The basic skills listed below summarize the more mechanical types of problems encountered in the exercise set for this section.

Basic Skills

- set up and solve DEs modeling free undamped motion of a spring/mass system

- set up and solve DEs modeling free damped motion of a spring/mass system

- set up and solve DEs modeling driven (forced) motion of a spring/mass system

- set up and solve DEs modeling an LRC series circuit

Phase Angle   As pointed out in the text, be careful in computing the phase angle ϕ in a solution of the form $x(t) = A\sin(\omega t + \phi)$. The problem with using the inverse tangent $\tan^{-1} X$ is that all calculators are programmed for this particular inverse trigonometric function to give angular values that are either in the first or fourth quadrants. In other words, $-\pi/2 < \tan^{-1} X < \pi/2$. For example, because of periodicity there are infinitely many angles for which $\tan\phi = 1$ and for which $\tan\phi = -1/\sqrt{3}$. But there is only *one* answer when we use the notation $\tan^{-1} 1$ and $\tan^{-1}(-1/\sqrt{3})$. The exact values are:

$$\tan^{-1} 1 = \frac{\pi}{4} \quad \text{(first quadrant) because } \tan\frac{\pi}{4} = 1,$$

$$\tan^{-1}\left(-\frac{1}{\sqrt{3}}\right) = -\frac{\pi}{6} \quad \text{(fourth quadrant) because } \tan\left(-\frac{\pi}{6}\right) = -\frac{1}{\sqrt{3}}.$$

Of course, if you use a calculator you will either get

$$\tan^{-1} 1 = 45° \qquad \text{or} \qquad \tan^{-1} 1 = 0.78539816\ldots$$

depending on whether your calculator is set in degree or radian mode. The decimal number $0.78539816\ldots$ is the number π divided by 4. Now in order to compute ϕ in $x(t) = A\sin(\omega t + \phi)$ we must take into consideration all three components given in (7) of Section 5.1 in the text. Since $A > 0$, the signs of $\sin\phi = c_1/A$ and $\cos\phi = c_2/A$ are determined from the algebraic signs of c_1 and c_2 and these signs, in turn, indicate the quadrant in which ϕ lies. We summarize the four possibilities:

$$\phi = \tan^{-1}\left(\frac{c_1}{c_2}\right) \quad \text{if} \quad \overbrace{c_1 > 0,\ c_2 > 0}^{\phi \text{ in 1st quadrant}} \quad \text{or if} \quad \overbrace{c_1 < 0,\ c_2 > 0}^{\phi \text{ in 4th quadrant}}$$

$$\phi = \pi + \tan^{-1}\left(\frac{c_1}{c_2}\right) \quad \text{if} \quad \overbrace{c_1 > 0,\ c_2 < 0}^{\phi \text{ in 2nd quadrant}} \quad \text{or if} \quad \overbrace{c_1 < 0,\ c_2 < 0}^{\phi \text{ in 3rd quadrant}}.$$

Damping   In Examples 4 and 5 in the text, the statements

"...a damping force numerically equal to 2 times the instantaneous velocity..."

and

"...surrounding medium offers a resistance numerically equal to the instantaneous velocity..."

are to be interpreted: In Example 4, the damping constant is $\beta = 2$ in equation (10) and in Example 5, the damping constant is $\beta = 1$ in equation (10).

5.1.1 Spring/Mass Systems: Free Undamped Motion

3. From $\frac{3}{4}x'' + 72x = 0$, $x(0) = -1/4$, and $x'(0) = 0$ we obtain $x = -\frac{1}{4}\cos 4\sqrt{6}\,t$.

6. From $50x'' + 200x = 0$, $x(0) = 0$, and $x'(0) = -10$ we obtain $x = -5\sin 2t$ and $x' = -10\cos 2t$.

9. (a) Since the weight is 8 pounds the mass is $8/32 = 1/4$ slug. The spring constant is $k = 1$ so $\omega^2 = 1/\frac{1}{4} = 4$ and the equation of motion is

$$\frac{1}{4}x'' + x = 0 \qquad \text{or} \qquad x'' + 4x = 0.$$

From the initial conditions we have $x(0) = 6$ in $= \frac{1}{2}$ ft and $x'(0) = \frac{3}{2}$ ft/sec. By (3) in the text the form of the equation of motion is

$$x(t) = c_1 \cos 2t + c_2 \sin 2t.$$

The initial conditions imply that $c_1 = \frac{1}{2}$ and $c_2 = \frac{3}{4}$. Thus the equation of motion of the system is

$$x(t) = \frac{1}{2} \cos 2t + \frac{3}{4} \sin 2t.$$

(b) To write the solution in the form $x(t) = A \sin(\omega t + \phi)$ we note that

$$A = \sqrt{\left(\frac{1}{2}\right)^2 + \left(\frac{3}{4}\right)^2} = \frac{1}{4}\sqrt{4+9} = \frac{\sqrt{13}}{4},$$

and, initially at least,

$$\phi = \tan^{-1}\frac{c_1}{c_2} = \tan^{-1}\frac{1/2}{3/4} = \tan^{-1}\frac{2}{3} \approx 0.5880 \text{ rad.}$$

Since $\sin\phi > 0$ and $\cos\phi > 0$, ϕ is in the first quadrant and

$$x(t) = \frac{\sqrt{13}}{4} \sin(2t + 0.5880).$$

(c) To write the system in the form $x(t) = A \cos(\omega t - \phi)$ we note that $A = \sqrt{13}/4$ as in part **(b)** and

$$\phi = \tan^{-1}\frac{3/4}{1/2} = \frac{3}{2} \approx 0.9828.$$

Since $\cos\phi > 0$ and $\sin\phi > 0$, ϕ is in the first quadrant and

$$x(t) = \frac{\sqrt{13}}{4} \cos(2t - 0.9828).$$

12. From $x'' + 9x = 0$, $x(0) = -1$, and $x'(0) = -\sqrt{3}$ we obtain

$$x = -\cos 3t - \frac{\sqrt{3}}{3} \sin 3t = \frac{2}{\sqrt{3}} \sin\left(3t + \frac{4\pi}{3}\right)$$

and $x' = 2\sqrt{3} \cos(3t + 4\pi/3)$. If $x' = 3$ then $t = -7\pi/18 + 2n\pi/3$ and $t = -\pi/2 + 2n\pi/3$ for $n = 1, 2, 3, \ldots$.

15. For large values of t the differential equation is approximated by $x'' = 0$. The solution of this equation is the linear function $x = c_1 t + c_2$. Thus, for large time, the restoring force will have decayed to the point where the spring is incapable of returning the mass, and the spring will simply keep on stretching.

5.1.2 Spring/Mass Systems: Free Damped Motion

18. (a) below   (b) from rest

21. From $\frac{1}{8}x'' + x' + 2x = 0$, $x(0) = -1$, and $x'(0) = 8$ we obtain $x = 4te^{-4t} - e^{-4t}$ and $x' = 8e^{-4t} - 16te^{-4t}$. If $x = 0$ then $t = 1/4$ second. If $x' = 0$ then $t = 1/2$ second and the extreme displacement is $x = e^{-2}$ feet.

24. (a) $x = \frac{1}{3}e^{-8t}\left(4e^{6t} - 1\right)$ is not zero for $t \geq 0$; the extreme displacement is $x(0) = 1$ meter.

 (b) $x = \frac{1}{3}e^{-8t}\left(5 - 2e^{6t}\right) = 0$ when $t = \frac{1}{6}\ln\frac{5}{2} \approx 0.153$ second; if $x' = \frac{4}{3}e^{-8t}\left(e^{6t} - 10\right) = 0$ then $t = \frac{1}{6}\ln 10 \approx 0.384$ second and the extreme displacement is $x = -0.232$ meter.

27. From $\frac{5}{16}x'' + \beta x' + 5x = 0$ we find that the roots of the auxiliary equation are

$$m = -\frac{8}{5}\beta \pm \frac{4}{5}\sqrt{4\beta^2 - 25}\,.$$

 (a) If $4\beta^2 - 25 > 0$ then $\beta > 5/2$.

 (b) If $4\beta^2 - 25 = 0$ then $\beta = 5/2$.

 (c) If $4\beta^2 - 25 < 0$ then $0 < \beta < 5/2$.

5.1.3 Spring/Mass Systems: Driven Motion

30. (a) If $x'' + 2x' + 5x = 12\cos 2t + 3\sin 2t$, $x(0) = 1$, and $x'(0) = 5$ then

$$x_c = e^{-t}(c_1\cos 2t + c_2\sin 2t) \qquad\text{and}\qquad x_p = 3\sin 2t$$

so that the equation of motion is $x = e^{-t}\cos 2t + 3\sin 2t$.

(b)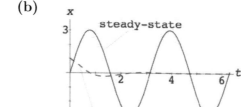

(c)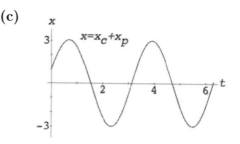

33. From $2x'' + 32x = 68e^{-2t}\cos 4t$, $x(0) = 0$, and $x'(0) = 0$ we obtain $x_c = c_1\cos 4t + c_2\sin 4t$ and $x_p = \frac{1}{2}e^{-2t}\cos 4t - 2e^{-2t}\sin 4t$ so that

$$x = -\frac{1}{2}\cos 4t + \frac{9}{4}\sin 4t + \frac{1}{2}e^{-2t}\cos 4t - 2e^{-2t}\sin 4t.$$

36. (a) From $100x'' + 1600x = 1600\sin 8t$, $x(0) = 0$, and $x'(0) = 0$ we obtain $x_c = c_1\cos 4t + c_2\sin 4t$ and $x_p = -\frac{1}{3}\sin 8t$ so that by a trig identity

$$x = \frac{2}{3}\sin 4t - \frac{1}{3}\sin 8t = \frac{2}{3}\sin 4t - \frac{2}{3}\sin 4t\cos 4t.$$

(b) If $x = \frac{1}{3}\sin 4t(2 - 2\cos 4t) = 0$ then $t = n\pi/4$ for $n = 0, 1, 2, \dots$.

(c) If $x' = \frac{8}{3}\cos 4t - \frac{8}{3}\cos 8t = \frac{8}{3}(1 - \cos 4t)(1 + 2\cos 4t) = 0$ then $t = \pi/3 + n\pi/2$ and $t = \pi/6 + n\pi/2$ for $n = 0, 1, 2, \dots$ at the extreme values. There are many other values of t for which $x' = 0$.

(d) $x(\pi/6 + n\pi/2) = \sqrt{3}/2$ cm and $x(\pi/3 + n\pi/2) = -\sqrt{3}/2$ cm

(e)

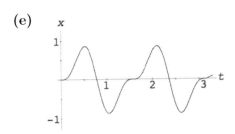

39. (a) From $x'' + \omega^2 x = F_0\cos\gamma t$, $x(0) = 0$, and $x'(0) = 0$ we obtain $x_c = c_1\cos\omega t + c_2\sin\omega t$ and $x_p = (F_0\cos\gamma t)/(\omega^2 - \gamma^2)$ so that

$$x = -\frac{F_0}{\omega^2 - \gamma^2}\cos\omega t + \frac{F_0}{\omega^2 - \gamma^2}\cos\gamma t.$$

(b) $\displaystyle\lim_{\gamma\to\omega}\frac{F_0}{\omega^2 - \gamma^2}(\cos\gamma t - \cos\omega t) = \lim_{\gamma\to\omega}\frac{-F_0 t\sin\gamma t}{-2\gamma} = \frac{F_0}{2\omega}t\sin\omega t.$

5.1.4 Series Circuit Analogue

45. Solving $\frac{1}{20}q'' + 2q' + 100q = 0$ we obtain $q(t) = e^{-20t}(c_1\cos 40t + c_2\sin 40t)$. The initial conditions $q(0) = 5$ and $q'(0) = 0$ imply $c_1 = 5$ and $c_2 = 5/2$. Thus

$$q(t) = e^{-20t}\left(5\cos 40t + \frac{5}{2}\sin 40t\right) = \sqrt{25 + 25/4}\,e^{-20t}\sin(40t + 1.1071)$$

and $q(0.01) \approx 4.5676$ coulombs. The charge is zero for the first time when $40t + 1.1071 = \pi$ or $t \approx 0.0509$ second.

48. Solving $q'' + 100q' + 2500q = 30$ we obtain $q(t) = c_1 e^{-50t} + c_2 t e^{-50t} + 0.012$. The initial conditions $q(0) = 0$ and $q'(0) = 2$ imply $c_1 = -0.012$ and $c_2 = 1.4$. Thus, using $i(t) = q'(t)$ we get

$$q(t) = -0.012 e^{-50t} + 1.4 t e^{-50t} + 0.012 \quad \text{and} \quad i(t) = 2e^{-50t} - 70 t e^{-50t}.$$

Solving $i(t) = 0$ we see that the maximum charge occurs when $t = 1/35$ second and $q(1/35) \approx 0.01871$ coulomb.

51. The differential equation is $\frac{1}{2}q'' + 20q' + 1000q = 100 \sin 60t$. To use Example 10 in the text we identify $E_0 = 100$ and $\gamma = 60$. Then

$$X = L\gamma - \frac{1}{c\gamma} = \frac{1}{2}(60) - \frac{1}{0.001(60)} \approx 13.3333,$$

$$Z = \sqrt{X^2 + R^2} = \sqrt{X^2 + 400} \approx 24.0370,$$

and

$$\frac{E_0}{Z} = \frac{100}{Z} \approx 4.1603.$$

From Problem 50, then

$$i_p(t) \approx 4.1603 \sin(60t + \phi)$$

where $\sin\phi = -X/Z$ and $\cos\phi = R/Z$. Thus $\tan\phi = -X/R \approx -0.6667$ and ϕ is a fourth quadrant angle. Now $\phi \approx -0.5880$ and

$$i_p(t) = 4.1603 \sin(60t - 0.5880).$$

54. In Problem 50 it is shown that the amplitude of the steady-state current is E_0/Z, where $Z = \sqrt{X^2 + R^2}$ and $X = L\gamma - 1/C\gamma$. Since E_0 is constant the amplitude will be a maximum when Z is a minimum. Since R is constant, Z will be a minimum when $X = 0$. Solving $L\gamma - 1/C\gamma = 0$ for γ we obtain $\gamma = 1/\sqrt{LC}$. The maximum amplitude will be E_0/R.

57. In an LC-series circuit there is no resistor, so the differential equation is

$$L\frac{d^2 q}{dt^2} + \frac{1}{C}q = E(t).$$

Then $q(t) = c_1 \cos\left(t/\sqrt{LC}\right) + c_2 \sin\left(t/\sqrt{LC}\right) + q_p(t)$ where $q_p(t) = A \sin\gamma t + B\cos\gamma t$. Substituting $q_p(t)$ into the differential equation we find

$$\left(\frac{1}{C} - L\gamma^2\right)A\sin\gamma t + \left(\frac{1}{C} - L\gamma^2\right)B\cos\gamma t = E_0\cos\gamma t.$$

Equating coefficients we obtain $A = 0$ and $B = E_0 C/(1 - LC\gamma^2)$. Thus, the charge is

$$q(t) = c_1 \cos\frac{1}{\sqrt{LC}}t + c_2 \sin\frac{1}{\sqrt{LC}}t + \frac{E_0 C}{1 - LC\gamma^2}\cos\gamma t.$$

The initial conditions $q(0) = q_0$ and $q'(0) = i_0$ imply $c_1 = q_0 - E_0C/(1 - LC\gamma^2)$ and $c_2 = i_0\sqrt{LC}$. The current is $i(t) = q'(t)$ or

$$i(t) = -\frac{c_1}{\sqrt{LC}}\sin\frac{1}{\sqrt{LC}}t + \frac{c_2}{\sqrt{LC}}\cos\frac{1}{\sqrt{LC}}t - \frac{E_0C\gamma}{1 - LC\gamma^2}\sin\gamma t$$

$$= i_0\cos\frac{1}{\sqrt{LC}}t - \frac{1}{\sqrt{LC}}\left(q_0 - \frac{E_0C}{1 - LC\gamma^2}\right)\sin\frac{1}{\sqrt{LC}}t - \frac{E_0C\gamma}{1 - LC\gamma^2}\sin\gamma t.$$

5.2 Linear Models: Boundary-Value Problems

The terminology and concepts listed below provide an outline of the main ideas encountered in this section. These can be useful when preparing for a quiz or test.

Terminology and Concepts

- axis of symmetry of a beam
- deflection curve or elastic curve of a beam
- cantilever beam
- eigenvalues and corresponding eigenfunctions of a boundary-value problem

The basic skills listed below summarize the more mechanical types of problems encountered in the exercise set for this section.

Basic Skills

- solve problems involving beams satisfying various end (boundary) conditions such as embedded, free, or simply supported
- find the eigenvalues and corresponding eigenfunctions for second-order boundary-value problems

3. **(a)** The general solution is

$$y(x) = c_1 + c_2x + c_3x^2 + c_4x^3 + \frac{w_0}{24EI}x^4.$$

The boundary conditions are $y(0) = 0$, $y'(0) = 0$, $y(L) = 0$, $y''(L) = 0$. The first two conditions give $c_1 = 0$ and $c_2 = 0$. The conditions at $x = L$ give the system

$$c_3L^2 + c_4L^3 + \frac{w_0}{24EI}L^4 = 0$$

$$2c_3 + 6c_4 L + \frac{w_0}{2EI} L^2 = 0.$$

Solving, we obtain $c_3 = w_0 L^2/16EI$ and $c_4 = -5w_0 L/48EI$. The deflection is

$$y(x) = \frac{w_0}{48EI}(3L^2 x^2 - 5Lx^3 + 2x^4).$$

6. (a) $y_{\max} = y(L) = w_0 L^4/8EI$

(b) Replacing both L and x by $L/2$ in $y(x)$ we obtain $w_0 L^4/128EI$, which is $1/16$ of the maximum deflection when the length of the beam is L.

(c) $y_{\max} = y(L/2) = 5w_0 L^4/384EI$

(d) The maximum deflection in Example 1 is $y(L/2) = (w_0/24EI)L^4/16 = w_0 L^4/384EI$, which is $1/5$ of the maximum displacement of the beam in part **(c)**.

9. This is Example 2 in the text with $L = \pi$. The eigenvalues are $\lambda_n = n^2\pi^2/\pi^2 = n^2$, $n = 1, 2, 3, \ldots$ and the corresponding eigenfunctions are $y_n = \sin(n\pi x/\pi) = \sin nx$, $n = 1, 2, 3, \ldots$.

12. For $\lambda \le 0$ the only solution of the boundary-value problem is $y = 0$. For $\lambda = \alpha^2 > 0$ we have

$$y = c_1 \cos \alpha x + c_2 \sin \alpha x.$$

Since $y(0) = 0$ implies $c_1 = 0$, $y = c_2 \sin x\, dx$. Now

$$y'\left(\frac{\pi}{2}\right) = c_2 \alpha \cos \alpha \frac{\pi}{2} = 0$$

gives

$$\alpha \frac{\pi}{2} = \frac{(2n-1)\pi}{2} \quad \text{or} \quad \lambda = \alpha^2 = (2n-1)^2, \ n = 1, 2, 3, \ldots.$$

The eigenvalues $\lambda_n = (2n-1)^2$ correspond to the eigenfunctions $y_n = \sin(2n-1)x$.

15. The auxiliary equation has solutions

$$m = \frac{1}{2}\left(-2 \pm \sqrt{4 - 4(\lambda + 1)}\right) = -1 \pm \alpha.$$

For $\lambda = -\alpha^2 < 0$ we have

$$y = e^{-x}\left(c_1 \cosh \alpha x + c_2 \sinh \alpha x\right).$$

The boundary conditions imply

$$y(0) = c_1 = 0$$

$$y(5) = c_2 e^{-5} \sinh 5\alpha = 0$$

so $c_1 = c_2 = 0$ and the only solution of the boundary-value problem is $y = 0$.

For $\lambda = 0$ we have

$$y = c_1 e^{-x} + c_2 x e^{-x}$$

and the only solution of the boundary-value problem is $y = 0$.

For $\lambda = \alpha^2 > 0$ we have

$$y = e^{-x}\left(c_1 \cos \alpha x + c_2 \sin \alpha x\right).$$

Now $y(0) = 0$ implies $c_1 = 0$, so

$$y(5) = c_2 e^{-5} \sin 5\alpha = 0$$

gives

$$5\alpha = n\pi \quad \text{or} \quad \lambda = \alpha^2 = \frac{n^2 \pi^2}{25}, \quad n = 1, 2, 3, \ldots.$$

The eigenvalues $\lambda_n = \dfrac{n^2 \pi^2}{25}$ correspond to the eigenfunctions $y_n = e^{-x} \sin \dfrac{n\pi}{5} x$ for $n = 1, 2, 3, \ldots$.

18. For $\lambda = 0$ the general solution is $y = c_1 + c_2 \ln x$. Now $y' = c_2/x$, so $y'(e^{-1}) = c_2 e = 0$ implies $c_2 = 0$. Then $y = c_1$ and $y(1) = 0$ gives $c_1 = 0$. Thus $y(x) = 0$.

For $\lambda = -\alpha^2 < 0$, $y = c_1 x^{-\alpha} + c_2 x^{\alpha}$. The boundary conditions give $c_2 = c_1 e^{2\alpha}$ and $c_1 = 0$, so that $c_2 = 0$ and $y(x) = 0$.

For $\lambda = \alpha^2 > 0$, $y = c_1 \cos(\alpha \ln x) + c_2 \sin(\alpha \ln x)$. From $y(1) = 0$ we obtain $c_1 = 0$ and $y = c_2 \sin(\alpha \ln x)$. Now $y' = c_2(\alpha/x) \cos(\alpha \ln x)$, so $y'(e^{-1}) = c_2 e \alpha \cos \alpha = 0$ implies $\cos \alpha = 0$ or $\alpha = (2n-1)\pi/2$ and $\lambda = \alpha^2 = (2n-1)^2 \pi^2/4$ for $n = 1, 2, 3, \ldots$. The corresponding eigenfunctions are

$$y_n = \sin\left(\frac{2n-1}{2}\pi \ln x\right).$$

21. If restraints are put on the column at $x = L/4$, $x = L/2$, and $x = 3L/4$, then the critical load will be P_4.

24. (a) The boundary-value problem is

$$\frac{d^4y}{dx^4} + \lambda \frac{d^2y}{dx^2} = 0, \quad y(0) = 0, y''(0) = 0, \ y(L) = 0, y'(L) = 0,$$

where $\lambda = \alpha^2 = P/EI$. The solution of the differential equation is $y = c_1 \cos \alpha x + c_2 \sin \alpha x + c_3 x + c_4$ and the conditions $y(0) = 0$, $y''(0) = 0$ yield $c_1 = 0$ and $c_4 = 0$. Next, by applying $y(L) = 0$, $y'(L) = 0$ to $y = c_2 \sin \alpha x + c_3 x$ we get the system of equations

$$c_2 \sin \alpha L + c_3 L = 0$$

$$\alpha c_2 \cos \alpha L + c_3 \ = 0.$$

To obtain nontrivial solutions c_2, c_3, we must have the determinant of the coefficients equal to zero:

$$\begin{vmatrix} \sin \alpha L & L \\ \alpha \cos \alpha L & 1 \end{vmatrix} = 0 \quad \text{or} \quad \tan \beta = \beta,$$

where $\beta = \alpha L$. If β_n denotes the positive roots of the last equation, then the eigenvalues are found from $\beta_n = \alpha_n L = \sqrt{\lambda_n}\, L$ or $\lambda_n = (\beta_n/L)^2$. From $\lambda = P/EI$ we see that the critical loads are $P_n = \beta_n^2 EI/L^2$. With the aid of a CAS we find that the first positive root of $\tan \beta = \beta$ is (approximately) $\beta_1 = 4.4934$, and so the Euler load is (approximately) $P_1 = 20.1907 EI/L^2$. Finally, if we use $c_3 = -c_2 \alpha \cos \alpha L$, then the deflection curves are

$$y_n(x) = c_2 \sin \alpha_n x + c_3 x = c_2 \left[\sin \left(\frac{\beta_n}{L} x \right) - \left(\frac{\beta_n}{L} \cos \beta_n \right) x \right].$$

(b) With $L = 1$ and c_2 appropriately chosen, the general shape of the first buckling mode,

$$y_1(x) = c_2 \left[\sin \left(\frac{4.4934}{L} x \right) - \left(\frac{4.4934}{L} \cos(4.4934) \right) x \right],$$

is shown below.

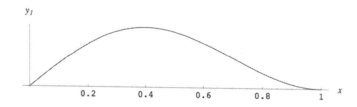

27. The auxiliary equation is $m^2 + m = m(m + 1) = 0$ so that $u(r) = c_1 r^{-1} + c_2$. The boundary conditions $u(a) = u_0$ and $u(b) = u_1$ yield the system $c_1 a^{-1} + c_2 = u_0$, $c_1 b^{-1} + c_2 = u_1$. Solving gives

$$c_1 = \left(\frac{u_0 - u_1}{b - a} \right) ab \quad \text{and} \quad c_2 = \frac{u_1 b - u_0 a}{b - a}.$$

Thus

$$u(r) = \left(\frac{u_0 - u_1}{b - a}\right)\frac{ab}{r} + \frac{u_1 b - u_0 a}{b - a}.$$

30. The initial-value problem is

$$x'' + \frac{2}{m}x' + \frac{k}{m}x = 0, \quad x(0) = 0, \ x'(0) = v_0.$$

With $k = 10$, the auxiliary equation has roots $\gamma = -1/m \pm \sqrt{1 - 10m}/m$. Consider the three cases:

(*i*) $m = \frac{1}{10}$. The roots are $\gamma_1 = \gamma_2 = 10$ and the solution of the differential equation is $x(t) = c_1 e^{-10t} + c_2 t e^{-10t}$. The initial conditions imply $c_1 = 0$ and $c_2 = v_0$ and so $x(t) = v_0 t e^{-10t}$. The condition $x(1) = 0$ implies $v_0 e^{-10} = 0$ which is impossible because $v_0 \neq 0$.

(*ii*) $1 - 10m > 0$ or $0 < m < \frac{1}{10}$. The roots are

$$\gamma_1 = -\frac{1}{m} - \frac{1}{m}\sqrt{1 - 10m} \quad \text{and} \quad \gamma_2 = -\frac{1}{m} + \frac{1}{m}\sqrt{1 - 10m}$$

and the solution of the differential equation is $x(t) = c_1 e^{\gamma_1 t} + c_2 e^{\gamma_2 t}$. The initial conditions imply

$$c_1 + c_2 = 0$$

$$\gamma_1 c_1 + \gamma_2 c_2 = v_0$$

so $c_1 = v_0/(\gamma_1 - \gamma_2)$, $c_2 = -v_0/(\gamma_1 - \gamma_2)$, and

$$x(t) = \frac{v_0}{\gamma_1 - \gamma_2}\left(e^{\gamma_1 t} - e^{\gamma_2 t}\right).$$

Again, $x(1) = 0$ is impossible because $v_0 \neq 0$.

(*iii*) $1 - 10m < 0$ or $m > \frac{1}{10}$. The roots of the auxiliary equation are

$$\gamma_1 = -\frac{1}{m} - \frac{1}{m}\sqrt{10m - 1}\,i \quad \text{and} \quad \gamma_2 = -\frac{1}{m} + \frac{1}{m}\sqrt{10m - 1}\,i$$

and the solution of the differential equation is

$$x(t) = c_1 e^{-t/m} \cos\frac{1}{m}\sqrt{10m - 1}\,t + c_2 e^{-t/m} \sin\frac{1}{m}\sqrt{10m - 1}\,t.$$

The initial conditions imply $c_1 = 0$ and $c_2 = mv_0/\sqrt{10m - 1}$, so that

$$x(t) = \frac{mv_0}{\sqrt{10m - 1}} e^{-t/m} \sin\left(\frac{1}{m}\sqrt{10m - 1}\,t\right),$$

The condition $x(1) = 0$ implies

$$\frac{mv_0}{\sqrt{10m - 1}} e^{-1/m} \sin\frac{1}{m}\sqrt{10m - 1} = 0$$

$$\sin\frac{1}{m}\sqrt{10m - 1} = 0$$

$$\frac{1}{m}\sqrt{10m - 1} = n\pi$$

$$\frac{10m - 1}{m^2} = n^2\pi^2, \quad n = 1, 2, 3, \ldots$$

$$(n^2\pi^2)m^2 - 10m + 1 = 0$$

$$m = \frac{10\sqrt{100 - 4n^2\pi^2}}{2n^2\pi^2} = \frac{5 \pm \sqrt{25 - n^2\pi^2}}{n^2\pi^2}.$$

Since m is real, $25 - n^2\pi^2 \geq 0$. If $25 - n^2\pi^2 = 0$, then $n^2 = 25/\pi^2$, and n is not an integer. Thus, $25 - n^2\pi^2 = (5 - n\pi)(5 + n\pi) > 0$ and since $n > 0$, $5 + n\pi > 0$, so $5 - n\pi > 0$ also. Then $n < 5/\pi$, and so $n = 1$. Therefore, the mass m will pass through the equilibrium position when $t = 1$ for

$$m_1 = \frac{5 + \sqrt{25 - \pi^2}}{\pi^2} \quad \text{and} \quad m_2 = \frac{5 - \sqrt{25 - \pi^2}}{\pi^2}.$$

33. (a) A solution curve has the same y-coordinate at both ends of the interval $[-\pi, \pi]$ and the tangent lines at the endpoints of the interval are parallel.

(b) For $\lambda = 0$ the solution of $y'' = 0$ is $y = c_1 x + c_2$. From the first boundary condition we have

$$y(-\pi) = -c_1\pi + c_2 = y(\pi) = c_1\pi + c_2$$

or $2c_1\pi = 0$. Thus, $c_1 = 0$ and $y = c_2$. This constant solution is seen to satisfy the boundary-value problem.

For $\lambda = -\alpha^2 < 0$ we have $y = c_1 \cosh \alpha x + c_2 \sinh \alpha x$. In this case the first boundary condition gives

$$y(-\pi) = c_1 \cosh(-\alpha\pi) + c_2 \sinh(-\alpha\pi)$$

$$= c_1 \cosh \alpha\pi - c_2 \sinh \alpha\pi$$

$$= y(\pi) = c_1 \cosh \alpha\pi + c_2 \sinh \alpha\pi$$

or $2c_2 \sinh \alpha\pi = 0$. Thus $c_2 = 0$ and $y = c_1 \cosh \alpha x$. The second boundary condition implies in a similar fashion that $c_1 = 0$. Thus, for $\lambda < 0$, the only solution of the boundary-value problem is $y = 0$.

For $\lambda = \alpha^2 > 0$ we have $y = c_1 \cos \alpha x + c_2 \sin \alpha x$. The first boundary condition implies

$$y(-\pi) = c_1 \cos(-\alpha\pi) + c_2 \sin(-\alpha\pi)$$

$$= c_1 \cos \alpha\pi - c_2 \sin \alpha\pi$$

$$= y(\pi) = c_1 \cos \alpha\pi + c_2 \sin \alpha\pi$$

or $2c_2 \sin \alpha\pi = 0$. Similarly, the second boundary condition implies $2c_1 \alpha \sin \alpha\pi = 0$. If $c_1 = c_2 = 0$ the solution is $y = 0$. However, if $c_1 \neq 0$ or $c_2 \neq 0$, then $\sin \alpha\pi = 0$, which implies that α must be an integer, n. Therefore, for c_1 and c_2 not both 0, $y = c_1 \cos nx + c_2 \sin nx$ is a nontrivial solution of the boundary-value problem. Since $\cos(-nx) = \cos nx$ and $\sin(-nx) = -\sin nx$, we may assume without loss of generality that the eigenvalues are $\lambda_n = \alpha^2 = n^2$, for n a positive integer. The corresponding eigenfunctions are $y_n = \cos nx$ and $y_n = \sin nx$.

(c)

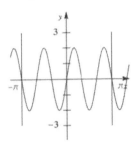

$y = 2\sin 3x$

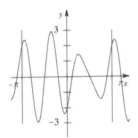

$y = \sin 4x - 2\cos 3x$

36. Using a CAS we find that the first four nonnegative roots of $\tan x = -x$ are approximately $2.02876, 4.91318, 7.97867$, and 11.0855. The corresponding eigenvalues are $4.11586, 24.1393,$ 63.6591, and 122.889, with eigenfunctions $\sin(2.02876x), \sin(4.91318x), \sin(7.97867x)$, and $\sin(11.0855x)$.

5.3 | Nonlinear Models

The terminology and concepts listed below provide an outline of the main ideas encountered in this section. These can be useful when preparing for a quiz or test.

Terminology and Concepts

- nonlinear spring models

- hard spring

- soft spring

- simple pendulum

- linearization of a DE

The basic skills listed below summarize the more mechanical types of problems encountered in the exercise set for this section.

Basic Skills

- use a numerical solver to plot solution curves of nonlinear DEs

Nonlinear Springs

3. The period corresponding to $x(0) = 1$, $x'(0) = 1$ is approximately 5.8. The second initial-value problem does not have a periodic solution.

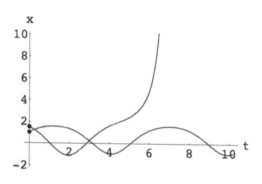

6. From the graphs we see that the interval is approximately $(-0.8, 1.1)$.

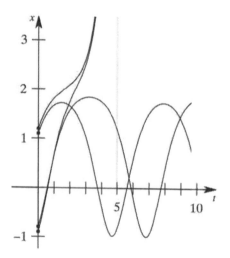

9. This is a damped hard spring, so x will approach 0 as t approaches ∞.

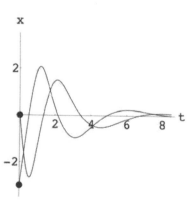

12. (a)

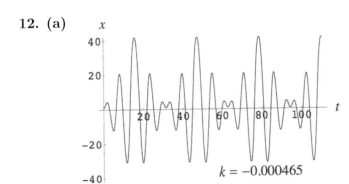

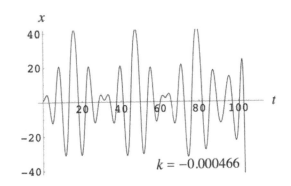

The system appears to be oscillatory for $-0.000465 \leq k_1 < 0$ and nonoscillatory for $k_1 \leq -0.000466$.

(b)

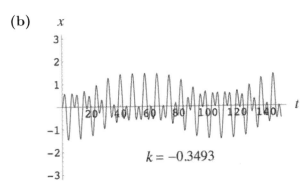

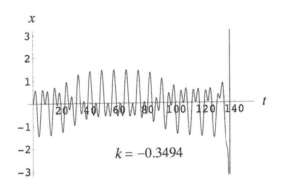

The system appears to be oscillatory for $-0.3493 \leq k_1 < 0$ and nonoscillatory for $k_1 \leq -0.3494$.

Variable Mass

15. (a) Intuitively, one might expect that only half of a 10-pound chain could be lifted by a 5-pound vertical force.

(b) Since $x = 0$ when $t = 0$, and $v = dx/dt = \sqrt{160 - 64x/3}$, we have $v(0) = \sqrt{160} \approx 12.65$ ft/s.

(c) Since x should always be positive, we solve $x(t) = 0$, getting $t = 0$ and $t = \frac{3}{2}\sqrt{5/2} \approx 2.3717$. Since the graph of $x(t)$ is a parabola, the maximum value occurs at $t_m = \frac{3}{4}\sqrt{5/2}$. This can also be obtained by solving $x'(t) = 0$. At this time the height of the chain is $x(t_m) \approx 7.5$ ft. This is higher than predicted because of the momentum generated by the force. When the chain is 5 feet high it still has a positive velocity of about 7.3 ft/s, which keeps it going higher for a while.

(d) As discussed in the solution to part **(c)** of this problem, the chain has momentum generated by the force applied to it that will cause it to go higher than expected. It will then fall back to below the expected maximum height, again due to momentum. This, in turn, will cause it to next go higher than expected, and so on.

Miscellaneous Mathematical Models

18. (a) Let (r, θ) denote the polar coordinates of the destroyer S_1. When S_1 travels the 6 miles from $(9, 0)$ to $(3, 0)$ it stands to reason, since S_2 travels half as fast as S_1, that the polar coordinates of S_2 are $(3, \theta_2)$, where θ_2 is unknown. In other words, the distances of the ships from $(0, 0)$ are the same and $r(t) = 15t$ then gives the radial distance of both ships. This is necessary if S_1 is to intercept S_2.

(b) The differential of arc length in polar coordinates is $(ds)^2 = (r\, d\theta)^2 + (dr)^2$, so that

$$\left(\frac{ds}{dt}\right)^2 = r^2 \left(\frac{d\theta}{dt}\right)^2 + \left(\frac{dr}{dt}\right)^2.$$

Using $ds/dt = 30$ and $dr/dt = 15$ then gives

$$900 = 225t^2 \left(\frac{d\theta}{dt}\right)^2 + 225$$

$$675 = 225t^2 \left(\frac{d\theta}{dt}\right)^2$$

$$\frac{d\theta}{dt} = \frac{\sqrt{3}}{t}$$

$$\theta(t) = \sqrt{3}\ln t + c = \sqrt{3}\ln\frac{r}{15} + c.$$

When $r = 3$, $\theta = 0$, so $c = -\sqrt{3}\ln\frac{1}{5}$ and

$$\theta(t) = \sqrt{3}\left(\ln\frac{r}{15} - \ln\frac{1}{5}\right) = \sqrt{3}\ln\frac{r}{3}.$$

Thus $r = 3e^{\theta/\sqrt{3}}$, whose graph is a logarithmic spiral.

(c) The time for S_1 to go from $(9, 0)$ to $(3, 0) = \frac{1}{5}$ hour. Now S_1 must intercept the path of S_2 for some angle β, where $0 < \beta < 2\pi$. At the time of interception t_2 we have $15t_2 = 3e^{\beta/\sqrt{3}}$ or $t = \frac{1}{5}e^{\beta/\sqrt{3}}$. The total time is then

$$t = \frac{1}{5} + \frac{1}{5}e^{\beta/\sqrt{3}} < \frac{1}{5}(1 + e^{2\pi/\sqrt{3}}).$$

5.R Chapter 5 in Review

3. $5/4$ m, since $x = -\cos 4t + \frac{3}{4}\sin 4t$

6. False; since the equation of motion in this case is $x(t) = e^{-\lambda t}(c_1 + c_2 t)$ and $x(t) = 0$ can have at most one real solution

9. $y = 0$ because $\lambda = 8$ is not an eigenvalue

12. **(a)** Solving $\frac{3}{8}x'' + 6x = 0$ subject to $x(0) = 1$ and $x'(0) = -4$ we obtain

$$x = \cos 4t - \sin 4t = \sqrt{2}\sin\left(4t + 3\pi/4\right).$$

   (b) The amplitude is $\sqrt{2}$, period is $\pi/2$, and frequency is $2/\pi$.

   (c) If $x = 1$ then $t = n\pi/2$ and $t = -\pi/8 + n\pi/2$ for $n = 1, 2, 3, \ldots$.

   (d) If $x = 0$ then $t = \pi/16 + n\pi/4$ for $n = 0, 1, 2, \ldots$. The motion is upward for n even and downward for n odd.

   (e) $x'(3\pi/16) = 0$

   (f) If $x' = 0$ then $4t + 3\pi/4 = \pi/2 + n\pi$ or $t = 3\pi/16 + n\pi$.

15. From $mx'' + 4x' + 2x = 0$ we see that nonoscillatory motion results if $16 - 8m \geq 0$ or $0 < m \leq 2$.

18. Clearly $x_p = A/\omega^2$ suffices.

21. From $q'' + 10^4 q = 100\sin 50t$, $q(0) = 0$, and $q'(0) = 0$ we obtain $q_c = c_1\cos 100t + c_2\sin 100t$, $q_p = \frac{1}{75}\sin 50t$, and

   (a) $q = -\frac{1}{150}\sin 100t + \frac{1}{75}\sin 50t$,

   (b) $i = -\frac{2}{3}\cos 100t + \frac{2}{3}\cos 50t$, and

   (c) $q = 0$ when $\sin 50t(1 - \cos 50t) = 0$ or $t = n\pi/50$ for $n = 0, 1, 2, \ldots$.

24. (a) The differential equation is $d^2r/dt^2 - \omega^2 r = -g\sin\omega t$. The auxiliary equation is $m^2 - \omega^2 = 0$, so $r_c = c_1 e^{\omega t} + c_2 e^{-\omega t}$. A particular solution has the form $r_p = A\sin\omega t + B\cos\omega t$. Substituting into the differential equation we find $-2A\omega^2 \sin\omega t - 2B\omega^2 \cos\omega t = -g\sin\omega t$. Thus, $B = 0$, $A = g/2\omega^2$, and $r_p = (g/2\omega^2)\sin\omega t$. The general solution of the differential equation is $r(t) = c_1 e^{\omega t} + c_2 e^{-\omega t} + (g/2\omega^2)\sin\omega t$. The initial conditions imply $c_1 + c_2 = r_0$ and $g/2\omega - \omega c_1 + \omega c_2 = v_0$. Solving for c_1 and c_2 we get

$$c_1 = (2\omega^2 r_0 + 2\omega v_0 - g)/4\omega^2 \quad \text{and} \quad c_2 = (2\omega^2 r_0 - 2\omega v_0 + g)/4\omega^2,$$

so that

$$r(t) = \frac{2\omega^2 r_0 + 2\omega v_0 - g}{4\omega^2} e^{\omega t} + \frac{2\omega^2 r_0 - 2\omega v_0 + g}{4\omega^2} e^{-\omega t} + \frac{g}{2\omega^2}\sin\omega t.$$

(b) The bead will exhibit simple harmonic motion when the exponential terms are missing. Solving $c_1 = 0$, $c_2 = 0$ for r_0 and v_0 we find $r_0 = 0$ and $v_0 = g/2\omega$.

To find the minimum length of rod that will accommodate simple harmonic motion we determine the amplitude of $r(t)$ and double it. Thus $L = g/\omega^2$.

(c) As t increases, $e^{\omega t}$ approaches infinity and $e^{-\omega t}$ approaches 0. Since $\sin\omega t$ is bounded, the distance, $r(t)$, of the bead from the pivot point increases without bound and the distance of the bead from P will eventually exceed $L/2$.

(d)

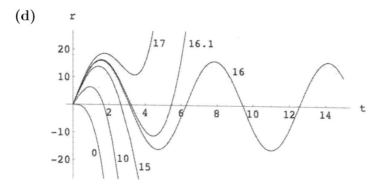

(e) For each v_0 we want to find the smallest value of t for which $r(t) = \pm 20$. Whether we look for $r(t) = -20$ or $r(t) = 20$ is determined by looking at the graphs in part (d). The total times that the bead stays on the rod is shown in the table below.

v_0	0	10	15	16.1	17
r	-20	-20	-20	20	20
t	1.55007	2.35494	3.43088	6.11627	4.22339

When $v_0 = 16$ the bead never leaves the rod.

27. The force of kinetic friction opposing the motion of the mass in μN, where μ is the coefficient of sliding friction and N is the normal component of the weight. Since friction is a force opposite to the direction of motion and since N is pointed directly downward (it is simply the weight of the mass), Newton's second law gives, for motion to the right $(x' > 0)$,

$$m \frac{d^2 x}{dt^2} = -kx - \mu mg,$$

and for motion to the left $(x' < 0)$,

$$m \frac{d^2 x}{dt^2} = -kx + \mu mg.$$

Traditionally, these two equations are written as one expression

$$m \frac{d^2 x}{dt^2} + f_k \, \text{sgn}(x') + kx = 0,$$

where $f_k = \mu mg$ and

$$\text{sgn}(x') = \begin{cases} 1, & x' > 0 \\ -1, & x' < 0. \end{cases}$$

6 Series Solutions of Linear Equations

6.1 | Review of Power Series

The terminology and concepts listed below provide an outline of the main ideas encountered in this section. These can be useful when preparing for a quiz or test.

Terminology and Concepts

- power series centered at a

- Maclaurin series

- interval and radius of convergence

- absolute convergence

- ratio test

- arithmetic of power series

- solution of differential equations using power series

Basic Skills

- find the interval and radius of convergence of a power series

- use known power series find new power series

- be able to shift the summation index of a power series

- be able to solve simple first-order, linear differential equations

140

3. $\lim\limits_{n\to\infty}\left|\dfrac{2^{n+1}x^{n+1}/(n+1)}{2^n x^n/n}\right| = \lim\limits_{n\to\infty}\dfrac{2n}{n+1}|x| = 2|x|$

The series is absolutely convergent for $2|x| < 1$ or $|x| < \frac{1}{2}$. The radius of convergence is $R = \frac{1}{2}$.

At $x = -\frac{1}{2}$, the series $\sum\limits_{n=1}^{\infty}(-1)^n/n$ converges by the alternating series test. At $x = \frac{1}{2}$, the series

$\sum\limits_{n=1}^{\infty}1/n$ is the harmonic series which diverges. Thus, the given series converges on $[-\frac{1}{2}, \frac{1}{2})$, and

the radius of convergence is $\frac{1}{2}$.

6. $\lim\limits_{k\to\infty}\left|\dfrac{(k+1)!(x-1)^{k+1}}{k!(x-1)^k}\right| = \lim\limits_{k\to\infty}(k+1)|x-1| = \begin{cases}\infty, & x\neq 1\\ 0, & x=1\end{cases}$

The radius of convergence is $R = 0$ and the series converges only for $x = 1$.

9. Write the series as $\sum\limits_{k=1}^{\infty}\left(\dfrac{32}{75}\right)^k x^k$. Then

$$\lim\limits_{n\to\infty}\left|\dfrac{a_{n+1}}{a_n}\right| = \lim\limits_{n\to\infty}\left|\dfrac{(32/75)^{n+1}x^{n+1}}{(32/75)^n x^n}\right| = \lim\limits_{n\to\infty}\left|\dfrac{32}{75}x\right| = \dfrac{32}{75}|x|.$$

The series is absolutely convergent for $\dfrac{32}{75}|x| < 1$, or on $(-75/32, 75/32)$. At $x = -75/32$ the

series $\sum\limits_{k=1}^{\infty}(-1)^k$ diverges by the n-th term test. At $x = 75/32$ the series $\sum\limits_{k=1}^{\infty}1$ diverges by

the n-th term test. Thus, the given series converges on $(-75/32, 75/32)$, and the radius of

convergence is $75/32$.

12. We replace x by $3x$ in the Maclaurin series of e^x and multiply the result by x.

$$xe^{3x} = x\cdot\sum\limits_{n=0}^{\infty}\dfrac{1}{n!}(3x)^n = \sum\limits_{n=0}^{\infty}\dfrac{3^n}{n!}x^{n+1}$$

15. We replace x by $-x$ in the Maclaurin series of $\ln(1+x)$.

$$\ln(1-x) = -x - \dfrac{(-x)^2}{2} + \dfrac{(-x)^3}{3} + \cdots = -x - \dfrac{x^2}{2} - \dfrac{x^3}{3} - \cdots = \sum\limits_{n=1}^{\infty}\dfrac{-1}{n}x^n$$

18. We first note that

$$\ln x = \ln\left[2\left(1 + \dfrac{x-2}{2}\right)\right] = \ln 2 + \ln\left(1 + \dfrac{x-2}{2}\right).$$

Next use the Maclaurin series of $\ln(1+x)$ with x replaced by $\dfrac{x-2}{2}$.

$$\ln x = \ln 2 + \frac{x-2}{2} - \frac{1}{2}\left(\frac{x-2}{2}\right)^2 + \frac{1}{3}\left(\frac{x-2}{2}\right)^3 - \frac{1}{4}\left(\frac{x-2}{2}\right)^4 + \cdots$$

$$= \ln 2 + \sum_{n=1}^{\infty} \frac{(-1)^{n+1}}{n2^n}(x-2)^n$$

21. $\sec x = \dfrac{1}{\cos x} = \dfrac{1}{1 - \dfrac{x^2}{2} + \dfrac{x^4}{4!} - \dfrac{x^6}{6!} + \cdots} = 1 + \dfrac{x^2}{2} + \dfrac{5x^4}{4!} + \dfrac{61x^6}{6!} + \cdots$

Since $\cos(\pi/2) = \cos(-\pi/2) = 0$, the series converges on $(-\pi/2, \pi/2)$.

24. Let $k = n - 3$ so that $n = k + 3$ and

$$\sum_{n=3}^{\infty}(2n-1)c_n x^{n-3} = \sum_{k=0}^{\infty}(2k+5)c_{k+3}x^k.$$

27. In the first summation let $k = n - 1$ and in the second summation let $k = n + 1$. Then $n = k + 1$ in the first summation, $n = k - 1$ in the second summation, and

$$\sum_{n=1}^{\infty}2nc_n x^{n-1} + \sum_{n=0}^{\infty}6c_n x^{n+1} = 2 \cdot 1 \cdot c_1 x^0 + \sum_{n=2}^{\infty}2nc_n x^{n-1} + \sum_{n=0}^{\infty}6c_n x^{n+1}$$

$$= 2c_1 + \sum_{k=1}^{\infty}2(k+1)c_{k+1}x^k + \sum_{k=1}^{\infty}6c_{k-1}x^k$$

$$= 2c_1 + \sum_{k=1}^{\infty}\left[2(k+1)c_{k+1} + 6c_{k-1}\right]x^k.$$

30. In the first and third summations let $k = n$ and in the second summation let $k = n - 2$. Then $n = k$ in the first and third summations, $n = k + 2$ in the second summation, and

$$\sum_{n=2}^{\infty}n(n-1)c_n x^n + 2\sum_{n=2}^{\infty}n(n-1)c_n x^{n-2} + 3\sum_{n=1}^{\infty}nc_n x^n$$

$$= 2 \cdot 2 \cdot 1c_2 x^0 + 2 \cdot 3 \cdot 2c_3 x^1 + 3 \cdot 1 \cdot c_1 x^1 + \sum_{n=2}^{\infty}n(n-1)c_n x^n + 2\sum_{n=4}^{\infty}n(n-1)c_n x^{n-2} + 3\sum_{n=2}^{\infty}nc_n x^n$$

$$= 4c_2 + (3c_1 + 12c_3)x + \sum_{k=2}^{\infty}k(k-1)c_k x^k + 2\sum_{k=2}^{\infty}(k+2)(k+1)c_{k+2}x^k + 3\sum_{k=2}^{\infty}kc_k x^k$$

$$= 4c_2 + (3c_1 + 12c_3)x + \sum_{k=2}^{\infty}\left[\left(k(k-1) + 3k\right)c_k + 2(k+2)(k+1)c_{k+2}\right]x^k$$

$$= 4c_2 + (3c_1 + 12c_3)x + \sum_{k=2}^{\infty}\left[k(k+2)c_k + 2(k+1)(k+2)c_{k+2}\right]x^k$$

33. In this problem we must take special care with starting values for the indices of summation. Normally when a power series is given in summation notation, successive derivatives of the power series of the unknown function start with an index that is one higher than the preceding one. In this case, the power series starts with $n = 1$ to avoid division by zero. From the first derivative on this is no longer necessary and the index of summation starts again with $n = 1$ for y'. To justify this to yourself you could simply write out the first few term of the power series for y.

Since

$$y' = \sum_{n=1}^{\infty} (-1)^{n+1} x^{n-1} \qquad \text{and} \qquad y'' = \sum_{n=2}^{\infty} (-1)^{n+1}(n-1)x^{n-2},$$

we have

$$(x+1)y'' + y' = (x+1)\sum_{n=2}^{\infty}(-1)^{n+1}(n-1)x^{n-2} + \sum_{n=1}^{\infty}(-1)^{n+1}x^{n-1}$$

$$= \sum_{n=2}^{\infty}(-1)^{n+1}(n-1)x^{n-1} + \sum_{n=2}^{\infty}(-1)^{n+1}(n-1)x^{n-2} + \sum_{n=1}^{\infty}(-1)^{n+1}x^{n-1}$$

$$= \underbrace{\sum_{n=2}^{\infty}(-1)^{n+1}(n-1)x^{n-1}}_{k=n-1} + \underbrace{\sum_{n=2}^{\infty}(-1)^{n+1}(n-1)x^{n-2}}_{k=n-2} + \underbrace{\sum_{n=1}^{\infty}(-1)^{n+1}x^{n-1}}_{k=n-1}$$

$$= \sum_{k=1}^{\infty}(-1)^{k+2}kx^k + \sum_{k=0}^{\infty}(-1)^{k+3}(k+1)x^k + \sum_{k=0}^{\infty}(-1)^{k+2}x^k$$

$$= -x^0 + x^0 + \sum_{k=2}^{\infty}\left[(-1)^{k+2}k - (-1)^{k+2}k - (-1)^{k+2} + (-1)^{k+2}\right]x^k = 0.$$

36. Substituting into the differential equation we have

$$4y' + y = 4\sum_{n=1}^{\infty} nc_n x^{n-1} + \sum_{n=0}^{\infty} c_n x^n = 4\sum_{k=0}^{\infty}(k+1)c_{k+1}x^k + \sum_{k=0}^{\infty} c_k x^k$$

$$= \sum_{k=0}^{\infty}\left[4(k+1)c_{k+1} + c_k\right]x^k = 0.$$

Thus $c_{k+1} = -\dfrac{1}{4(k+1)} c_k$, for $k = 0, 1, 2, \ldots$, and

$$c_1 = -\frac{1}{4 \cdot 1} c_0$$

$$c_2 = -\frac{1}{4 \cdot 2} c_1 = \frac{1}{4^2 \cdot 1 \cdot 2} c_0$$

$$c_3 = -\frac{1}{4 \cdot 3} c_2 = -\frac{1}{4^3 \cdot 1 \cdot 2 \cdot 3} c_0$$

$$c_4 = -\frac{1}{4 \cdot 4} c_3 = \frac{1}{4^4 \cdot 1 \cdot 2 \cdot 3 \cdot 4} c_0$$

$$\vdots$$

Hence,

$$y = c_0 - \frac{1}{4 \cdot 1} c_0 x + \frac{1}{4^2 \cdot 1 \cdot 2} c_0 x^2 - \frac{1}{4^3 \cdot 1 \cdot 2 \cdot 3} c_0 x^3 + \frac{1}{4^4 \cdot 1 \cdot 2 \cdot 3 \cdot 4} c_0 x^4 - \cdots$$

and

$$y = c_0 \sum_{k=0}^{\infty} \frac{(-1)^k}{4^k k!} x^k = c_0 \sum_{k=0}^{\infty} \frac{(-1^k}{k!} \left(\frac{x}{4}\right)^k = c_0 e^{-x/4}.$$

6.2 | Solutions About Ordinary Points

The terminology and concepts listed below provide an outline of the main ideas encountered in this section. These can be useful when preparing for a quiz or test.

Terminology and Concepts

- power series centered at a; $\sum_{n=0}^{\infty} c_n (x-a)^n$
- convergence of a power series
- interval of convergence
- radius of convergence
- absolute convergence
- ratio test
- a power series defines a function
- a function analytic at a point

- ordinary point

- singular point

- power series solution of a DE

- recurrence relation

The basic skills listed below summarize the more mechanical types of problems encountered in the exercise set for this section.

Basic Skills

- find the radius of convergence of a power series

- convert an expression like $\sum_{n=3}^{\infty}(n+1)x^{n-2} + \sum_{n=2}^{\infty} nx^n$ to an expression involving a single infinite summation: $4x + \sum_{k=2}^{\infty}(2k+3)x^k$

- find power series solutions of linear second-order DEs

Summation Notation   Without a doubt, the key to successfully solving a linear DE with variable coefficients by the power series method depends on your ability to recast a combination of power series centered at the same point in terms of a *single* summation. In Example 1 of Section 6.1 we gave an example involving two series; here is another example of how this is done using three series. We wish to combine the three sums

$$\sum_{n=2}^{\infty} c_n n(n-1)x^{n-2} + 2\sum_{n=1}^{\infty} c_n n x^n + 5\sum_{n=0}^{\infty} c_n x^{n+1}$$

into one term. Notice that the first series starts (for $n=2$) with x^0, the second series starts (for $n=1$) with x^1, and the third series also starts (for $n=0$) with x^1. To bring the first series in line with the last two, we write out the term corresponding to $n=2$ and observe that now all three series

$$2c_2 + \sum_{n=3}^{\infty} c_n n(n-1)x^{n-2} + \sum_{n=1}^{\infty} 2c_n n x^n + \sum_{n=0}^{\infty} 5c_n x^{n+1}$$

begin with the same power x^1. The next thing that must be done is to get a common summation index for the series:

- In the first series, we let $k = n - 2$. As $n = 3, 4, 5, \ldots$, k takes on the values $1, 2, 3, \ldots$.

- In the second series, we let $k = n$. Here $n = 1, 2, 3, \ldots$ corresponds to $k = 1, 2, 3, \ldots$.

- In the third series, we let $k = n + 1$. Then $n = 0, 1, 2, \ldots$ also yields $k = 1, 2, 3, \ldots$.

Next, we substitute $n = k + 2$ in the first series, $n = k$ in the second series, and then $n = k - 1$ in the third:

$$2c_2 + \sum_{k=1}^{\infty} c_{k+2}(k+2)(k+1)x^k + \sum_{k=1}^{\infty} 2c_k k x^k + \sum_{k=1}^{\infty} 5c_{k-1}x^k.$$

At this point, when all the series start with the same power of x and have the same summation index starting with the same value (in this case, $k = 1$), we can add the series term-by-term to get

$$2c_2 + \sum_{k=1}^{\infty} [c_{k+2}(k+2)(k+1) + 2c_k k + 5c_{k-1}]x^k.$$

Alternatively, we can make the change of summation variable first as illustrated in the solution of Problem 18 in this section of the manual.

Use of Computers   Both *Mathematica* and *Maple* can express the sum of several terms in a concise way without actually listing each term. The commands below can be used to express the first four terms of the infinite sum $\sum_{n=0}^{\infty} c_n(-1)^n/(n+1)^2$.

Clear[c]   (*Mathematica*)
Sum[c[n](-1)^n/(n+1)^2, {n, 0, 3}]

```
sum(c(n)*(-1)^n/(n+1)^2, n=0..3);
```
(*Maple*)

To obtain the power series representation about $x = 0$ of a function like $f(x) = e^x$ use

Series[Exp[x], {x, 0, 5}]   (*Mathematica*)

or

```
series(exp(x),x=0,5);
```
(*Maple*)

The *Mathematica* output corresponding to the above command is

$$1 + x + \frac{1}{2}x^2 + \frac{1}{6}x^3 + \frac{1}{24}x^4 + \frac{1}{120}x^5 + O[x]^6,$$

whereas for *Maple* the output is

$$1 + x + \frac{1}{2}x^2 + \frac{1}{6}x^3 + \frac{1}{24}x^4 + O(x^5).$$

Note that in the *Mathematica* command the 5 tells the program to compute the series out to, and including, the term containing x^5. On the other hand, in *Maple*, the 5 tells the program to compute the series out to, but excluding, the term containing x^5.

Depending on how you want to use the series representation of a function found using a CAS, it may be inconvenient to have the order of the series included in the output. For example, neither *Mathematica* nor *Maple* will plot the graph of an expression containing $O[x]^{\wedge}6$, in the case of *Mathematica*, or $O(x^5)$ in the case of *Maple*. To get around this, both programs have commands that truncate the series, converting it to a polynomial.

$$\textbf{Normal[Series[Exp[x], \{x, 0, 5\}]]} \qquad\qquad (Mathematica)$$

or

$$\textbf{Series[Exp[x], \{x, 0, 5\}]//Normal}$$

```
convert(series(exp(x),x=0,5),polynom);
```
$\qquad\qquad$ (*Maple*)

3. We have

$$y'' + y = \underbrace{\sum_{n=2}^{\infty} n(n-1)c_n x^{n-2}}_{k=n-2} + \underbrace{\sum_{n=0}^{\infty} c_n x^n}_{k=n}$$

$$= \sum_{k=0}^{\infty} (k+2)(k+1)c_{k+2} x^k + \sum_{k=0}^{\infty} c_k x^k$$

$$= \sum_{k=0}^{\infty} \left[(k+2)(k+1)c_{k+2} + c_k \right] x^k = 0.$$

Thus $c_{k+2} = -\dfrac{c_k}{(k+2)(k+1)}$, for $k = 0, 1, 2, \ldots,$ and for $k = 0, 2, 4, 6, \ldots$ we get

$$c_2 = -\frac{c_0}{2!}$$

$$c_4 = \frac{c_0}{4!}$$

$$c_6 = -\frac{c_0}{6!}$$

$$\vdots$$

For $k = 1, 3, 5, 7, \ldots$ we get

$$c_3 = -\frac{c_1}{3!}$$

$$c_5 = \frac{c_1}{5!}$$

$$c_7 = -\frac{c_1}{7!}$$

$$\vdots$$

Hence,

$$y_1(x) = c_0 \left[1 - \frac{1}{2!} x^2 + \frac{1}{4!} x^4 - \frac{1}{6!} x^6 + \cdots \right]$$

and

$$y_2(x) = c_1 \left[x - \frac{1}{3!} x^3 + \frac{1}{5!} x^5 - \frac{1}{7!} x^7 + \cdots \right].$$

The solution $y_1(x)$ is recognized as $y_1(x) = c_0 \cos x$, and the solution $y_2(x)$ is recognized as $y_2(x) = c_1 \sin x$

6. We have

$$y'' + 2y' = \underbrace{\sum_{n=2}^{\infty} n(n-1)c_n x^{n-2}}_{k=n-2} + 2 \underbrace{\sum_{n=1}^{\infty} n c_n x^{n-1}}_{k=n-1}$$

$$= \sum_{k=0}^{\infty} (k+2)(k+1)c_{k+2} x^k + \sum_{k=0}^{\infty} 2(k+1)c_{k+1} x^k$$

$$= \sum_{k=0}^{\infty} \left[(k+2)(k+1)c_{k+2} + 2(k+1)c_{k+1} \right] x^k = 0.$$

Thus $c_{k+2} = -\dfrac{2(k+1)c_{k+1}}{(k+2)(k+1)} = -\dfrac{2c_{k+1}}{k+2}$, for $k = 0, 1, 2, \ldots$, so

$$c_2 = -\frac{2c_1}{2!}$$

$$c_3 = \frac{2^2 c_2}{3} = \frac{2^2 c_1}{3!}$$

$$c_4 = -\frac{2^3 c_3}{4} = -\frac{2^3 c_1}{4!}$$

$$\vdots$$

Hence, the solution of the differential equation is

$$y(x) = y_1(x) + y_2(x) = c_0 + c_1 \left[x - \frac{2}{2!} x^2 + \frac{2^2}{3!} x^3 - \frac{2^3}{4!} x^4 + \cdots \right]$$

$$= c_0 + \frac{1}{2} c_1 \left[2x - \frac{2^2}{2!} x^2 + \frac{2^3}{3!} x^3 - \frac{2^4}{4!} x^4 + \cdots \right]$$

$$= c_0 + \frac{1}{2} c_1 \left[1 - 12x - \frac{2^2}{2!} x^2 + \frac{2^3}{3!} x^3 - \frac{2^4}{4!} x^4 + \cdots \right]$$

$$= c_0 + \frac{1}{2} c_1 - \frac{1}{2} c_1 \sum_{k=0}^{\infty} \frac{1}{k!} (2x)^k.$$

The solutions $y_1(x)$ and $y_2(x)$ are recognized as

$$y_1(x) = c_0 \qquad \text{and} \qquad y_2(x) = \frac{1}{2} c_1 - \frac{1}{2} c_1 e^{-2x}.$$

9. Substituting $y = \sum_{n=0}^{\infty} c_n x^n$ into the differential equation we have

$$y'' - 2xy' + y = \sum_{n=2}^{\infty} \underbrace{n(n-1)c_n x^{n-2}}_{k=n-2} - 2\sum_{n=1}^{\infty} \underbrace{nc_n x^n}_{k=n} + \sum_{n=0}^{\infty} \underbrace{c_n x^n}_{k=n}$$

$$= \sum_{k=0}^{\infty} (k+2)(k+1)c_{k+2} x^k - 2\sum_{k=1}^{\infty} kc_k x^k + \sum_{k=0}^{\infty} c_k x^k$$

$$= 2c_2 + c_0 + \sum_{k=1}^{\infty} [(k+2)(k+1)c_{k+2} - (2k-1)c_k]x^k = 0.$$

Thus

$$2c_2 + c_0 = 0$$

$$(k+2)(k+1)c_{k+2} - (2k-1)c_k = 0$$

and

$$c_2 = -\frac{1}{2}c_0$$

$$c_{k+2} = \frac{2k-1}{(k+2)(k+1)} c_k, \quad k = 1, 2, 3, \ldots.$$

Choosing $c_0 = 1$ and $c_1 = 0$ we find

$$c_2 = -\frac{1}{2}$$

$$c_3 = c_5 = c_7 = \cdots = 0$$

$$c_4 = -\frac{1}{8}$$

$$c_6 = -\frac{7}{240}$$

and so on. For $c_0 = 0$ and $c_1 = 1$ we obtain

$$c_2 = c_4 = c_6 = \cdots = 0$$

$$c_3 = \frac{1}{6}$$

$$c_5 = \frac{1}{24}$$

$$c_7 = \frac{1}{112}$$

and so on. Thus, two solutions are

$$y_1 = 1 - \frac{1}{2}x^2 - \frac{1}{8}x^4 - \frac{7}{240}x^6 - \cdots \qquad \text{and} \qquad y_2 = x + \frac{1}{6}x^3 + \frac{1}{24}x^5 + \frac{1}{112}x^7 + \cdots.$$

12. Substituting $y = \sum_{n=0}^{\infty} c_n x^n$ into the differential equation we have

$$y'' + 2xy' + 2y = \underbrace{\sum_{n=2}^{\infty} n(n-1)c_n x^{n-2}}_{k=n-2} + 2\underbrace{\sum_{n=1}^{\infty} nc_n x^n}_{k=n} + 2\underbrace{\sum_{n=0}^{\infty} c_n x^n}_{k=n}$$

$$= \sum_{k=0}^{\infty} (k+2)(k+1)c_{k+2} x^k + 2\sum_{k=1}^{\infty} kc_k x^k + 2\sum_{k=0}^{\infty} c_k x^k$$

$$= 2c_2 + 2c_0 + \sum_{k=1}^{\infty} [(k+2)(k+1)c_{k+2} + 2(k+1)c_k]x^k = 0.$$

Thus

$$2c_2 + 2c_0 = 0$$

$$(k+2)(k+1)c_{k+2} + 2(k+1)c_k = 0$$

and

$$c_2 = -c_0$$

$$c_{k+2} = -\frac{2}{k+2}c_k, \quad k = 1, 2, 3, \ldots.$$

Choosing $c_0 = 1$ and $c_1 = 0$ we find

$$c_2 = -1$$

$$c_3 = c_5 = c_7 = \cdots = 0$$

$$c_4 = \frac{1}{2}$$

$$c_6 = -\frac{1}{6}$$

and so on. For $c_0 = 0$ and $c_1 = 1$ we obtain

$$c_2 = c_4 = c_6 = \cdots = 0$$

$$c_3 = -\frac{2}{3}$$

$$c_5 = \frac{4}{15}$$

$$c_7 = -\frac{8}{105}$$

and so on. Thus, two solutions are

$$y_1 = 1 - x^2 + \frac{1}{2}x^4 - \frac{1}{6}x^6 + \cdots \quad \text{and} \quad y_2 = x - \frac{2}{3}x^3 + \frac{4}{15}x^5 - \frac{8}{105}x^7 + \cdots.$$

15. Substituting $y = \sum_{n=0}^{\infty} c_n x^n$ into the differential equation we have

$$y'' - (x+1)y' - y = \underbrace{\sum_{n=2}^{\infty} n(n-1)c_n x^{n-2}}_{k=n-2} - \underbrace{\sum_{n=1}^{\infty} nc_n x^n}_{k=n} - \underbrace{\sum_{n=1}^{\infty} nc_n x^{n-1}}_{k=n-1} - \underbrace{\sum_{n=0}^{\infty} c_n x^n}_{k=n}$$

$$= \sum_{k=0}^{\infty} (k+2)(k+1)c_{k+2} x^k - \sum_{k=1}^{\infty} kc_k x^k - \sum_{k=0}^{\infty} (k+1)c_{k+1} x^k - \sum_{k=0}^{\infty} c_k x^k$$

$$= 2c_2 - c_1 - c_0 + \sum_{k=1}^{\infty} [(k+2)(k+1)c_{k+2} - (k+1)c_{k+1} - (k+1)c_k] x^k = 0.$$

Thus

$$2c_2 - c_1 - c_0 = 0$$

$$(k+2)(k+1)c_{k+2} - (k+1)(c_{k+1} + c_k) = 0$$

and

$$c_2 = \frac{c_1 + c_0}{2}$$

$$c_{k+2} = \frac{c_{k+1} + c_k}{k+2}, \quad k = 1, 2, 3, \dots .$$

Choosing $c_0 = 1$ and $c_1 = 0$ we find

$$c_2 = \frac{1}{2}, \qquad c_3 = \frac{1}{6}, \qquad c_4 = \frac{1}{6},$$

and so on. For $c_0 = 0$ and $c_1 = 1$ we obtain

$$c_2 = \frac{1}{2}, \qquad c_3 = \frac{1}{2}, \qquad c_4 = \frac{1}{4},$$

and so on. Thus, two solutions are

$$y_1 = 1 + \frac{1}{2}x^2 + \frac{1}{6}x^3 + \frac{1}{6}x^4 + \cdots \qquad \text{and} \qquad y_2 = x + \frac{1}{2}x^2 + \frac{1}{2}x^3 + \frac{1}{4}x^4 + \cdots .$$

18. Substituting $y = \sum_{n=0}^{\infty} c_n x^n$ into the differential equation we have

$$\left(x^2 - 1\right) y'' + xy' - y = \underbrace{\sum_{n=2}^{\infty} n(n-1)c_n x^n}_{k=n} - \underbrace{\sum_{n=2}^{\infty} n(n-1)c_n x^{n-2}}_{k=n-2} + \underbrace{\sum_{n=1}^{\infty} nc_n x^n}_{k=n} - \underbrace{\sum_{n=0}^{\infty} c_n x^n}_{k=n}$$

$$= \sum_{k=2}^{\infty} k(k-1)c_k x^k - \sum_{k=0}^{\infty} (k+2)(k+1)c_{k+2} x^k + \sum_{k=1}^{\infty} kc_k x^k - \sum_{k=0}^{\infty} c_k x^k$$

$$= (-2c_2 - c_0) - 6c_3 x + \sum_{k=2}^{\infty} \left[-(k+2)(k+1)c_{k+2} + \left(k^2 - 1\right) c_k \right] x^k = 0.$$

Thus

$$-2c_2 - c_0 = 0$$

$$-6c_3 = 0$$

$$-(k+2)(k+1)c_{k+2} + (k-1)(k+1)c_k = 0$$

and

$$c_2 = -\frac{1}{2}c_0$$

$$c_3 = 0$$

$$c_{k+2} = \frac{k-1}{k+2}c_k, \quad k = 2, 3, 4, \ldots.$$

Choosing $c_0 = 1$ and $c_1 = 0$ we find

$$c_2 = -\frac{1}{2}$$

$$c_3 = c_5 = c_7 = \cdots = 0$$

$$c_4 = -\frac{1}{8}$$

and so on. For $c_0 = 0$ and $c_1 = 1$ we obtain

$$c_2 = c_4 = c_6 = \cdots = 0$$

$$c_3 = c_5 = c_7 = \cdots = 0.$$

Thus, two solutions are

$$y_1 = 1 - \frac{1}{2}x^2 - \frac{1}{8}x^4 - \cdots \qquad \text{and} \qquad y_2 = x.$$

21. Substituting $y = \sum_{n=0}^{\infty} c_n x^n$ into the differential equation we have

$$y'' - 2xy' + 8y = \underbrace{\sum_{n=2}^{\infty} n(n-1)c_n x^{n-2}}_{k=n-2} - 2\underbrace{\sum_{n=1}^{\infty} nc_n x^n}_{k=n} + 8\underbrace{\sum_{n=0}^{\infty} c_n x^n}_{k=n}$$

$$= \sum_{k=0}^{\infty} (k+2)(k+1)c_{k+2} x^k - 2\sum_{k=1}^{\infty} kc_k x^k + 8\sum_{k=0}^{\infty} c_k x^k$$

$$= 2c_2 + 8c_0 + \sum_{k=1}^{\infty} [(k+2)(k+1)c_{k+2} + (8-2k)c_k]x^k = 0.$$

Thus

$$2c_2 + 8c_0 = 0$$

$$(k+2)(k+1)c_{k+2} + (8 - 2k)c_k = 0$$

and

$$c_2 = -4c_0$$

$$c_{k+2} = \frac{2(k-4)}{(k+2)(k+1)} c_k, \quad k = 1, 2, 3, \ldots .$$

Choosing $c_0 = 1$ and $c_1 = 0$ we find

$$c_2 = -4$$

$$c_3 = c_5 = c_7 = \cdots = 0$$

$$c_4 = \frac{4}{3}$$

$$c_6 = c_8 = c_{10} = \cdots = 0.$$

For $c_0 = 0$ and $c_1 = 1$ we obtain

$$c_2 = c_4 = c_6 = \cdots = 0$$

$$c_3 = -1$$

$$c_5 = \frac{1}{10}$$

and so on. Thus,

$$y = C_1 \left(1 - 4x^2 + \frac{4}{3}x^4 \right) + C_2 \left(x - x^3 + \frac{1}{10}x^5 + \cdots \right)$$

and

$$y' = C_1 \left(-8x + \frac{16}{3}x^3 \right) + C_2 \left(1 - 3x^2 + \frac{1}{2}x^4 + \cdots \right).$$

The initial conditions imply $C_1 = 3$ and $C_2 = 0$, so

$$y = 3 \left(1 - 4x^2 + \frac{4}{3}x^4 \right) = 3 - 12x^2 + 4x^4.$$

24. Substituting $y = \sum_{n=0}^{\infty} c_n x^n$ into the differential equation we have

$$y'' + e^x y' - y = \sum_{n=2}^{\infty} n(n-1)c_n x^{n-2}$$

$$+ \left(1 + x + \frac{1}{2}x^2 + \frac{1}{6}x^3 + \cdots\right)\left(c_1 + 2c_2 x + 3c_3 x^2 + 4c_4 x^3 + \cdots\right) - \sum_{n=0}^{\infty} c_n x^n$$

$$= \left[2c_2 + 6c_3 x + 12c_4 x^2 + 20c_5 x^3 + \cdots\right]$$

$$+ \left[c_1 + (2c_2 + c_1)x + \left(3c_3 + 2c_2 + \frac{1}{2}c_1\right)x^2 + \cdots\right] - \left[c_0 + c_1 x + c_2 x^2 + \cdots\right]$$

$$= (2c_2 + c_1 - c_0) + (6c_3 + 2c_2)x + \left(12c_4 + 3c_3 + c_2 + \frac{1}{2}c_1\right)x^2 + \cdots = 0.$$

Thus

$$2c_2 + c_1 - c_0 = 0$$

$$6c_3 + 2c_2 = 0$$

$$12c_4 + 3c_3 + c_2 + \frac{1}{2}c_1 = 0$$

and

$$c_2 = \frac{1}{2}c_0 - \frac{1}{2}c_1$$

$$c_3 = -\frac{1}{3}c_2$$

$$c_4 = -\frac{1}{4}c_3 + \frac{1}{12}c_2 - \frac{1}{24}c_1.$$

Choosing $c_0 = 1$ and $c_1 = 0$ we find

$$c_2 = \frac{1}{2}, \qquad c_3 = -\frac{1}{6}, \qquad c_4 = 0$$

and so on. For $c_0 = 0$ and $c_1 = 1$ we obtain

$$c_2 = -\frac{1}{2}, \qquad c_3 = \frac{1}{6}, \qquad c_4 = -\frac{1}{24}$$

and so on. Thus, two solutions are

$$y_1 = 1 + \frac{1}{2}x^2 - \frac{1}{6}x^3 + \cdots \qquad \text{and} \qquad y_2 = x - \frac{1}{2}x^2 + \frac{1}{6}x^3 - \frac{1}{24}x^4 + \cdots.$$

6.3 | Solutions About Singular Points

The terminology and concepts listed below provide an outline of the main ideas encountered in this section. These can be useful when preparing for a quiz or test.

Terminology and Concepts

- regular singular point of a DE

- irregular singular point of a DE

- method of Frobenius for finding a series solution about a regular singular point of a DE

- indicial equation

The basic skills listed below summarize the more mechanical types of problems encountered in the exercise set for this section.

Basic Skills

- determine the singular points of a linear DE and classify them as regular or irregular

- use the method of Frobenius to obtain two linearly independent solutions about $x = 0$ of a homogeneous linear DE with regular singular point $x = 0$ when the indicial roots

   1. do not differ by an integer, or
   2. differ by an integer.

3. Irregular singular point: $x = 3$; regular singular point: $x = -3$

6. Irregular singular point: $x = 5$; regular singular point: $x = 0$

9. Irregular singular point: $x = 0$; regular singular points: $x = 2, \pm 5$

12. Writing the differential equation in the form

$$y'' + \frac{x+3}{x} y' + 7xy = 0$$

we see that $x_0 = 0$ is a regular singular point. Multiplying by x^2, the differential equation can be put in the form

$$x^2 y'' + x(x+3)y' + 7x^3 y = 0.$$

We identify $p(x) = x + 3$ and $q(x) = 7x^3$.

15. Substituting $y = \sum_{n=0}^{\infty} c_n x^{n+r}$ into the differential equation and collecting terms, we obtain

$$2xy'' - y' + 2y = \left(2r^2 - 3r\right) c_0 x^{r-1} + \sum_{k=1}^{\infty} [2(k+r-1)(k+r)c_k - (k+r)c_k + 2c_{k-1}]x^{k+r-1} = 0,$$

which implies

$$2r^2 - 3r = r(2r - 3) = 0$$

and

$$(k+r)(2k+2r-3)c_k + 2c_{k-1} = 0.$$

The indicial roots are $r = 0$ and $r = 3/2$. For $r = 0$ the recurrence relation is

$$c_k = -\frac{2c_{k-1}}{k(2k-3)}, \quad k = 1, 2, 3, \ldots,$$

and

$$c_1 = 2c_0, \qquad c_2 = -2c_0, \qquad c_3 = \frac{4}{9}c_0,$$

and so on. For $r = 3/2$ the recurrence relation is

$$c_k = -\frac{2c_{k-1}}{(2k+3)k}, \quad k = 1, 2, 3, \ldots,$$

and

$$c_1 = -\frac{2}{5}c_0, \qquad c_2 = \frac{2}{35}c_0, \qquad c_3 = -\frac{4}{945}c_0,$$

and so on. The general solution on $(0, \infty)$ is

$$y = C_1 \left(1 + 2x - 2x^2 + \frac{4}{9}x^3 + \cdots \right) + C_2 x^{3/2} \left(1 - \frac{2}{5}x + \frac{2}{35}x^2 - \frac{4}{945}x^3 + \cdots \right).$$

18. Substituting $y = \sum_{n=0}^{\infty} c_n x^{n+r}$ into the differential equation and collecting terms, we obtain

$$2x^2 y'' - xy' + (x^2 + 1)\, y = (2r^2 - 3r + 1)\, c_0 x^r + (2r^2 + r)\, c_1 x^{r+1}$$

$$+ \sum_{k=2}^{\infty} [2(k+r)(k+r-1)c_k - (k+r)c_k + c_k + c_{k-2}]x^{k+r}$$

$$= 0,$$

which implies

$$2r^2 - 3r + 1 = (2r-1)(r-1) = 0,$$
$$(2r^2 + r)\, c_1 = 0,$$

and

$$[(k+r)(2k+2r-3) + 1]c_k + c_{k-2} = 0.$$

The indicial roots are $r = 1/2$ and $r = 1$, so $c_1 = 0$. For $r = 1/2$ the recurrence relation is

$$c_k = -\frac{c_{k-2}}{k(2k-1)}, \quad k = 2, 3, 4, \ldots,$$

and

$$c_2 = -\frac{1}{6}c_0, \qquad c_3 = 0, \qquad c_4 = \frac{1}{168}c_0,$$

and so on. For $r = 1$ the recurrence relation is

$$c_k = -\frac{c_{k-2}}{k(2k+1)}, \quad k = 2, 3, 4, \ldots,$$

and

$$c_2 = -\frac{1}{10}c_0, \qquad c_3 = 0, \qquad c_4 = \frac{1}{360}c_0,$$

and so on. The general solution on $(0, \infty)$ is

$$y = C_1 x^{1/2} \left(1 - \frac{1}{6}x^2 + \frac{1}{168}x^4 + \cdots \right) + C_2 x \left(1 - \frac{1}{10}x^2 + \frac{1}{360}x^4 + \cdots \right).$$

21. Substituting $y = \sum_{n=0}^{\infty} c_n x^{n+r}$ into the differential equation and collecting terms, we obtain

$$2xy'' - (3 + 2x)y' + y = (2r^2 - 5r)\, c_0 x^{r-1} + \sum_{k=1}^{\infty} [2(k+r)(k+r-1)c_k$$

$$- 3(k+r)c_k - 2(k+r-1)c_{k-1} + c_{k-1}]x^{k+r-1}$$

$$= 0,$$

which implies

$$2r^2 - 5r = r(2r - 5) = 0$$

and

$$(k+r)(2k+2r-5)c_k - (2k+2r-3)c_{k-1} = 0.$$

The indicial roots are $r = 0$ and $r = 5/2$. For $r = 0$ the recurrence relation is

$$c_k = \frac{(2k-3)c_{k-1}}{k(2k-5)}, \quad k = 1, 2, 3, \ldots,$$

and

$$c_1 = \frac{1}{3}c_0, \qquad c_2 = -\frac{1}{6}c_0, \qquad c_3 = -\frac{1}{6}c_0,$$

and so on. For $r = 5/2$ the recurrence relation is

$$c_k = \frac{2(k+1)c_{k-1}}{k(2k+5)}, \quad k = 1, 2, 3, \ldots,$$

and

$$c_1 = \frac{4}{7}c_0, \qquad c_2 = \frac{4}{21}c_0, \qquad c_3 = \frac{32}{693}c_0,$$

and so on. The general solution on $(0, \infty)$ is

$$y = C_1\left(1 + \frac{1}{3}x - \frac{1}{6}x^2 - \frac{1}{6}x^3 + \cdots\right) + C_2 x^{5/2}\left(1 + \frac{4}{7}x + \frac{4}{21}x^2 + \frac{32}{693}x^3 + \cdots\right).$$

24. Substituting $y = \sum_{n=0}^{\infty} c_n x^{n+r}$ into the differential equation and collecting terms, we obtain

$$2x^2 y'' + 3xy' + (2x-1)y = \left(2r^2 + r - 1\right)c_0 x^r$$

$$+ \sum_{k=1}^{\infty}[2(k+r)(k+r-1)c_k + 3(k+r)c_k - c_k + 2c_{k-1}]x^{k+r}$$

$$= 0,$$

which implies

$$2r^2 + r - 1 = (2r-1)(r+1) = 0$$

and

$$[(k+r)(2k+2r+1) - 1]c_k + 2c_{k-1} = 0.$$

The indicial roots are $r = -1$ and $r = 1/2$. For $r = -1$ the recurrence relation is

$$c_k = -\frac{2c_{k-1}}{k(2k-3)}, \quad k = 1, 2, 3, \ldots,$$

and

$$c_1 = 2c_0, \qquad c_2 = -2c_0, \qquad c_3 = \frac{4}{9}c_0,$$

and so on. For $r = 1/2$ the recurrence relation is

$$c_k = -\frac{2c_{k-1}}{k(2k+3)}, \quad k = 1, 2, 3, \ldots,$$

and

$$c_1 = -\frac{2}{5}c_0, \qquad c_2 = \frac{2}{35}c_0, \qquad c_3 = -\frac{4}{945}c_0,$$

and so on. The general solution on $(0, \infty)$ is

$$y = C_1 x^{-1}\left(1 + 2x - 2x^2 + \frac{4}{9}x^3 + \cdots\right) + C_2 x^{1/2}\left(1 - \frac{2}{5}x + \frac{2}{35}x^2 - \frac{4}{945}x^3 + \cdots\right).$$

27. Substituting $y = \sum_{n=0}^{\infty} c_n x^{n+r}$ into the differential equation and collecting terms, we obtain

$$xy'' - xy' + y = \left(r^2 - r\right)c_0 x^{r-1} + \sum_{k=0}^{\infty}[(k+r+1)(k+r)c_{k+1} - (k+r)c_k + c_k]x^{k+r} = 0$$

which implies

$$r^2 - r = r(r-1) = 0$$

and

$$(k+r+1)(k+r)c_{k+1} - (k+r-1)c_k = 0.$$

The indicial roots are $r_1 = 1$ and $r_2 = 0$. For $r_1 = 1$ the recurrence relation is

$$c_{k+1} = \frac{kc_k}{(k+2)(k+1)}, \quad k = 0, 1, 2, \ldots,$$

so $c_1 = c_2 = c_3 = \cdots = 0$ and one solution is $y_1 = c_0 x$. A second solution is

$$y_2 = x\int \frac{e^{-\int(-1)\,dx}}{x^2}\,dx = x\int \frac{e^x}{x^2}\,dx = x\int \frac{1}{x^2}\left(1 + x + \frac{1}{2}x^2 + \frac{1}{3!}x^3 + \cdots\right)dx$$

$$= x\int\left(\frac{1}{x^2} + \frac{1}{x} + \frac{1}{2} + \frac{1}{3!}x + \frac{1}{4!}x^2 + \cdots\right)dx = x\left[-\frac{1}{x} + \ln x + \frac{1}{2}x + \frac{1}{12}x^2 + \frac{1}{72}x^3 + \cdots\right]$$

$$= x\ln x - 1 + \frac{1}{2}x^2 + \frac{1}{12}x^3 + \frac{1}{72}x^4 + \cdots.$$

The general solution on $(0, \infty)$ is

$$y = C_1 x + C_2 y_2(x).$$

30. Substituting $y = \sum_{n=0}^{\infty} c_n x^{n+r}$ into the differential equation and collecting terms, we obtain

$$xy'' + y' + y = r^2 c_0 x^{r-1} + \sum_{k=1}^{\infty}[(k+r)(k+r-1)c_k + (k+r)c_k + c_{k-1}]x^{k+r-1} = 0$$

which implies $r^2 = 0$ and

$$(k+r)^2 c_k + c_{k-1} = 0.$$

The indicial roots are $r_1 = r_2 = 0$ and the recurrence relation is

$$c_k = -\frac{c_{k-1}}{k^2}, \quad k = 1, 2, 3, \ldots .$$

One solution is

$$y_1 = c_0\left(1 - x + \frac{1}{2^2}x^2 - \frac{1}{(3!)^2}x^3 + \frac{1}{(4!)^2}x^4 - \cdots\right) = c_0 \sum_{n=0}^{\infty} \frac{(-1)^n}{(n!)^2}x^n.$$

A second solution is

$$y_2 = y_1 \int \frac{e^{-\int(1/x)dx}}{y_1^2}\, dx = y_1 \int \frac{dx}{x\left(1 - x + \frac{1}{4}x^2 - \frac{1}{36}x^3 + \cdots\right)^2}$$

$$= y_1 \int \frac{dx}{x\left(1 - 2x + \frac{3}{2}x^2 - \frac{5}{9}x^3 + \frac{35}{288}x^4 - \cdots\right)}$$

$$= y_1 \int \frac{1}{x}\left(1 + 2x + \frac{5}{2}x^2 + \frac{23}{9}x^3 + \frac{677}{288}x^4 + \cdots\right) dx$$

$$= y_1 \int \left(\frac{1}{x} + 2 + \frac{5}{2}x + \frac{23}{9}x^2 + \frac{677}{288}x^3 + \cdots\right) dx$$

$$= y_1 \left[\ln x + 2x + \frac{5}{4}x^2 + \frac{23}{27}x^3 + \frac{677}{1{,}152}x^4 + \cdots\right]$$

$$= y_1 \ln x + y_1\left(2x + \frac{5}{4}x^2 + \frac{23}{27}x^3 + \frac{677}{1{,}152}x^4 + \cdots\right).$$

The general solution on $(0, \infty)$ is

$$y = C_1 y_1(x) + C_2 y_2(x).$$

33. (a) From $t = 1/x$ we have $dt/dx = -1/x^2 = -t^2$. Then

$$\frac{dy}{dx} = \frac{dy}{dt}\frac{dt}{dx} = -t^2\frac{dy}{dt}$$

and

$$\frac{d^2y}{dx^2} = \frac{d}{dx}\left(\frac{dy}{dx}\right) = \frac{d}{dx}\left(-t^2\frac{dy}{dt}\right) = -t^2\frac{d^2y}{dt^2}\frac{dt}{dx} - \frac{dy}{dt}\left(2t\frac{dt}{dx}\right) = t^4\frac{d^2y}{dt^2} + 2t^3\frac{dy}{dt}.$$

Now

$$x^4\frac{d^2y}{dx^2} + \lambda y = \frac{1}{t^4}\left(t^4\frac{d^2y}{dt^2} + 2t^3\frac{dy}{dt}\right) + \lambda y = \frac{d^2y}{dt^2} + \frac{2}{t}\frac{dy}{dt} + \lambda y = 0$$

becomes

$$t\frac{d^2y}{dt^2} + 2\frac{dy}{dt} + \lambda t y = 0.$$

(b) Substituting $y = \sum_{n=0}^{\infty} c_n t^{n+r}$ into the differential equation and collecting terms, we obtain

$$t\frac{d^2y}{dt^2} + 2\frac{dy}{dt} + \lambda t y = (r^2 + r)c_0 t^{r-1} + (r^2 + 3r + 2)c_1 t^r$$

$$+ \sum_{k=2}^{\infty}[(k+r)(k+r-1)c_k + 2(k+r)c_k + \lambda c_{k-2}]t^{k+r-1}$$

$$= 0,$$

which implies

$$r^2 + r = r(r+1) = 0,$$
$$\left(r^2 + 3r + 2\right)c_1 = 0,$$

and

$$(k+r)(k+r+1)c_k + \lambda c_{k-2} = 0.$$

The indicial roots are $r_1 = 0$ and $r_2 = -1$, so $c_1 = 0$. For $r_1 = 0$ the recurrence relation is

$$c_k = -\frac{\lambda c_{k-2}}{k(k+1)}, \quad k = 2, 3, 4, \ldots,$$

and

$$c_2 = -\frac{\lambda}{3!}c_0$$

$$c_3 = c_5 = c_7 = \cdots = 0$$

$$c_4 = \frac{\lambda^2}{5!}c_0$$

$$\vdots$$

$$c_{2n} = (-1)^n\frac{\lambda^n}{(2n+1)!}c_0.$$

For $r_2 = -1$ the recurrence relation is

$$c_k = -\frac{\lambda c_{k-2}}{k(k-1)}, \quad k = 2, 3, 4, \ldots,$$

and

$$c_2 = -\frac{\lambda}{2!} c_0$$

$$c_3 = c_5 = c_7 = \cdots = 0$$

$$c_4 = \frac{\lambda^2}{4!} c_0$$

$$\vdots$$

$$c_{2n} = (-1)^n \frac{\lambda^n}{(2n)!} c_0.$$

The general solution on $(0, \infty)$ is

$$y(t) = c_1 \sum_{n=0}^{\infty} \frac{(-1)^n}{(2n+1)!} (\sqrt{\lambda}\, t)^{2n} + c_2 t^{-1} \sum_{n=0}^{\infty} \frac{(-1)^n}{(2n)!} (\sqrt{\lambda}\, t)^{2n}$$

$$= \frac{1}{t} \left[C_1 \sum_{n=0}^{\infty} \frac{(-1)^n}{(2n+1)!} (\sqrt{\lambda}\, t)^{2n+1} + C_2 \sum_{n=0}^{\infty} \frac{(-1)^n}{(2n)!} (\sqrt{\lambda}\, t)^{2n} \right]$$

$$= \frac{1}{t} [C_1 \sin \sqrt{\lambda}\, t + C_2 \cos \sqrt{\lambda}\, t].$$

(c) Using $t = 1/x$, the solution of the original equation is

$$y(x) = C_1 x \sin \frac{\sqrt{\lambda}}{x} + C_2 x \cos \frac{\sqrt{\lambda}}{x}.$$

6.4 | Special Functions

The terminology and concepts listed below provide an outline of the main ideas encountered in this section. These can be useful when preparing for a quiz or test.

Terminology and Concepts

- Bessel's DE of order ν

- Bessel functions of the first kind: $J_\nu(x)$ and $J_{-\nu}(x)$

- Bessel functions of the second kind: $Y_\nu(x)$

- parametric Bessel equation of order ν

- modified Bessel equation of order ν

- modified Bessel function of the first kind: $I_\nu(x)$
- modified Bessel function of the second kind: $K_\nu(x)$
- aging spring
- Euler's constant: γ
- differential recurrence relation
- spherical Bessel functions
- Legendre's DE of order n
- Legendre polynomials
- Rodrigues' formula

The basic skills listed below summarize the more mechanical types of problems encountered in the exercise set for this section.

Basic Skills

- solve Bessel's DE of order ν
- solve DEs of the form $x^2 y'' + xy' + (\alpha^2 x^2 - \nu^2)y = 0$
- solve DEs of the form $y'' + [(1 - 2a)/x]y' + [b^2 c^2 x^{2c-2} + (a^2 - p^2 c^2)/x^2]y = 0, \ p \geq 0$
- verify various recurrence relations involving Bessel functions
- use a recurrence relation to generate Legendre polynomials
- use Rodrigues' formula to generate Legendre polynomials

Use of Computers See Section 2.3 in this manual for the syntax used to obtain Bessel functions and Legendre polynomials in *Mathematica* and *Maple*.

Bessel's Equation

3. Since $\nu^2 = 25/4$ the general solution is $y = c_1 J_{5/2}(x) + c_2 J_{-5/2}(x)$.

6. Since $\nu^2 = 4$ the general solution is $y = c_1 J_2(x) + c_2 Y_2(x)$.

9. We identify $\alpha = 5$ and $\nu = \frac{2}{3}$. Then the general solution is $y = c_1 J_{2/3}(5x) + c_2 J_{-2/3}(5x)$.

12. If $y = \sqrt{x}\, v(x)$ then

$$y' = x^{1/2} v'(x) + \frac{1}{2}x^{-1/2} v(x)$$

$$y'' = x^{1/2} v''(x) + x^{-1/2} v'(x) - \frac{1}{4}x^{-3/2} v(x)$$

and

$$x^2 y'' + \left(\alpha^2 x^2 - \nu^2 + \frac{1}{4}\right) y = x^{5/2} v''(x) + x^{3/2} v'(x) - \frac{1}{4} x^{1/2} v(x) + \left(\alpha^2 x^2 - \nu^2 + \frac{1}{4}\right) x^{1/2} v(x)$$

$$= x^{5/2} v''(x) + x^{3/2} v'(x) + (\alpha^2 x^{5/2} - \nu^2 x^{1/2}) v(x) = 0.$$

Multiplying by $x^{-1/2}$ we obtain

$$x^2 v''(x) + x v'(x) + (\alpha^2 x^2 - \nu^2) v(x) = 0,$$

whose solution is $v(x) = c_1 J_\nu(\alpha x) + c_2 Y_\nu(\alpha x)$. Then $y = c_1 \sqrt{x}\, J_\nu(\alpha x) + c_2 \sqrt{x}\, Y_\nu(\alpha x)$.

15. Write the differential equation in the form $y'' - (1/x)y' + y = 0$. This is the form of (18) in the text with $a = 1$, $c = 1$, $b = 1$, and $p = 1$, so, by (19) in the text, the general solution is

$$y = x[c_1 J_1(x) + c_2 Y_1(x)].$$

18. Write the differential equation in the form $y'' + (4 + 1/4x^2)y = 0$. This is the form of (18) in the text with $a = \frac{1}{2}$, $c = 1$, $b = 2$, and $p = 0$, so, by (19) in the text, the general solution is

$$y = x^{1/2}[c_1 J_0(2x) + c_2 Y_0(2x)].$$

21. Using the fact that $i^2 = -1$, along with the definition of $J_\nu(x)$ in (7) in the text, we have

$$I_\nu(x) = i^{-\nu} J_\nu(ix) = i^{-\nu} \sum_{n=0}^{\infty} \frac{(-1)^n}{n!\Gamma(1 + \nu + n)} \left(\frac{ix}{2}\right)^{2n+\nu}$$

$$= \sum_{n=0}^{\infty} \frac{(-1)^n}{n!\Gamma(1 + \nu + n)} i^{2n+\nu-\nu} \left(\frac{x}{2}\right)^{2n+\nu}$$

$$= \sum_{n=0}^{\infty} \frac{(-1)^n}{n!\Gamma(1 + \nu + n)} (i^2)^n \left(\frac{x}{2}\right)^{2n+\nu}$$

$$= \sum_{n=0}^{\infty} \frac{(-1)^{2n}}{n!\Gamma(1 + \nu + n)} \left(\frac{x}{2}\right)^{2n+\nu}$$

$$= \sum_{n=0}^{\infty} \frac{1}{n!\Gamma(1 + \nu + n)} \left(\frac{x}{2}\right)^{2n+\nu},$$

which is a real function.

24. Write the differential equation in the form $y'' + (4/x)y' + (1 + 2/x^2)y = 0$. This is the form of (18) in the text with

$$1 - 2a = 4 \implies a = -\frac{3}{2}$$

$$2c - 2 = 0 \implies c = 1$$

$$b^2 c^2 = 1 \implies b = 1$$

$$a^2 - p^2 c^2 = 2 \implies p = \frac{1}{2}.$$

Then, by (19), (23), and (24) in the text,

$$y = x^{-3/2}[c_1 J_{1/2}(x) + c_2 J_{-1/2}(x)] = x^{-3/2}\left[c_1 \sqrt{\frac{2}{\pi x}} \sin x + c_2 \sqrt{\frac{2}{\pi x}} \cos x\right]$$

$$= C_1 \frac{1}{x^2} \sin x + C_2 \frac{1}{x^2} \cos x.$$

27. (a) The recurrence relation follows from

$$-\nu J_\nu(x) + x J_{\nu-1}(x) = -\sum_{n=0}^{\infty} \frac{(-1)^n \nu}{n!\Gamma(1+\nu+n)} \left(\frac{x}{2}\right)^{2n+\nu} + x \sum_{n=0}^{\infty} \frac{(-1)^n}{n!\Gamma(\nu+n)} \left(\frac{x}{2}\right)^{2n+\nu-1}$$

$$= -\sum_{n=0}^{\infty} \frac{(-1)^n \nu}{n!\Gamma(1+\nu+n)} \left(\frac{x}{2}\right)^{2n+\nu} + \sum_{n=0}^{\infty} \frac{(-1)^n (\nu+n)}{n!\Gamma(1+\nu+n)} \cdot 2 \left(\frac{x}{2}\right) \left(\frac{x}{2}\right)^{2n+\nu-1}$$

$$= \sum_{n=0}^{\infty} \frac{(-1)^n (2n+\nu)}{n!\Gamma(1+\nu+n)} \left(\frac{x}{2}\right)^{2n+\nu} = x J_\nu'(x).$$

(b) The formula in part **(a)** is a linear first-order differential equation in $J_\nu(x)$. An integrating factor for this equation is x^ν, so

$$\frac{d}{dx}[x^\nu J_\nu(x)] = x^\nu J_{\nu-1}(x).$$

30. From (20) we obtain $J_0'(x) = -J_1(x)$, and from (21) we obtain $J_0'(x) = J_{-1}(x)$. Thus $J_0'(x) = J_{-1}(x) = -J_1(x)$.

33. Letting

$$s = \frac{2}{\alpha} \sqrt{\frac{k}{m}} e^{-\alpha t/2},$$

we have

$$\frac{dx}{dt} = \frac{dx}{ds}\frac{ds}{dt} = \frac{dx}{dt}\left[\frac{2}{\alpha}\sqrt{\frac{k}{m}}\left(-\frac{\alpha}{2}\right)e^{-\alpha t/2}\right] = \frac{dx}{ds}\left(-\sqrt{\frac{k}{m}}e^{-\alpha t/2}\right)$$

and

$$\frac{d^2x}{dt^2} = \frac{d}{dt}\left(\frac{dx}{dt}\right) = \frac{dx}{ds}\left(\frac{\alpha}{2}\sqrt{\frac{k}{m}}e^{-\alpha t/2}\right) + \frac{d}{dt}\left(\frac{dx}{ds}\right)\left(-\sqrt{\frac{k}{m}}e^{-\alpha t/2}\right)$$

$$= \frac{dx}{ds}\left(\frac{\alpha}{2}\sqrt{\frac{k}{m}}e^{-\alpha t/2}\right) + \frac{d^2x}{ds^2}\frac{ds}{dt}\left(-\sqrt{\frac{k}{m}}e^{-\alpha t/2}\right)$$

$$= \frac{dx}{ds}\left(\frac{\alpha}{2}\sqrt{\frac{k}{m}}e^{-\alpha t/2}\right) + \frac{d^2x}{ds^2}\left(\frac{k}{m}e^{-\alpha t}\right).$$

Then

$$m \frac{d^2x}{dt^2} + ke^{-\alpha t} x = ke^{-\alpha t} \frac{d^2x}{ds^2} + \frac{m\alpha}{2} \sqrt{\frac{k}{m}} \, e^{-\alpha t/2} \frac{dx}{ds} + ke^{-\alpha t} x = 0.$$

Multiplying by $2^2/\alpha^2 m$ we have

$$\frac{2^2}{\alpha^2} \frac{k}{m} e^{-\alpha t} \frac{d^2x}{ds^2} + \frac{2}{\alpha} \sqrt{\frac{k}{m}} \, e^{-\alpha t/2} \frac{dx}{ds} + \frac{2^2}{\alpha^2} \frac{k}{m} e^{-\alpha t} x = 0$$

or, since $s = (2/\alpha)\sqrt{k/m}\,e^{-\alpha t/2}$,

$$s^2 \frac{d^2x}{ds^2} + s \frac{dx}{ds} + s^2 x = 0.$$

36. The general solution of the differential equation is

$$y(x) = c_1 J_0(\alpha x) + c_2 Y_0(\alpha x).$$

In order to satisfy the conditions that $\lim_{x \to 0^+} y(x)$ and $\lim_{x \to 0^+} y'(x)$ are finite we are forced to define $c_2 = 0$. Thus, $y(x) = c_1 J_0(\alpha x)$. The second boundary condition, $y(2) = 0$, implies $c_1 = 0$ or $J_0(2\alpha) = 0$. In order to have a nontrivial solution we require that $J_0(2\alpha) = 0$. From Table 6.1, the first three positive zeros of J_0 are found to be

$$2\alpha_1 = 2.4048, \quad 2\alpha_2 = 5.5201, \quad 2\alpha_3 = 8.6537$$

and so $\alpha_1 = 1.2024$, $\alpha_2 = 2.7601$, $\alpha_3 = 4.3269$. The eigenfunctions corresponding to the eigenvalues $\lambda_1 = \alpha_1^2$, $\lambda_2 = \alpha_2^2$, $\lambda_3 = \alpha_3^2$ are $J_0(1.2024x)$, $J_0(2.7601x)$, and $J_0(4.3269x)$.

Legendre's Equation

45. The recurrence relation can be written

$$P_{k+1}(x) = \frac{2k+1}{k+1} x P_k(x) - \frac{k}{k+1} P_{k-1}(x), \qquad k = 2, \, 3, \, 4, \, \ldots \,.$$

$k = 1$: $P_2(x) = \frac{3}{2}x^2 - \frac{1}{2}$

$k = 2$: $P_3(x) = \frac{5}{3}x \left(\frac{3}{2}x^2 - \frac{1}{2} \right) - \frac{2}{3}x = \frac{5}{2}x^3 - \frac{3}{2}x$

$k = 3$: $P_4(x) = \frac{7}{4}x \left(\frac{5}{2}x^3 - \frac{3}{2}x \right) - \frac{3}{4} \left(\frac{3}{2}x^2 - \frac{1}{2} \right) = \frac{35}{8}x^4 - \frac{30}{8}x^2 + \frac{3}{8}$

$k = 4$: $P_5(x) = \frac{9}{5}x \left(\frac{35}{8}x^4 - \frac{30}{8}x^2 + \frac{3}{8} \right) - \frac{4}{5} \left(\frac{5}{2}x^3 - \frac{3}{2}x \right) = \frac{63}{8}x^5 - \frac{35}{4}x^3 + \frac{15}{8}x$

$$k = 5: \quad P_6(x) = \frac{11}{6}x\left(\frac{63}{8}x^5 - \frac{35}{4}x^3 + \frac{15}{8}x\right) - \frac{5}{6}\left(\frac{35}{8}x^4 - \frac{30}{8}x^2 + \frac{3}{8}\right)$$

$$= \frac{231}{16}x^6 - \frac{315}{16}x^4 + \frac{105}{16}x^2 - \frac{5}{16}$$

$$k = 6: \quad P_7(x) = \frac{13}{7}x\left(\frac{231}{16}x^6 - \frac{315}{16}x^4 + \frac{105}{16}x^2 - \frac{5}{16}\right) - \frac{6}{7}\left(\frac{63}{8}x^5 - \frac{35}{4}x^3 + \frac{15}{8}x\right)$$

$$= \frac{429}{16}x^7 - \frac{693}{16}x^5 + \frac{315}{16}x^3 - \frac{35}{16}x$$

Miscellaneous Differential Equations

51. Letting $y = \displaystyle\sum_{n=0}^{\infty} c_n x^n$ we have

$$y' = \sum_{n=1}^{\infty} n c_n x^{n-1} \quad \text{and} \quad y'' = \sum_{n=2}^{\infty} n(n+1)c_n x^{n-2}.$$

Then, with appropriate substitutions, we have

$$y'' - 2xy' + 2\alpha y = \sum_{k=0}^{\infty} \left[(k+2)(k+1)c_{k+2} + (2\alpha - 2k)c_k\right]x^k = 0.$$

This leads to the recurrence relation

$$c_{k+2} = -\frac{2\alpha - 2k}{(k+2)(k+1)}c_k, \quad \text{for} \quad k = 0, 1, 2, 3, \ldots.$$

Thus

$$c_2 = -\frac{2\alpha}{2!}c_0$$

$$c_3 = -\frac{2\alpha - 2}{3!}c_1$$

$$c_4 = -\frac{2\alpha - 4}{4 \cdot 3}c_2 = \frac{2^2\alpha(\alpha - 2)}{4!}c_0$$

$$c_5 = -\frac{2\alpha - 6}{5 \cdot 4}c_3 = \frac{2^2(\alpha - 1)(\alpha - 3)}{5!}c_1$$

$$\vdots$$

Then two solutions are

$$y_1(x) = 1 - \frac{2\alpha}{2!}x^2 + \frac{2^2\alpha(\alpha - 2)}{4!}x^4 - \cdots = 1 + \sum_{k=1}^{\infty} \frac{(-1)^k 2^k \alpha(\alpha - 2)\cdots(\alpha - 2k + 2)}{(2k)!}x^{2k}$$

and

$$y_2(x) = x - \frac{2\alpha - 2}{3!}x^3 + \frac{2^2(\alpha - 1)(\alpha - 3)}{5!}x^5 - \cdots$$

$$= 1 + \sum_{k=1}^{\infty} \frac{(-1)^k 2^k (\alpha - 1)(\alpha - 3)\cdots(\alpha - 2k + 1)}{(2k+1)!}x^{2k+1},$$

and the general solution is $y(x) = c_0 y_1(x) + c_1 y_2(x)$.

54. With $R(x) = (\alpha x)^{-1/2}Z(z)$ the product rule gives:

and

$$R' = \alpha^{-1/2}\left(x^{-1/2}Z' - \frac{1}{2}x^{-3/2}Z\right)$$

$$R'' = \alpha^{-1/2}\left(x^{-1/2}Z'' - \frac{1}{2}x^{-3/2}Z' - \frac{1}{2}x^{-3/2}Z' + \frac{3}{4}x^{-5/2}Z\right)$$

$$= \alpha^{-1/2}\left(x^{-1/2}Z'' - x^{-3/2}Z' + \frac{3}{4}x^{-5/2}Z\right).$$

The differential equation then becomes

$$\alpha^{-1/2}x^2\left(x^{-1/2}Z'' - x^{-3/2}Z' + \frac{3}{4}x^{-5/2}Z\right) + \alpha^{-1/2}2x\left(x^{-1/2}Z' - \frac{1}{2}x^{-3/2}Z\right)$$

$$+ \left[\alpha^2 x^2 - n(n+1)\right]\alpha^{-1/2}x^{-1/2}Z(x) = 0$$

or

$$x^2 Z'' + xZ' + \left[\alpha^2 x^2 - \left(n^2 + n + \frac{1}{4}\right)\right]Z = 0.$$

This is equivalent to

$$x^2 Z'' + xZ' + \left[\alpha^2 x^2 - \left(n + \frac{1}{2}\right)^2\right]Z = 0,$$

which is the parametric Bessel equation, so

$$Z(x) = C_1 J_{n+1/2}(\alpha x) + C_2 Y_{n+1/2}(\alpha x),$$

and

$$R(x) = \alpha^{-1/2}x^{1/2}Z(x) = \alpha^{-1/2}x^{-1/2}\left[C_1 J_{n+1/2}(\alpha x) + C_2 Y_{n+1/2}(\alpha x)\right].$$

Renaming C_1 and C_2 this becomes

$$R(x) = \left(c_1\sqrt{\frac{\pi}{2}}\right)\frac{J_{n+1/2}(\alpha x)}{\sqrt{\alpha x}} + \left(c_2\sqrt{\frac{\pi}{2}}\right)\frac{Y_{n+1/2}(\alpha x)}{\sqrt{\alpha x}}$$

$$= c_1\sqrt{\frac{\pi}{2\alpha x}}J_{n+1/2}(\alpha x) + c_2\sqrt{\frac{\pi}{2\alpha x}}Y_{n+1/2}(\alpha x)$$

$$= c_1 j_n(\alpha x) + c_2 y_n(\alpha x),$$

where $j_n(\alpha x)$ and $y_n(\alpha x)$ are the spherical Bessel functions of the first and second kind defined in the text.

6.R | Chapter 6 in Review

3. $x = -1$ is the nearest singular point to the ordinary point $x = 0$. Theorem 6.2.1 guarantees the existence of two power series solutions $y = \sum_{n=1}^{\infty} c_n x^n$ of the differential equation that converge at least for $-1 < x < 1$. Since $-\frac{1}{2} \le x \le \frac{1}{2}$ is properly contained in $-1 < x < 1$, both power series must converge for all points contained in $-\frac{1}{2} \le x \le \frac{1}{2}$.

6. We have

$$f(x) = \frac{\sin x}{\cos x} = \frac{x - \dfrac{x^3}{6} + \dfrac{x^5}{120} - \cdots}{1 - \dfrac{x^2}{2} + \dfrac{x^4}{24} - \cdots} = x + \frac{x^3}{3} + \frac{2x^5}{15} + \cdots .$$

9. Substituting $y = \sum_{n=0}^{\infty} c_n x^{n+r}$ into the differential equation we obtain

$$2xy'' + y' + y = \left(2r^2 - r\right) c_0 x^{r-1} + \sum_{k=1}^{\infty} [2(k+r)(k+r-1)c_k + (k+r)c_k + c_{k-1}] x^{k+r-1} = 0$$

which implies

$$2r^2 - r = r(2r - 1) = 0$$

and

$$(k+r)(2k+2r-1)c_k + c_{k-1} = 0.$$

The indicial roots are $r = 0$ and $r = 1/2$. For $r = 0$ the recurrence relation is

$$c_k = -\frac{c_{k-1}}{k(2k-1)}, \quad k = 1, 2, 3, \ldots,$$

so

$$c_1 = -c_0, \qquad c_2 = \frac{1}{6}c_0, \qquad c_3 = -\frac{1}{90}c_0.$$

For $r = 1/2$ the recurrence relation is

$$c_k = -\frac{c_{k-1}}{k(2k+1)}, \quad k = 1, 2, 3, \ldots,$$

so

$$c_1 = -\frac{1}{3}c_0, \qquad c_2 = \frac{1}{30}c_0, \qquad c_3 = -\frac{1}{630}c_0.$$

Two linearly independent solutions are

$$y_1 = 1 - x + \frac{1}{6}x^2 - \frac{1}{90}x^3 + \cdots$$

and

$$y_2 = x^{1/2}\left(1 - \frac{1}{3}x + \frac{1}{30}x^2 - \frac{1}{630}x^3 + \cdots\right).$$

12. Substituting $y = \sum_{n=0}^{\infty} c_n x^n$ into the differential equation we obtain

$$y'' - x^2 y' + xy = 2c_2 + (6c_3 + c_0)x + \sum_{k=1}^{\infty}[(k+3)(k+2)c_{k+3} - (k-1)c_k] x^{k+1} = 0$$

which implies $c_2 = 0$, $c_3 = -c_0/6$, and

$$c_{k+3} = \frac{k-1}{(k+3)(k+2)} c_k, \quad k = 1, 2, 3, \ldots .$$

Choosing $c_0 = 1$ and $c_1 = 0$ we find

$$c_3 = -\frac{1}{6}$$

$$c_4 = c_7 = c_{10} = \cdots = 0$$

$$c_5 = c_8 = c_{11} = \cdots = 0$$

$$c_6 = -\frac{1}{90}$$

and so on. For $c_0 = 0$ and $c_1 = 1$ we obtain

$$c_3 = c_6 = c_9 = \cdots = 0$$

$$c_4 = c_7 = c_{10} = \cdots = 0$$

$$c_5 = c_8 = c_{11} = \cdots = 0$$

and so on. Thus, two solutions are

$$y_1 = 1 - \frac{1}{6}x^3 - \frac{1}{90}x^6 - \cdots \quad \text{and} \quad y_2 = x.$$

15. Substituting $y = \sum_{n=0}^{\infty} c_n x^n$ into the differential equation we have

$$y'' + xy' + 2y = \underbrace{\sum_{n=2}^{\infty} n(n-1)c_n x^{n-2}}_{k=n-2} + \underbrace{\sum_{n=1}^{\infty} nc_n x^n}_{k=n} + 2\underbrace{\sum_{n=0}^{\infty} c_n x^n}_{k=n}$$

$$= \sum_{k=0}^{\infty} (k+2)(k+1)c_{k+2}x^k + \sum_{k=1}^{\infty} kc_k x^k + 2\sum_{k=0}^{\infty} c_k x^k$$

$$= 2c_2 + 2c_0 + \sum_{k=1}^{\infty} [(k+2)(k+1)c_{k+2} + (k+2)c_k]x^k = 0$$

Thus

$$2c_2 + 2c_0 = 0$$

$$(k+2)(k+1)c_{k+2} + (k+2)c_k = 0$$

and

$$c_2 = -c_0$$

$$c_{k+2} = -\frac{1}{k+1} c_k, \quad k = 1, 2, 3, \ldots .$$

Choosing $c_0 = 1$ and $c_1 = 0$ we find

$$c_2 = -1$$

$$c_3 = c_5 = c_7 = \cdots = 0$$

$$c_4 = \frac{1}{3}$$

$$c_6 = -\frac{1}{15}$$

and so on. For $c_0 = 0$ and $c_1 = 1$ we obtain

$$c_2 = c_4 = c_6 = \cdots = 0$$

$$c_3 = -\frac{1}{2}$$

$$c_5 = \frac{1}{8}$$

$$c_7 = -\frac{1}{48}$$

and so on. Thus, the general solution is

$$y = C_0 \left(1 - x^2 + \frac{1}{3}x^4 - \frac{1}{15}x^6 + \cdots \right) + C_1 \left(x - \frac{1}{2}x^3 + \frac{1}{8}x^5 - \frac{1}{48}x^7 + \cdots \right)$$

and

$$y' = C_0 \left(-2x + \frac{4}{3}x^3 - \frac{2}{5}x^5 + \cdots \right) + C_1 \left(1 - \frac{3}{2}x^2 + \frac{5}{8}x^4 - \frac{7}{48}x^6 + \cdots \right).$$

Setting $y(0) = 3$ and $y'(0) = -2$ we find $c_0 = 3$ and $c_1 = -2$. Therefore, the solution of the initial-value problem is

$$y = 3 - 2x - 3x^2 + x^3 + x^4 - \frac{1}{4}x^5 - \frac{1}{5}x^6 + \frac{1}{24}x^7 + \cdots .$$

18. While we can find two solutions of the form

$$y_1 = c_0[1 + \cdots] \quad \text{and} \quad y_2 = c_1[x + \cdots],$$

the initial conditions at $x = 1$ give solutions for c_0 and c_1 in terms of infinite series. Letting $t = x - 1$ the initial-value problem becomes

$$\frac{d^2 y}{dt^2} + (t+1)\frac{dy}{dt} + y = 0, \qquad y(0) = -6, \ y'(0) = 3.$$

Substituting $y = \sum_{n=0}^{\infty} c_n t^n$ into the differential equation we have

$$\frac{d^2 y}{dt^2} + (t+1)\frac{dy}{dt} + y = \underbrace{\sum_{n=2}^{\infty} n(n-1)c_n t^{n-2}}_{k=n-2} + \underbrace{\sum_{n=1}^{\infty} nc_n t^{n}}_{k=n} + \underbrace{\sum_{n=1}^{\infty} nc_n t^{n-1}}_{k=n-1} + \underbrace{\sum_{n=0}^{\infty} c_n t^{n}}_{k=n}$$

$$= \sum_{k=0}^{\infty}(k+2)(k+1)c_{k+2}t^k + \sum_{k=1}^{\infty}kc_kt^k + \sum_{k=0}^{\infty}(k+1)c_{k+1}t^k + \sum_{k=0}^{\infty}c_kt^k$$

$$= 2c_2 + c_1 + c_0 + \sum_{k=1}^{\infty}[(k+2)(k+1)c_{k+2} + (k+1)c_{k+1} + (k+1)c_k]t^k = 0.$$

Thus

$$2c_2 + c_1 + c_0 = 0$$

$$(k+2)(k+1)c_{k+2} + (k+1)c_{k+1} + (k+1)c_k = 0$$

and

$$c_2 = -\frac{c_1 + c_0}{2}$$

$$c_{k+2} = -\frac{c_{k+1} + c_k}{k+2}, \quad k = 1, 2, 3, \ldots.$$

Choosing $c_0 = 1$ and $c_1 = 0$ we find

$$c_2 = -\frac{1}{2}, \quad c_3 = \frac{1}{6}, \quad c_4 = \frac{1}{12},$$

and so on. For $c_0 = 0$ and $c_1 = 1$ we find

$$c_2 = -\frac{1}{2}, \quad c_3 = -\frac{1}{6}, \quad c_4 = \frac{1}{6},$$

and so on. Thus, the general solution is

$$y = c_0\left[1 - \frac{1}{2}t^2 + \frac{1}{6}t^3 + \frac{1}{12}t^4 + \cdots\right] + c_1\left[t - \frac{1}{2}t^2 - \frac{1}{6}t^3 + \frac{1}{6}t^4 + \cdots\right].$$

The initial conditions then imply $c_0 = -6$ and $c_1 = 3$. Thus the solution of the initial-value problem is

$$y = -6\left[1 - \frac{1}{2}(x-1)^2 + \frac{1}{6}(x-1)^3 + \frac{1}{12}(x-1)^4 + \cdots\right]$$

$$+ 3\left[(x-1) - \frac{1}{2}(x-1)^2 - \frac{1}{6}(x-1)^3 + \frac{1}{6}(x-1)^4 + \cdots\right].$$

21. Substituting $y = \sum_{n=0}^{\infty}c_nx^n$ into the differential equation we have

$$y'' + x^2y' + 2xy = \underbrace{\sum_{n=2}^{\infty}n(n-1)c_nx^{n-2}}_{k=n-2} + \underbrace{\sum_{n=1}^{\infty}nc_nx^{n+1}}_{k=n+1} + 2\underbrace{\sum_{n=0}^{\infty}c_nx^{n+1}}_{k=n+1}$$

$$= \sum_{k=0}^{\infty}(k+2)(k+1)c_{k+2}x^k + \sum_{k=2}^{\infty}(k-1)c_{k-1}x^k + 2\sum_{k=1}^{\infty}c_{k-1}x^k$$

$$= 2c_2 + (6c_3 + 2c_0)x + \sum_{k=2}^{\infty}[(k+2)(k+1)c_{k+2} + (k+1)c_{k-1}]x^k$$

$$= 5 - 2x + 10x^3.$$

Thus, equating coefficients of like powers of x gives

$$2c_2 = 5$$

$$6c_3 + 2c_0 = -2$$

$$12c_4 + 3c_1 = 0$$

$$20c_5 + 4c_2 = 10$$

$$(k+2)(k+1)c_{k+2} + (k+1)c_{k-1} = 0, \quad k = 4, 5, 6, \ldots,$$

and

$$c_2 = \frac{5}{2}$$

$$c_3 = -\frac{1}{3}c_0 - \frac{1}{3}$$

$$c_4 = -\frac{1}{4}c_1$$

$$c_5 = \frac{1}{2} - \frac{1}{5}c_2 = \frac{1}{2} - \frac{1}{5}\left(\frac{5}{2}\right) = 0$$

$$c_{k+2} = -\frac{1}{k+2}c_{k-1}.$$

Using the recurrence relation, we find

$$c_6 = -\frac{1}{6}c_3 = \frac{1}{3\cdot 6}(c_0 + 1) = \frac{1}{3^2\cdot 2!}c_0 + \frac{1}{3^2\cdot 2!}$$

$$c_7 = -\frac{1}{7}c_4 = \frac{1}{4\cdot 7}c_1$$

$$c_8 = c_{11} = c_{14} = \cdots = 0$$

$$c_9 = -\frac{1}{9}c_6 = -\frac{1}{3^3\cdot 3!}c_0 - \frac{1}{3^3\cdot 3!}$$

$$c_{10} = -\frac{1}{10}c_7 = -\frac{1}{4\cdot 7\cdot 10}c_1$$

$$c_{12} = -\frac{1}{12}c_9 = \frac{1}{3^4\cdot 4!}c_0 + \frac{1}{3^4\cdot 4!}$$

$$c_{13} = -\frac{1}{13}c_0 = \frac{1}{4\cdot 7\cdot 10\cdot 13}c_1$$

and so on. Thus

$$y = c_0 \left[1 - \frac{1}{3}x^3 + \frac{1}{3^2 \cdot 2!}x^6 - \frac{1}{3^3 \cdot 3!}x^9 + \frac{1}{3^4 \cdot 4!}x^{12} - \cdots \right]$$

$$+ c_1 \left[x - \frac{1}{4}x^4 + \frac{1}{4 \cdot 7}x^7 - \frac{1}{4 \cdot 7 \cdot 10}x^{10} + \frac{1}{4 \cdot 7 \cdot 10 \cdot 13}x^{13} - \cdots \right]$$

$$+ \left[\frac{5}{2}x^2 - \frac{1}{3}x^3 + \frac{1}{3^2 \cdot 2!}x^6 - \frac{1}{3^3 \cdot 3!}x^9 + \frac{1}{3^4 \cdot 4!}x^{12} - \cdots \right].$$

24. **(a)** Using formula (5) of Section 4.2 in the text, we find that a second solution of $(1 - x^2)y'' - 2xy' = 0$ is

$$y_2(x) = 1 \cdot \int \frac{e^{\int 2x\,dx/(1-x^2)}}{1^2}\,dx = \int e^{-\ln(1-x^2)}\,dx$$

$$= \int \frac{dx}{1 - x^2} = \frac{1}{2}\ln\left(\frac{1+x}{1-x}\right),$$

where partial fractions was used to obtain the last integral.

(b) Using formula (5) of Section 4.2 in the text, we find that a second solution of $(1 - x^2)y'' - 2xy' + 2y = 0$ is

$$y_2(x) = x \cdot \int \frac{e^{\int 2x\,dx/(1-x^2)}}{x^2}\,dx = x \int \frac{e^{-\ln(1-x^2)}}{x^2}\,dx$$

$$= x \int \frac{dx}{x^2(1 - x^2)}\,dx = x \left[\frac{1}{2}\ln\left(\frac{1+x}{1-x}\right) - \frac{1}{x} \right]$$

$$= \frac{x}{2}\ln\left(\frac{1+x}{1-x}\right) - 1,$$

where partial fractions was used to obtain the last integral.

(c)

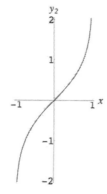

$$y_2(x) = \frac{1}{2}\ln\left(\frac{1+x}{1-x}\right)$$

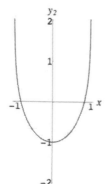

$$y_2 = \frac{x}{2}\ln\left(\frac{1+x}{1-x}\right) - 1$$

7 The Laplace Transform

The terminology and concepts listed below provide an outline of the main ideas encountered in this section. These can be useful when preparing for a quiz or test.

Terminology and Concepts

- integral transform
- definition of the Laplace transform
- linearity of the Laplace transform
- exponential order
- piecewise-continuous function
- transform of a piecewise-continuous function
- behavior of the Laplace transform $F(s)$ as $s \to \infty$

The basic skills listed below summarize the more mechanical types of problems encountered in the exercise set for this section.

Basic Skills

- find the Laplace transform of a function using the definition
- find the Laplace transform of a function using linearity and known transforms of some basic functions

Piecewise Continuity A function f that is piecewise continuous on $[0, \infty)$ has only a finite number of *finite* discontinuities on any finite interval of $[0, \infty)$. Another name for a **finite discontinuity** is a **jump discontinuity**; this means that at any point of discontinuity t_1, the left- and right-hand limits, $\lim_{t \to t_1^-} f(t)$ and $\lim_{t \to t_1^+} f(t)$, respectively, exist but are different. Put another way, f cannot have any vertical asymptotes on the interval $[0, \infty)$.

175

Use of Computers   The Laplace transform function is built into version 5 of *Mathematica*. In *Maple* it is contained in the integral transforms package `inttrans`. The routines below define the Laplace transform of $f(t) = t^2 \cos t$ as a function $F(s)$.

$$\textbf{Clear}[\textbf{f}] \qquad\qquad (\textit{Mathematica})$$

F[s_]:=Evaluate[LaplaceTransform[t^2 Cos[t], t, s]]

F[s]

```
with(inttrans);                          (Maple)
F:=s->laplace(t^2*cos(t),t,s);
F(s);
```

3. $\mathscr{L}\{f(t)\} = \displaystyle\int_0^1 te^{-st}dt + \int_1^\infty e^{-st}dt = \left(-\frac{1}{s}te^{-st} - \frac{1}{s^2}e^{-st}\right)\bigg|_0^1 - \frac{1}{s}e^{-st}\bigg|_1^\infty$

$$= \left(-\frac{1}{s}e^{-s} - \frac{1}{s^2}e^{-s}\right) - \left(0 - \frac{1}{s^2}\right) - \frac{1}{s}(0 - e^{-s}) = \frac{1}{s^2}(1 - e^{-s}), \quad s > 0$$

6. $\mathscr{L}\{f(t)\} = \displaystyle\int_{\pi/2}^\infty (\cos t)e^{-st}dt = \left(-\frac{s}{s^2+1}e^{-st}\cos t + \frac{1}{s^2+1}e^{-st}\sin t\right)\bigg|_{\pi/2}^\infty$

$$= 0 - \left(0 + \frac{1}{s^2+1}e^{-\pi s/2}\right) = -\frac{1}{s^2+1}e^{-\pi s/2}, \quad s > 0$$

9. The function is $f(t) = \begin{cases} 1 - t, & 0 < t < 1 \\ 0, & t > 1 \end{cases}$ so

$$\mathscr{L}\{f(t)\} = \int_0^1 (1-t)e^{-st}\,dt + \int_1^\infty 0e^{-st}\,dt = \int_0^1 (1-t)e^{-st}\,dt = \left(-\frac{1}{s}(1-t)e^{-st} + \frac{1}{s^2}e^{-st}\right)\bigg|_0^1$$

$$= \frac{1}{s^2}e^{-s} + \frac{1}{s} - \frac{1}{s^2}, \quad s > 0$$

12. $\mathscr{L}\{f(t)\} = \displaystyle\int_0^\infty e^{-2t-5}e^{-st}dt = e^{-5}\int_0^\infty e^{-(s+2)t}dt = -\frac{e^{-5}}{s+2}e^{-(s+2)t}\bigg|_0^\infty = \frac{e^{-5}}{s+2}, \quad s > -2$

15. $\mathscr{L}\{f(t)\} = \displaystyle\int_0^\infty e^{-t}(\sin t)e^{-st}dt = \int_0^\infty (\sin t)e^{-(s+1)t}dt$

$$= \left(\frac{-(s+1)}{(s+1)^2+1}e^{-(s+1)t}\sin t - \frac{1}{(s+1)^2+1}e^{-(s+1)t}\cos t\right)\bigg|_0^\infty$$

$$= \frac{1}{(s+1)^2+1} = \frac{1}{s^2+2s+2}, \quad s > -1$$

18. $\mathscr{L}\{f(t)\} = \displaystyle\int_0^\infty t(\sin t)e^{-st}dt$

$$= \left[\left(-\frac{t}{s^2+1} - \frac{2s}{(s^2+1)^2}\right)(\cos t)e^{-st} - \left(\frac{st}{s^2+1} + \frac{s^2-1}{(s^2+1)^2}\right)(\sin t)e^{-st}\right]_0^\infty$$

$$= \frac{2s}{(s^2+1)^2}, \quad s > 0$$

21. $\mathscr{L}\{4t - 10\} = \dfrac{4}{s^2} - \dfrac{10}{s}$

24. $\mathscr{L}\{-4t^2 + 16t + 9\} = -4\dfrac{2}{s^3} + \dfrac{16}{s^2} + \dfrac{9}{s}$

27. $\mathscr{L}\{1 + e^{4t}\} = \dfrac{1}{s} + \dfrac{1}{s-4}$

30. $\mathscr{L}\{e^{2t} - 2 + e^{-2t}\} = \dfrac{1}{s-2} - \dfrac{2}{s} + \dfrac{1}{s+2}$

33. $\mathscr{L}\{\sinh kt\} = \dfrac{1}{2}\mathscr{L}\left\{e^{kt} - e^{-kt}\right\} = \dfrac{1}{2}\left[\dfrac{1}{s-k} - \dfrac{1}{s+k}\right] = \dfrac{k}{s^2-k^2}$

36. $\mathscr{L}\{e^{-t}\cosh t\} = \mathscr{L}\left\{e^{-t}\dfrac{e^t + e^{-t}}{2}\right\} = \mathscr{L}\left\{\dfrac{1}{2} + \dfrac{1}{2}e^{-2t}\right\} = \dfrac{1}{2s} + \dfrac{1}{2(s+2)}$

39. From the addition formula for the sine function, $\sin(4t + 5) = (\sin 4t)(\cos 5) + (\cos 4t)(\sin 5)$ so

$$\mathscr{L}\{\sin(4t+5)\} = (\cos 5)\mathscr{L}\{\sin 4t\} + (\sin 5)\mathscr{L}\{\cos 4t\}$$

$$= (\cos 5)\frac{4}{s^2+16} + (\sin 5)\frac{s}{s^2+16}$$

$$= \frac{4\cos 5 + (\sin 5)s}{s^2+16}.$$

42. Let $u = st$ so that $du = s\,dt$. Then

$$\mathscr{L}\{t^\alpha\} = \int_0^\infty e^{-st}t^\alpha dt = \int_0^\infty e^{-u}\left(\frac{u}{s}\right)^\alpha \frac{1}{s}\,du = \frac{1}{s^{\alpha+1}}\Gamma(\alpha+1), \quad \alpha > -1.$$

45. $\mathscr{L}\left\{t^{3/2}\right\} = \dfrac{\Gamma(5/2)}{s^{5/2}} = \dfrac{3\sqrt{\pi}}{4s^{5/2}}$

7.2 | Inverse Transforms and Transforms of Derivatives

The terminology and concepts listed below provide an outline of the main ideas encountered in this section. These can be useful when preparing for a quiz or test.

Terminology and Concepts

- definition of the inverse Laplace transform
- linearity of the inverse Laplace transform
- Laplace transform of the derivative of a function
- use of the Laplace transform to solve an IVP

The basic skills listed below summarize the more mechanical types of problems encountered in the exercise set for this section.

Basic Skills

- find the inverse Laplace transform of a function (frequently using partial fractions)
- solve a linear IVP using the Laplace transform

Partial Fractions Again

The key to success in working with the Laplace transform is the ability to do algebra quickly and accurately. Partial fraction decomposition is used almost every time you have to solve some sort of equation. In this manual, check out the review of partial fractions given in Section 2.2.

Use of Computers

As can be seen in this section, partial fractions are used extensively in the computation of inverse Laplace transforms. This process can be very computationally intensive, and as such, is ideally suited to a CAS. The examples below show how to find the partial fraction decomposition of $(s^2 + 1)/(s^3 + 4s)$.

> **Apart**[(s^2+1)/(s^3+4s)]   (*Mathematica*)
>
> convert((s^2+1)/(s^3+4*s),parfrac,s);   (*Maple*)

The routines below define the inverse Laplace transform of $F(s) = (2s^3 - 6s)/(1 + s^2)^3$ as a function $f(t)$.

> **Clear**[f]   (*Mathematica*)
>
> f[t_]:=**Evaluate**[**InverseLaplaceTransform**[(2s^3 - 6s)/(s^2 + 1)^3, s, t]]
>
> f[t]

```
with(inttrans);
f:=t->invlaplace((2*s^3-6*s)/(s^2+1)^3,s,t);
f(t);
```

3. $\mathcal{L}^{-1}\left\{\dfrac{1}{s^2} - \dfrac{48}{s^5}\right\} = \mathcal{L}^{-1}\left\{\dfrac{1}{s^2} - \dfrac{48}{24}\cdot\dfrac{4!}{s^5}\right\} = t - 2t^4$

6. $\mathcal{L}^{-1}\left\{\dfrac{(s+2)^2}{s^3}\right\} = \mathcal{L}^{-1}\left\{\dfrac{1}{s} + 4\cdot\dfrac{1}{s^2} + 2\cdot\dfrac{2}{s^3}\right\} = 1 + 4t + 2t^2$

9. $\mathcal{L}^{-1}\left\{\dfrac{1}{4s+1}\right\} = \dfrac{1}{4}\mathcal{L}^{-1}\left\{\dfrac{1}{s+1/4}\right\} = \dfrac{1}{4}e^{-t/4}$

12. $\mathcal{L}^{-1}\left\{\dfrac{10s}{s^2+16}\right\} = 10\cos 4t$

15. $\mathcal{L}^{-1}\left\{\dfrac{2s-6}{s^2+9}\right\} = \mathcal{L}^{-1}\left\{2\cdot\dfrac{s}{s^2+9} - 2\cdot\dfrac{3}{s^2+9}\right\} = 2\cos 3t - 2\sin 3t$

18. $\mathcal{L}^{-1}\left\{\dfrac{s+1}{s^2-4s}\right\} = \mathcal{L}^{-1}\left\{-\dfrac{1}{4}\cdot\dfrac{1}{s} + \dfrac{5}{4}\cdot\dfrac{1}{s-4}\right\} = -\dfrac{1}{4} + \dfrac{5}{4}e^{4t}$

21. $\mathcal{L}^{-1}\left\{\dfrac{0.9s}{(s-0.1)(s+0.2)}\right\} = \mathcal{L}^{-1}\left\{(0.3)\cdot\dfrac{1}{s-0.1} + (0.6)\cdot\dfrac{1}{s+0.2}\right\} = 0.3e^{0.1t} + 0.6e^{-0.2t}$

24. $\mathcal{L}^{-1}\left\{\dfrac{s^2+1}{s(s-1)(s+1)(s-2)}\right\} = \mathcal{L}^{-1}\left\{\dfrac{1}{2}\cdot\dfrac{1}{s} - \dfrac{1}{s-1} - \dfrac{1}{3}\cdot\dfrac{1}{s+1} + \dfrac{5}{6}\cdot\dfrac{1}{s-2}\right\}$

$$= \dfrac{1}{2} - e^t - \dfrac{1}{3}e^{-t} + \dfrac{5}{6}e^{2t}$$

27. $\mathcal{L}^{-1}\left\{\dfrac{2s-4}{(s^2+s)(s^2+1)}\right\} = \mathcal{L}^{-1}\left\{\dfrac{2s-4}{s(s+1)(s^2+1)}\right\} = \mathcal{L}^{-1}\left\{-\dfrac{4}{s} + \dfrac{3}{s+1} + \dfrac{s}{s^2+1} + \dfrac{3}{s^2+1}\right\}$

$$= -4 + 3e^{-t} + \cos t + 3\sin t$$

30. $\mathcal{L}^{-1}\left\{\dfrac{6s+3}{(s^2+1)(s^2+4)}\right\} = \mathcal{L}^{-1}\left\{2\cdot\dfrac{s}{s^2+1} + \dfrac{1}{s^2+1} - 2\cdot\dfrac{s}{s^2+4} - \dfrac{1}{2}\cdot\dfrac{2}{s^2+4}\right\}$

$$= 2\cos t + \sin t - 2\cos 2t - \dfrac{1}{2}\sin 2t$$

33. The Laplace transform of the initial-value problem is

$$s\mathcal{L}\{y\} - y(0) + 6\mathcal{L}\{y\} = \dfrac{1}{s-4}.$$

Solving for $\mathscr{L}\{y\}$ we obtain

$$\mathscr{L}\{y\} = \frac{1}{(s-4)(s+6)} + \frac{2}{s+6} = \frac{1}{10} \cdot \frac{1}{s-4} + \frac{19}{10} \cdot \frac{1}{s+6}.$$

Thus

$$y = \frac{1}{10}e^{4t} + \frac{19}{10}e^{-6t}.$$

36. The Laplace transform of the initial-value problem is

$$s^2 \mathscr{L}\{y\} - sy(0) - y'(0) - 4[s\mathscr{L}\{y\} - y(0)] = \frac{6}{s-3} - \frac{3}{s+1}.$$

Solving for $\mathscr{L}\{y\}$ we obtain

$$\mathscr{L}\{y\} = \frac{6}{(s-3)(s^2-4s)} - \frac{3}{(s+1)(s^2-4s)} + \frac{s-5}{s^2-4s}$$

$$= \frac{5}{2} \cdot \frac{1}{s} - \frac{2}{s-3} - \frac{3}{5} \cdot \frac{1}{s+1} + \frac{11}{10} \cdot \frac{1}{s-4}.$$

Thus

$$y = \frac{5}{2} - 2e^{3t} - \frac{3}{5}e^{-t} + \frac{11}{10}e^{4t}.$$

39. The Laplace transform of the initial-value problem is

$$2\left[s^3 \mathscr{L}\{y\} - s^2 y(0) - sy'(0) - y''(0)\right] + 3[s^2 \mathscr{L}\{y\} - sy(0) - y'(0)] - 3[s\mathscr{L}\{y\} - y(0)] - 2\mathscr{L}\{y\} = \frac{1}{s+1}.$$

Solving for $\mathscr{L}\{y\}$ we obtain

$$\mathscr{L}\{y\} = \frac{2s+3}{(s+1)(s-1)(2s+1)(s+2)} = \frac{1}{2}\frac{1}{s+1} + \frac{5}{18}\frac{1}{s-1} - \frac{8}{9}\frac{1}{s+1/2} + \frac{1}{9}\frac{1}{s+2}.$$

Thus

$$y = \frac{1}{2}e^{-t} + \frac{5}{18}e^{t} - \frac{8}{9}e^{-t/2} + \frac{1}{9}e^{-2t}.$$

42. The Laplace transform of the initial-value problem is

$$s^2 \mathscr{L}\{y\} - s \cdot 1 - 3 - 2[s\mathscr{L}\{y\} - 1] + 5\mathscr{L}\{y\} = (s^2 - 2s + 5)\mathscr{L}\{y\} - s - 1 = 0.$$

Solving for $\mathscr{L}\{y\}$ we obtain

$$\mathscr{L}\{y\} = \frac{s+1}{s^2-2s+5} = \frac{s-1+2}{(s-1)^2+2^2} = \frac{s-1}{(s-1)^2+2^2} + \frac{2}{(s-1)^2+2^2}.$$

Thus

$$y = e^t \cos 2t + e^t \sin 2t.$$

45. For $y'' - 4y' = 6e^{3t} - 3e^{-t}$ the transfer function is $W(s) = 1/(s^2 - 4s)$. The zero-input response is

$$y_0(t) = \mathscr{L}^{-1}\left\{\frac{s-5}{s^2-4s}\right\} = \mathscr{L}^{-1}\left\{\frac{5}{4}\cdot\frac{1}{s} - \frac{1}{4}\cdot\frac{1}{s-4}\right\} = \frac{5}{4} - \frac{1}{4}e^{4t},$$

and the zero-state response is

$$y_1(t) = \mathscr{L}^{-1}\left\{\frac{6}{(s-3)(s^2-4s)} - \frac{3}{(s+1)(s^2-4s)}\right\}$$

$$= \mathscr{L}^{-1}\left\{\frac{27}{20}\cdot\frac{1}{s-4} - \frac{2}{s-3} + \frac{5}{4}\cdot\frac{1}{s} - \frac{3}{5}\cdot\frac{1}{s+1}\right\}$$

$$= \frac{27}{20}e^{4t} - 2e^{3t} + \frac{5}{4} - \frac{3}{5}e^{-t}.$$

7.3 | Operational Properties I

The terminology and concepts listed below provide an outline of the main ideas encountered in this section. These can be useful when preparing for a quiz or test.

Terminology and Concepts

- first translation (or shifting) theorem
- unit step function (Heaviside function)
- second translation (or shifting) theorem

The basic skills listed below summarize the more mechanical types of problems encountered in the exercise set for this section.

Basic Skills

- use the first translation theorem to find the Laplace transform of a function that is a multiple of e^{at}.
- use the first translation theorem to find the inverse Laplace transform of the shifted function $F(s-a)$
- use the Laplace transform in conjunction with the first translation theorem to solve an IVP

- express a piecewise-defined function in terms of unit step functions
- find the Laplace transform of a piecewise-defined function
- find the inverse Laplace transform of a function that is a multiple of e^{-as}
- use the Laplace transform to solve an IVP involving piecewise-defined functions

Use of Computers   See Section 2.3 in this manual for *Mathematica* and *Maple* syntax for the unit step function.

3. $\mathscr{L}\left\{t^3 e^{-2t}\right\} = \dfrac{3!}{(s+2)^4}$

6. $\mathscr{L}\left\{e^{2t}(t-1)^2\right\} = \mathscr{L}\left\{t^2 e^{2t} - 2te^{2t} + e^{2t}\right\} = \dfrac{2}{(s-2)^3} - \dfrac{2}{(s-2)^2} + \dfrac{1}{s-2}$

9. $\mathscr{L}\left\{(1 - e^t + 3e^{-4t})\cos 5t\right\} = \mathscr{L}\left\{\cos 5t - e^t \cos 5t + 3e^{-4t}\cos 5t\right\}$

$$= \dfrac{s}{s^2 + 25} - \dfrac{s-1}{(s-1)^2 + 25} + \dfrac{3(s+4)}{(s+4)^2 + 25}$$

12. $\mathscr{L}^{-1}\left\{\dfrac{1}{(s-1)^4}\right\} = \dfrac{1}{6}\mathscr{L}^{-1}\left\{\dfrac{3!}{(s-1)^4}\right\} = \dfrac{1}{6}t^3 e^t$

15. $\mathscr{L}^{-1}\left\{\dfrac{s}{s^2 + 4s + 5}\right\} = \mathscr{L}^{-1}\left\{\dfrac{s+2}{(s+2)^2 + 1^2} - 2\dfrac{1}{(s+2)^2 + 1^2}\right\} = e^{-2t}\cos t - 2e^{-2t}\sin t$

18. $\mathscr{L}^{-1}\left\{\dfrac{5s}{(s-2)^2}\right\} = \mathscr{L}^{-1}\left\{\dfrac{5(s-2) + 10}{(s-2)^2}\right\} = \mathscr{L}^{-1}\left\{\dfrac{5}{s-2} + \dfrac{10}{(s-2)^2}\right\} = 5e^{2t} + 10te^{2t}$

21. The Laplace transform of the differential equation is

$$s\mathscr{L}\{y\} - y(0) + 4\mathscr{L}\{y\} = \dfrac{1}{s+4}.$$

Solving for $\mathscr{L}\{y\}$ we obtain

$$\mathscr{L}\{y\} = \dfrac{1}{(s+4)^2} + \dfrac{2}{s+4}.$$

Thus

$$y = te^{-4t} + 2e^{-4t}.$$

24. The Laplace transform of the differential equation is

$$s^2 \mathcal{L}\{y\} - sy(0) - y'(0) - 4\left[s\,\mathcal{L}\{y\} - y(0)\right] + 4\,\mathcal{L}\{y\} = \frac{6}{(s-2)^4}.$$

Solving for $\mathcal{L}\{y\}$ we obtain $\mathcal{L}\{y\} = \dfrac{1}{20}\dfrac{5!}{(s-2)^6}$. Thus, $y = \dfrac{1}{20}t^5 e^{2t}$.

27. The Laplace transform of the differential equation is

$$s^2 \mathcal{L}\{y\} - sy(0) - y'(0) - 6\left[s\,\mathcal{L}\{y\} - y(0)\right] + 13\,\mathcal{L}\{y\} = 0.$$

Solving for $\mathcal{L}\{y\}$ we obtain

$$\mathcal{L}\{y\} = -\frac{3}{s^2 - 6s + 13} = -\frac{3}{2}\frac{2}{(s-3)^2 + 2^2}.$$

Thus

$$y = -\frac{3}{2}e^{3t}\sin 2t.$$

30. The Laplace transform of the differential equation is

$$s^2 \mathcal{L}\{y\} - sy(0) - y'(0) - 2\left[s\,\mathcal{L}\{y\} - y(0)\right] + 5\,\mathcal{L}\{y\} = \frac{1}{s} + \frac{1}{s^2}.$$

Solving for $\mathcal{L}\{y\}$ we obtain

$$\mathcal{L}\{y\} = \frac{4s^2 + s + 1}{s^2(s^2 - 2s + 5)} = \frac{7}{25}\frac{1}{s} + \frac{1}{5}\frac{1}{s^2} + \frac{-7s/25 + 109/25}{s^2 - 2s + 5}$$

$$= \frac{7}{25}\frac{1}{s} + \frac{1}{5}\frac{1}{s^2} - \frac{7}{25}\frac{s-1}{(s-1)^2 + 2^2} + \frac{51}{25}\frac{2}{(s-1)^2 + 2^2}.$$

Thus

$$y = \frac{7}{25} + \frac{1}{5}t - \frac{7}{25}e^t \cos 2t + \frac{51}{25}e^t \sin 2t.$$

33. Recall from Section 5.1 that $mx'' = -kx - \beta x'$. Now $m = W/g = 4/32 = \frac{1}{8}$ slug, and $4 = 2k$ so that $k = 2$ lb/ft. Thus, the differential equation is $x'' + 7x' + 16x = 0$. The initial conditions are $x(0) = -3/2$ and $x'(0) = 0$. The Laplace transform of the differential equation is

$$s^2\mathcal{L}\{x\} + \frac{3}{2}s + 7s\mathcal{L}\{x\} + \frac{21}{2} + 16\mathcal{L}\{x\} = 0.$$

Solving for $\mathcal{L}\{x\}$ we obtain

$$\mathcal{L}\{x\} = \frac{-3s/2 - 21/2}{s^2 + 7s + 16} = -\frac{3}{2}\frac{s + 7/2}{(s+7/2)^2 + (\sqrt{15}/2)^2} - \frac{7\sqrt{15}}{10}\frac{\sqrt{15}/2}{(s+7/2)^2 + (\sqrt{15}/2)^2}.$$

Thus

$$x = -\frac{3}{2}e^{-7t/2}\cos\frac{\sqrt{15}}{2}t - \frac{7\sqrt{15}}{10}e^{-7t/2}\sin\frac{\sqrt{15}}{2}t.$$

36. The differential equation is

$$R\frac{dq}{dt} + \frac{1}{C}q = E_0 e^{-kt}, \quad q(0) = 0.$$

The Laplace transform of this equation is

$$Rs\,\mathscr{L}\{q\} + \frac{1}{C}\,\mathscr{L}\{q\} = E_0\,\frac{1}{s+k}\,.$$

Solving for $\mathscr{L}\{q\}$ we obtain

$$\mathscr{L}\{q\} = \frac{E_0 C}{(s+k)(RCs+1)} = \frac{E_0/R}{(s+k)(s+1/RC)}\,.$$

When $1/RC \neq k$ we have by partial fractions

$$\mathscr{L}\{q\} = \frac{E_0}{R}\left(\frac{1/(1/RC - k)}{s+k} - \frac{1/(1/RC - k)}{s+1/RC}\right) = \frac{E_0}{R}\,\frac{1}{1/RC - k}\left(\frac{1}{s+k} - \frac{1}{s+1/RC}\right).$$

Thus

$$q(t) = \frac{E_0 C}{1 - kRC}\left(e^{-kt} - e^{-t/RC}\right).$$

When $1/RC = k$ we have

$$\mathscr{L}\{q\} = \frac{E_0}{R}\,\frac{1}{(s+k)^2}\,.$$

Thus

$$q(t) = \frac{E_0}{R}\,te^{-kt} = \frac{E_0}{R}\,te^{-t/RC}.$$

39. $\mathscr{L}\{t\,\mathscr{U}(t-2)\} = \mathscr{L}\{(t-2)\,\mathscr{U}(t-2) + 2\,\mathscr{U}(t-2)\} = \dfrac{e^{-2s}}{s^2} + \dfrac{2e^{-2s}}{s}$

Alternatively, (16) of this section in the text could be used:

$$\mathscr{L}\{t\,\mathscr{U}(t-2)\} = e^{-2s}\mathscr{L}\{t+2\} = e^{-2s}\left(\frac{1}{s^2} + \frac{2}{s}\right).$$

42. $\mathscr{L}\left\{\sin t\,\mathscr{U}\left(t - \dfrac{\pi}{2}\right)\right\} = \mathscr{L}\left\{\cos\left(t - \dfrac{\pi}{2}\right)\mathscr{U}\left(t - \dfrac{\pi}{2}\right)\right\} = \dfrac{se^{-\pi s/2}}{s^2 + 1}$

Alternatively, (16) of this section in the text could be used:

$$\mathscr{L}\left\{\sin t\,\mathscr{U}\left(t - \frac{\pi}{2}\right)\right\} = e^{-\pi s/2}\mathscr{L}\left\{\sin\left(t + \frac{\pi}{2}\right)\right\} = e^{-\pi s/2}\mathscr{L}\{\cos t\} = e^{-\pi s/2}\,\frac{s}{s^2 + 1}\,.$$

45. $\mathscr{L}^{-1}\left\{\dfrac{e^{-\pi s}}{s^2+1}\right\} = \sin(t-\pi)\,\mathscr{U}(t-\pi) = -\sin t\,\mathscr{U}(t-\pi)$

48. $\mathscr{L}^{-1}\left\{\dfrac{e^{-2s}}{s^2(s-1)}\right\} = \mathscr{L}^{-1}\left\{-\dfrac{e^{-2s}}{s} - \dfrac{e^{-2s}}{s^2} + \dfrac{e^{-2s}}{s-1}\right\} = -\mathscr{U}(t-2) - (t-2)\mathscr{U}(t-2) + e^{t-2}\mathscr{U}(t-2)$

51. (f)   **54. (d)**

57. $\mathscr{L}\left\{t^2\,\mathscr{U}(t-1)\right\} = \mathscr{L}\left\{\left[(t-1)^2 + 2t - 1\right]\mathscr{U}(t-1)\right\} = \mathscr{L}\left\{\left[(t-1)^2 + 2(t-1) + 1\right]\mathscr{U}(t-1)\right\}$

$$= \left(\dfrac{2}{s^3} + \dfrac{2}{s^2} + \dfrac{1}{s}\right)e^{-s}$$

Alternatively, by (16) of this section in the text,

$$\mathscr{L}\left\{t^2\,\mathscr{U}(t-1)\right\} = e^{-s}\,\mathscr{L}\left\{t^2 + 2t + 1\right\} = e^{-s}\left(\dfrac{2}{s^3} + \dfrac{2}{s^2} + \dfrac{1}{s}\right).$$

60. $\mathscr{L}\left\{\sin t - \sin t\,\mathscr{U}(t-2\pi)\right\} = \mathscr{L}\left\{\sin t - \sin(t-2\pi)\,\mathscr{U}(t-2\pi)\right\} = \dfrac{1}{s^2+1} - \dfrac{e^{-2\pi s}}{s^2+1}$

63. The Laplace transform of the differential equation is

$$s\,\mathscr{L}\{y\} - y(0) + \mathscr{L}\{y\} = \dfrac{5}{s}e^{-s}.$$

Solving for $\mathscr{L}\{y\}$ we obtain

$$\mathscr{L}\{y\} = \dfrac{5e^{-s}}{s(s+1)} = 5e^{-s}\left[\dfrac{1}{s} - \dfrac{1}{s+1}\right].$$

Thus

$$y = 5\,\mathscr{U}(t-1) - 5e^{-(t-1)}\,\mathscr{U}(t-1).$$

66. The Laplace transform of the differential equation is

$$s^2\,\mathscr{L}\{y\} - sy(0) - y'(0) + 4\,\mathscr{L}\{y\} = \dfrac{1}{s} - \dfrac{e^{-s}}{s}.$$

Solving for $\mathscr{L}\{y\}$ we obtain

$$\mathscr{L}\{y\} = \dfrac{1-s}{s(s^2+4)} - e^{-s}\dfrac{1}{s(s^2+4)} = \dfrac{1}{4}\dfrac{1}{s} - \dfrac{1}{4}\dfrac{s}{s^2+4} - \dfrac{1}{2}\dfrac{2}{s^2+4} - e^{-s}\left[\dfrac{1}{4}\dfrac{1}{s} - \dfrac{1}{4}\dfrac{s}{s^2+4}\right].$$

Thus

$$y = \dfrac{1}{4} - \dfrac{1}{4}\cos 2t - \dfrac{1}{2}\sin 2t - \left[\dfrac{1}{4} - \dfrac{1}{4}\cos 2(t-1)\right]\mathscr{U}(t-1).$$

69. The Laplace transform of the differential equation is

$$s^2 \mathscr{L}\{y\} - sy(0) - y'(0) + \mathscr{L}\{y\} = \frac{e^{-\pi s}}{s} - \frac{e^{-2\pi s}}{s}.$$

Solving for $\mathscr{L}\{y\}$ we obtain

$$\mathscr{L}\{y\} = e^{-\pi s}\left[\frac{1}{s} - \frac{s}{s^2 + 1}\right] - e^{-2\pi s}\left[\frac{1}{s} - \frac{s}{s^2 + 1}\right] + \frac{1}{s^2 + 1}.$$

Thus

$$y = [1 - \cos(t - \pi)]\,\mathscr{U}(t - \pi) - [1 - \cos(t - 2\pi)]\,\mathscr{U}(t - 2\pi) + \sin t.$$

72. Recall from Section 5.1 that $mx'' = -kx + f(t)$. Now $m = W/g = 32/32 = 1$ slug, and $32 = 2k$ so that $k = 16$ lb/ft. Thus, the differential equation is $x'' + 16x = f(t)$. The initial conditions are $x(0) = 0$, $x'(0) = 0$. Also, since

$$f(t) = \begin{cases} \sin t, & 0 \le t < 2\pi \\ 0, & t \ge 2\pi \end{cases}$$

and $\sin t = \sin(t - 2\pi)$ we can write

$$f(t) = \sin t - \sin(t - 2\pi)\,\mathscr{U}(t - 2\pi).$$

The Laplace transform of the differential equation is

$$s^2 \mathscr{L}\{x\} + 16\,\mathscr{L}\{x\} = \frac{1}{s^2 + 1} - \frac{1}{s^2 + 1}e^{-2\pi s}.$$

Solving for $\mathscr{L}\{x\}$ we obtain

$$\mathscr{L}\{x\} = \frac{1}{(s^2 + 16)(s^2 + 1)} - \frac{1}{(s^2 + 16)(s^2 + 1)}e^{-2\pi s}$$

$$= \frac{-1/15}{s^2 + 16} + \frac{1/15}{s^2 + 1} - \left[\frac{-1/15}{s^2 + 16} + \frac{1/15}{s^2 + 1}\right]e^{-2\pi s}.$$

Thus

$$x(t) = -\frac{1}{60}\sin 4t + \frac{1}{15}\sin t + \frac{1}{60}\sin 4(t - 2\pi)\,\mathscr{U}(t - 2\pi) - \frac{1}{15}\sin(t - 2\pi)\,\mathscr{U}(t - 2\pi)$$

$$= \begin{cases} -\frac{1}{60}\sin 4t + \frac{1}{15}\sin t, & 0 \le t < 2\pi \\ 0, & t \ge 2\pi. \end{cases}$$

75. (a) The differential equation is

$$\frac{di}{dt} + 10i = \sin t + \cos\left(t - \frac{3\pi}{2}\right)\mathscr{U}\left(t - \frac{3\pi}{2}\right), \quad i(0) = 0.$$

The Laplace transform of this equation is

$$s\mathscr{L}\{i\} + 10\mathscr{L}\{i\} = \frac{1}{s^2 + 1} + \frac{se^{-3\pi s/2}}{s^2 + 1}.$$

Solving for $\mathscr{L}\{i\}$ we obtain

$$\mathscr{L}\{i\} = \frac{1}{(s^2+1)(s+10)} + \frac{s}{(s^2+1)(s+10)}e^{-3\pi s/2}$$

$$= \frac{1}{101}\left(\frac{1}{s+10} - \frac{s}{s^2+1} + \frac{10}{s^2+1}\right) + \frac{1}{101}\left(\frac{-10}{s+10} + \frac{10s}{s^2+1} + \frac{1}{s^2+1}\right)e^{-3\pi s/2}.$$

Thus

$$i(t) = \frac{1}{101}\left(e^{-10t} - \cos t + 10\sin t\right)$$

$$+ \frac{1}{101}\left[-10e^{-10(t-3\pi/2)} + 10\cos\left(t - \frac{3\pi}{2}\right) + \sin\left(t - \frac{3\pi}{2}\right)\right]\mathscr{U}\left(t - \frac{3\pi}{2}\right).$$

(b)

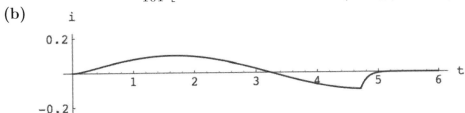

The maximum value of $i(t)$ is approximately 0.1 at $t = 1.7$, the minimum is approximately -0.1 at 4.7. [Using *Mathematica* we see that the maximum value is $i(t)$ is 0.0995037 at $t = 1.670465$, and the mininum value is $i(3\pi/2) \approx -0.0990099$ at $t = 3\pi/2$.]

78. The differential equation is

$$EI\frac{d^4y}{dx^4} = w_0[\mathscr{U}(x - L/3) - \mathscr{U}(x - 2L/3)].$$

Taking the Laplace transform of both sides and using $y(0) = y'(0) = 0$ we obtain

$$s^4\mathscr{L}\{y\} - sy''(0) - y'''(0) = \frac{w_0}{EI}\frac{1}{s}\left(e^{-Ls/3} - e^{-2Ls/3}\right).$$

Letting $y''(0) = c_1$ and $y'''(0) = c_2$ we have

$$\mathscr{L}\{y\} = \frac{c_1}{s^3} + \frac{c_2}{s^4} + \frac{w_0}{EI}\frac{1}{s^5}\left(e^{-Ls/3} - e^{-2Ls/3}\right)$$

so that

$$y(x) = \frac{1}{2}c_1x^2 + \frac{1}{6}c_2x^3 + \frac{1}{24}\frac{w_0}{EI}\left[\left(x - \frac{L}{3}\right)^4\mathscr{U}\left(x - \frac{L}{3}\right) - \left(x - \frac{2L}{3}\right)^4\mathscr{U}\left(x - \frac{2L}{3}\right)\right].$$

To find c_1 and c_2 we compute

$$y''(x) = c_1 + c_2x + \frac{1}{2}\frac{w_0}{EI}\left[\left(x - \frac{L}{3}\right)^2\mathscr{U}\left(x - \frac{L}{3}\right) - \left(x - \frac{2L}{3}\right)^2\mathscr{U}\left(x - \frac{2L}{3}\right)\right]$$

and

$$y'''(x) = c_2 + \frac{w_0}{EI}\left[\left(x - \frac{L}{3}\right)\mathscr{U}\left(x - \frac{L}{3}\right) - \left(x - \frac{2L}{3}\right)\mathscr{U}\left(x - \frac{2L}{3}\right)\right].$$

Then $y''(L) = y'''(L) = 0$ yields the system

$$c_1 + c_2 L + \frac{1}{2}\frac{w_0}{EI}\left[\left(\frac{2L}{3}\right)^2 - \left(\frac{L}{3}\right)^2\right] = c_1 + c_2 L + \frac{1}{6}\frac{w_0 L^2}{EI} = 0$$

$$c_2 + \frac{w_0}{EI}\left[\frac{2L}{3} - \frac{L}{3}\right] = c_2 + \frac{1}{3}\frac{w_0 L}{EI} = 0.$$

Solving for c_1 and c_2 we obtain $c_1 = \frac{1}{6}w_0 L^2/EI$ and $c_2 = -\frac{1}{3}w_0 L/EI$. Thus

$$y(x) = \frac{w_0}{EI}\left(\frac{1}{12}L^2 x^2 - \frac{1}{18}Lx^3 + \frac{1}{24}\left[\left(x - \frac{L}{3}\right)^4 \mathcal{U}\left(x - \frac{L}{3}\right) - \left(x - \frac{2L}{3}\right)^4 \mathcal{U}\left(x - \frac{2L}{3}\right)\right]\right).$$

81. (a) The temperature T of the cake inside the oven is modeled by

$$\frac{dT}{dt} = k(T - T_m)$$

where T_m is the ambient temperature of the oven. For $0 \le t \le 4$, we have

$$T_m = 70 + \frac{300 - 70}{4 - 0}t = 70 + 57.5t.$$

Hence for $t \ge 0$,

$$T_m = \begin{cases} 70 + 57.5t, & 0 \le t < 4 \\ 300, & t \ge 4. \end{cases}$$

In terms of the unit step function,

$$T_m = (70 + 57.5t)[1 - \mathcal{U}(t - 4)] + 300\,\mathcal{U}(t - 4) = 70 + 57.5t + (230 - 57.5t)\,\mathcal{U}(t - 4).$$

The initial-value problem is then

$$\frac{dT}{dt} = k[T - 70 - 57.5t - (230 - 57.5t)\,\mathcal{U}(t - 4)], \qquad T(0) = 70.$$

(b) Let $t(s) = \mathcal{L}\{T(t)\}$. Transforming the equation, using $230 - 57.5t = -57.5(t - 4)$ and Theorem 7.3.2, gives

$$st(s) - 70 = k\left(t(s) - \frac{70}{s} - \frac{57.5}{s^2} + \frac{57.5}{s^2}e^{-4s}\right)$$

or

$$t(s) = \frac{70}{s - k} - \frac{70k}{s(s - k)} - \frac{57.5k}{s^2(s - k)} + \frac{57.5k}{s^2(s - k)}e^{-4s}.$$

After using partial functions, the inverse transform is then

$$T(t) = 70 + 57.5\left(\frac{1}{k} + t - \frac{1}{k}e^{kt}\right) - 57.5\left(\frac{1}{k} + t - 4 - \frac{1}{k}e^{k(t-4)}\right)\mathcal{U}(t - 4).$$

Of course, the obvious question is: What is k? If the cake is supposed to bake for, say, 20 minutes, then $T(20) = 300$. That is,

$$300 = 70 + 57.5\left(\frac{1}{k} + 20 - \frac{1}{k}e^{20k}\right) - 57.5\left(\frac{1}{k} + 16 - \frac{1}{k}e^{16k}\right).$$

But this equation has no physically meaningful solution. This should be no surprise since the model predicts the asymptotic behavior $T(t) \to 300$ as t increases. Using $T(20) = 299$ instead, we find, with the help of a CAS, that $k \approx -0.3$.

7.4 | Operational Properties II

The terminology and concepts listed below provide an outline of the main ideas encountered in this section. These can be useful when preparing for a quiz or test.

Terminology and Concepts

- the relationship between $t^n f(t)$ and the derivative of the Laplace transform of $f(t)$
- convolution of two functions
- the Laplace transform of the convolution f*g of two functions
- the Laplace transform of the integral of a function
- Volterra integral equation
- integrodifferential differential equation
- the Laplace transform of a periodic function

The basic skills listed below summarize the more mechanical types of problems encountered in the exercise set for this section.

Basic Skills

- find the Laplace transform of the product $t^n f(t)$ in terms of the Laplace transform of $f(t)$
- find the Laplace transform of the convolution of two functions
- find the Laplace transform of the integral of a function
- solve an integral equation
- solve an integrodifferential equation
- find the Laplace transform of a periodic function

DEs With Variable Coefficients

The Laplace transform is not well-suited to solving linear differential equations with variable coefficients. Nevertheless, it can be used in *some* instances. If the DE has coefficients that are at most linear polynomials in t, then its transform is a linear first-order DE in the transformed function $Y(s)$. For example, let's find a solution of $ty'' + 2ty' + 2y = t$ subject to $y(0) = 0$. From Theorem 7.4.1 observe

$$\mathscr{L}\{ty''\} = -\frac{d}{ds}\mathscr{L}\{y''\} = -\frac{d}{ds}[s^2 Y(s) - sy(0) - y'(0)] = -s^2 Y'(s) - 2sY(s) + y(0) \qquad (1)$$

$$\mathscr{L}\{ty'\} = -\frac{d}{ds}\mathscr{L}\{y'\} = -\frac{d}{ds}[sY(s) - y(0)] = -sY'(s) - Y(s). \qquad (2)$$

Note here that we have used the product rule to differentiate $s^2Y(s)$ and $sY(s)$. Hence the transform of the second-order DE is the first-order DE,

$$-s^2Y'(s) - 2sY(s) - 2sY'(s) - 2Y(s) + 2Y(s) = \frac{1}{s^2}$$

or

$$\frac{dY}{ds} + \frac{2}{s+2}Y = -\frac{1}{s^3(s+2)},$$

where we have used $y(0) = 0$. Since this last equation is recognized as a linear first-order DE in Y, we find from Section 2.3 in the text that the integrating factor is $(s+2)^2$. Multiplying the last equation by this factor gives

$$\frac{d}{ds}[(s+2)^2 Y] = -\frac{s+2}{s^3} = -\frac{1}{s^2} - \frac{2}{s^3}.$$

Integrating, we obtain

$$(s+2)^2 Y = \frac{1}{s} + \frac{1}{s^2} + c = \frac{s+1}{s^2} + c.$$

Now solve for Y and expand into partial fractions:

$$Y = \frac{s+1}{s^2(s+2)^2} + \frac{c}{(s+2)^2} = \frac{1}{4}\frac{1}{s^2} - \frac{1}{4}\frac{1}{(s+2)^2} + \frac{c}{(s+2)^2}.$$

The inverse of the preceding expression is then

$$y(t) = \frac{1}{4}t - \frac{1}{4}te^{-2t} + cte^{-2t} \qquad \text{or} \qquad y(t) = \frac{1}{4}t + c_1te^{-2t},$$

where we have replaced the term $c - \frac{1}{4}$ by c_1.

7.4.1   DERIVATIVES OF A TRANSFORM

3. $\mathscr{L}\{t\cos 2t\} = -\dfrac{d}{ds}\left(\dfrac{s}{s^2+4}\right) = \dfrac{s^2-4}{(s^2+4)^2}$

6. $\mathscr{L}\{t^2\cos t\} = \dfrac{d^2}{ds^2}\left(\dfrac{s}{s^2+1}\right) = \dfrac{d}{ds}\left(\dfrac{1-s^2}{(s^2+1)^2}\right) = \dfrac{2s(s^2-3)}{(s^2+1)^3}$

9. The Laplace transform of the differential equation is

$$s\mathscr{L}\{y\} + \mathscr{L}\{y\} = \frac{2s}{(s^2+1)^2}.$$

Solving for $\mathscr{L}\{y\}$ we obtain

$$\mathscr{L}\{y\} = \frac{2s}{(s+1)(s^2+1)^2} = -\frac{1}{2}\frac{1}{s+1} - \frac{1}{2}\frac{1}{s^2+1} + \frac{1}{2}\frac{s}{s^2+1} + \frac{1}{(s^2+1)^2} + \frac{s}{(s^2+1)^2}.$$

Thus

$$y(t) = -\frac{1}{2}e^{-t} - \frac{1}{2}\sin t + \frac{1}{2}\cos t + \frac{1}{2}(\sin t - t\cos t) + \frac{1}{2}t\sin t$$

$$= -\frac{1}{2}e^{-t} + \frac{1}{2}\cos t - \frac{1}{2}t\cos t + \frac{1}{2}t\sin t.$$

12. The Laplace transform of the differential equation is

$$s^2 \mathscr{L}\{y\} - sy(0) - y'(0) + \mathscr{L}\{y\} = \frac{1}{s^2+1}.$$

Solving for $\mathscr{L}\{y\}$ we obtain

$$\mathscr{L}\{y\} = \frac{s^3 - s^2 + s}{(s^2+1)^2} = \frac{s}{s^2+1} - \frac{1}{s^2+1} + \frac{1}{(s^2+1)^2}.$$

Thus

$$y = \cos t - \sin t + \left(\frac{1}{2}\sin t - \frac{1}{2}t\cos t\right) = \cos t - \frac{1}{2}\sin t - \frac{1}{2}t\cos t.$$

15.

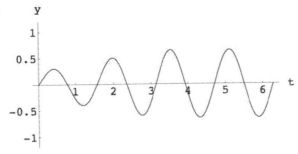

18. From Theorem 7.4.1 in the text

$$\mathscr{L}\{ty'\} = -\frac{d}{ds}\mathscr{L}\{y'\} = -\frac{d}{ds}[sY(s) - y(0)] = -s\frac{dY}{ds} - Y$$

so that the transform of the given second-order differential equation is the linear first-order differential equation in $Y(s)$:

$$Y' + \left(\frac{3}{s} - 2s\right)Y = -\frac{10}{s}.$$

Using the integrating factor $s^3 e^{-s^2}$, the last equation yields

$$Y(s) = \frac{5}{s^3} + \frac{c}{s^3}e^{s^2}.$$

But if $Y(s)$ is the Laplace transform of a piecewise-continuous function of exponential order, we must have, in view of Theorem 7.1.3, $\lim_{s\to\infty} Y(s) = 0$. In order to obtain this condition we require $c = 0$. Hence

$$y(t) = \mathscr{L}^{-1}\left\{\frac{5}{s^3}\right\} = \frac{5}{2}t^2.$$

7.4.2 TRANSFORMS OF INTEGRALS

21. $\mathscr{L}\{e^{-t} * e^t \cos t\} = \dfrac{s-1}{(s+1)\left[(s-1)^2+1\right]}$

24. $\mathscr{L}\left\{\displaystyle\int_0^t \cos\tau\, d\tau\right\} = \dfrac{1}{s}\mathscr{L}\{\cos t\} = \dfrac{s}{s(s^2+1)} = \dfrac{1}{s^2+1}$

27. $\mathscr{L}\left\{\displaystyle\int_0^t \tau e^{t-\tau}\, d\tau\right\} = \mathscr{L}\{t\}\,\mathscr{L}\{e^t\} = \dfrac{1}{s^2(s-1)}$

30. $\mathscr{L}\left\{t\displaystyle\int_0^t \tau e^{-\tau} d\tau\right\} = -\dfrac{d}{ds}\mathscr{L}\left\{\displaystyle\int_0^t \tau e^{-\tau} d\tau\right\} = -\dfrac{d}{ds}\left(\dfrac{1}{s}\dfrac{1}{(s+1)^2}\right) = \dfrac{3s+1}{s^2(s+1)^3}$

33. $\mathscr{L}^{-1}\left\{\dfrac{1}{s^3(s-1)}\right\} = \mathscr{L}^{-1}\left\{\dfrac{1/s^2(s-1)}{s}\right\} = \displaystyle\int_0^t (e^\tau - \tau - 1)d\tau = e^t - \dfrac{1}{2}t^2 - t - 1$

36. The Laplace transform of the differential equation is

$$s^2\mathscr{L}\{y\} + \mathscr{L}\{y\} = \dfrac{1}{(s^2+1)} + \dfrac{2s}{(s^2+1)^2}.$$

Thus

$$\mathscr{L}\{y\} = \dfrac{1}{(s^2+1)^2} + \dfrac{2s}{(s^2+1)^3}$$

and, using Problem 35 with $k=1$,

$$y = \dfrac{1}{2}(\sin t - t\cos t) + \dfrac{1}{4}(t\sin t - t^2\cos t).$$

39. The Laplace transform of the given equation is

$$\mathscr{L}\{f\} = \mathscr{L}\{te^t\} + \mathscr{L}\{t\}\,\mathscr{L}\{f\}.$$

Solving for $\mathscr{L}\{f\}$ we obtain

$$\mathscr{L}\{f\} = \dfrac{s^2}{(s-1)^3(s+1)} = \dfrac{1}{8}\dfrac{1}{s-1} + \dfrac{3}{4}\dfrac{1}{(s-1)^2} + \dfrac{1}{4}\dfrac{2}{(s-1)^3} - \dfrac{1}{8}\dfrac{1}{s+1}.$$

Thus

$$f(t) = \dfrac{1}{8}e^t + \dfrac{3}{4}te^t + \dfrac{1}{4}t^2 e^t - \dfrac{1}{8}e^{-t}$$

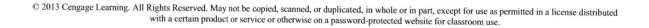

42. The Laplace transform of the given equation is

$$\mathscr{L}\{f\} = \mathscr{L}\{\cos t\} + \mathscr{L}\{e^{-t}\}\,\mathscr{L}\{f\}\,.$$

Solving for $\mathscr{L}\{f\}$ we obtain

$$\mathscr{L}\{f\} = \frac{s}{s^2+1} + \frac{1}{s^2+1}\,.$$

Thus

$$f(t) = \cos t + \sin t.$$

45. The Laplace transform of the given equation is

$$s\mathscr{L}\{y\} - y(0) = \mathscr{L}\{1\} - \mathscr{L}\{\sin t\} - \mathscr{L}\{1\}\,\mathscr{L}\{y\}\,.$$

Solving for $\mathscr{L}\{f\}$ we obtain

$$\mathscr{L}\{y\} = \frac{s^2 - s + 1}{(s^2+1)^2} = \frac{1}{s^2+1} - \frac{1}{2}\frac{2s}{(s^2+1)^2}\,.$$

Thus

$$y = \sin t - \frac{1}{2}t\sin t.$$

48. The differential equation is

$$0.005\frac{di}{dt} + i + \frac{1}{0.02}\int_0^t i(\tau)d\tau = 100\big[t - (t-1)\mathscr{U}(t-1)\big]$$

or

$$\frac{di}{dt} + 200i + 10{,}000\int_0^t i(\tau)d\tau = 20{,}000\big[t - (t-1)\mathscr{U}(t-1)\big],$$

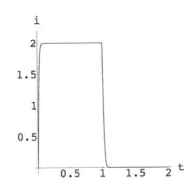

where $i(0) = 0$. The Laplace transform of the differential

equation is

$$s\mathscr{L}\{i\} + 200\mathscr{L}\{i\} + \frac{10{,}000}{s}\mathscr{L}\{i\} = 20{,}000\left(\frac{1}{s^2} - \frac{1}{s^2}e^{-s}\right).$$

Solving for $\mathscr{L}\{i\}$ we obtain

$$\mathscr{L}\{i\} = \frac{20{,}000}{s(s+100)^2}(1-e^{-s}) = \left[\frac{2}{s} - \frac{2}{s+100} - \frac{200}{(s+100)^2}\right](1-e^{-s}).$$

Thus

$$i(t) = 2 - 2e^{-100t} - 200te^{-100t} - 2\mathscr{U}(t-1)) + 2e^{-100(t-1)}\mathscr{U}(t-1))$$
$$+ 200(t-1)e^{-100(t-1)}\mathscr{U}(t-1)\,.$$

7.4.2 TRANSFORM OF A PERIODIC FUNCTION

51. Using integration by parts,

$$\mathscr{L}\{f(t)\} = \frac{1}{1 - e^{-bs}} \int_0^b \frac{a}{b} t e^{-st} dt = \frac{a}{s}\left(\frac{1}{bs} - \frac{1}{e^{bs} - 1}\right).$$

54. $\mathscr{L}\{f(t)\} = \dfrac{1}{1 - e^{-2\pi s}} \displaystyle\int_0^{\pi} e^{-st} \sin t \, dt = \dfrac{1}{s^2 + 1} \cdot \dfrac{1}{1 - e^{-\pi s}}$

57. The differential equation is $x'' + 2x' + 10x = 20f(t)$, where $f(t)$ is the meander function in Problem 49 with $a = \pi$. Using the initial conditions $x(0) = x'(0) = 0$ and taking the Laplace transform we obtain

$$(s^2 + 2s + 10)\mathscr{L}\{x(t)\} = \frac{20}{s}(1 - e^{-\pi s})\frac{1}{1 + e^{-\pi s}}$$

$$= \frac{20}{s}(1 - e^{-\pi s})(1 - e^{-\pi s} + e^{-2\pi s} - e^{-3\pi s} + \cdots)$$

$$= \frac{20}{s}(1 - 2e^{-\pi s} + 2e^{-2\pi s} - 2e^{-3\pi s} + \cdots)$$

$$= \frac{20}{s} + \frac{40}{s}\sum_{n=1}^{\infty}(-1)^n e^{-n\pi s}.$$

Then

$$\mathscr{L}\{x(t)\} = \frac{20}{s(s^2 + 2s + 10)} + \frac{40}{s(s^2 + 2s + 10)}\sum_{n=1}^{\infty}(-1)^n e^{-n\pi s}$$

$$= \frac{2}{s} - \frac{2s + 4}{s^2 + 2s + 10} + \sum_{n=1}^{\infty}(-1)^n\left[\frac{4}{s} - \frac{4s + 8}{s^2 + 2s + 10}\right]e^{-n\pi s}$$

$$= \frac{2}{s} - \frac{2(s + 1) + 2}{(s + 1)^2 + 9} + 4\sum_{n=1}^{\infty}(-1)^n\left[\frac{1}{s} - \frac{(s + 1) + 1}{(s + 1)^2 + 9}\right]e^{-n\pi s}$$

and

$$x(t) = 2\left(1 - e^{-t}\cos 3t - \frac{1}{3}e^{-t}\sin 3t\right)$$

$$+ 4\sum_{n=1}^{\infty}(-1)^n\left[1 - e^{-(t-n\pi)}\cos 3(t - n\pi) - \frac{1}{3}e^{-(t-n\pi)}\sin 3(t - n\pi)\right]\mathscr{U}(t - n\pi).$$

The graph of $x(t)$ on the interval $[0, 2\pi)$ is shown below.

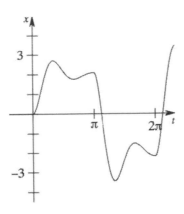

7.5 | The Dirac Delta Function

The terminology and concepts listed below provide an outline of the main ideas encountered in this section. These can be useful when preparing for a quiz or test.

Terminology and Concepts

- unit impulse function
- Dirac delta function
- Laplace transform of the Dirac delta function

The basic skills listed below summarize the more mechanical types of problems encountered in the exercise set for this section.

Basic Skills

- solve a linear second-order DE whose input function involves the Dirac delta function

3. The Laplace transform of the differential equation yields

$$\mathscr{L}\{y\} = \frac{1}{s^2 + 1}\left(1 + e^{-2\pi s}\right)$$

so that

$$y = \sin t + \sin t \, \mathscr{U}(t - 2\pi).$$

6. The Laplace transform of the differential equation yields

$$\mathscr{L}\{y\} = \frac{s}{s^2+1} + \frac{1}{s^2+1}\left(e^{-2\pi s} + e^{-4\pi s}\right)$$

so that

$$y = \cos t + \sin t [\mathscr{U}(t-2\pi) + \mathscr{U}(t-4\pi)].$$

9. The Laplace transform of the differential equation yields

$$\mathscr{L}\{y\} = \frac{1}{(s+2)^2+1}e^{-2\pi s}$$

so that

$$y = e^{-2(t-2\pi)}\sin t\,\mathscr{U}(t-2\pi).$$

12. The Laplace transform of the differential equation yields

$$\mathscr{L}\{y\} = \frac{1}{(s-1)^2(s-6)} + \frac{e^{-2s}+e^{-4s}}{(s-1)(s-6)}$$

$$= -\frac{1}{25}\frac{1}{s-1} - \frac{1}{5}\frac{1}{(s-1)^2} + \frac{1}{25}\frac{1}{s-6} + \left[-\frac{1}{5}\frac{1}{s-1} + \frac{1}{5}\frac{1}{s-6}\right]\left(e^{-2s}+e^{-4s}\right)$$

so that

$$y = -\frac{1}{25}e^t - \frac{1}{5}te^t + \frac{1}{25}e^{6t} + \left[-\frac{1}{5}e^{t-2} + \frac{1}{5}e^{6(t-2)}\right]\mathscr{U}(t-2)$$

$$+ \left[-\frac{1}{5}e^{t-4} + \frac{1}{5}e^{6(t-4)}\right]\mathscr{U}(t-4).$$

7.6 Systems of Linear Differential Equations

The terminology and concepts listed below provide an outline of the main ideas encountered in this section. These can be useful when preparing for a quiz or test.

Terminology and Concepts

- coupled spring systems
- doubled pendulum

The basic skills listed below summarize the more mechanical types of problems encountered in the exercise set for this section.

Basic Skills

- use Laplace transform methods to solve a linear system of differential equations
- solve algebraic systems of equations by elimination

3. Taking the Laplace transform of the system gives

$$s\mathscr{L}\{x\} + 1 = \mathscr{L}\{x\} - 2\mathscr{L}\{y\}$$

$$s\mathscr{L}\{y\} - 2 = 5\mathscr{L}\{x\} - \mathscr{L}\{y\}$$

so that

$$\mathscr{L}\{x\} = \frac{-s-5}{s^2+9} = -\frac{s}{s^2+9} - \frac{5}{3}\frac{3}{s^2+9}$$

and

$$x = -\cos 3t - \frac{5}{3}\sin 3t.$$

Then

$$y = \frac{1}{2}x - \frac{1}{2}x' = 2\cos 3t - \frac{7}{3}\sin 3t.$$

6. Taking the Laplace transform of the system gives

$$(s+1)\,\mathscr{L}\{x\} - (s-1)\mathscr{L}\{y\} = -1$$

$$s\mathscr{L}\{x\} + (s+2)\,\mathscr{L}\{y\} = 1$$

so that

$$\mathscr{L}\{y\} = \frac{s+1/2}{s^2+s+1} = \frac{s+1/2}{(s+1/2)^2+(\sqrt{3}/2)^2}$$

and

$$\mathscr{L}\{x\} = \frac{-3/2}{s^2+s+1} = -\sqrt{3}\,\frac{\sqrt{3}/2}{(s+1/2)^2+(\sqrt{3}/2)^2}.$$

Then

$$y = e^{-t/2}\cos\frac{\sqrt{3}}{2}t \qquad \text{and} \qquad x = -\sqrt{3}\,e^{-t/2}\sin\frac{\sqrt{3}}{2}t.$$

9. Adding the equations and then subtracting them gives

$$\frac{d^2x}{dt^2} = \frac{1}{2}t^2 + 2t$$

$$\frac{d^2y}{dt^2} = \frac{1}{2}t^2 - 2t.$$

Taking the Laplace transform of the system gives

$$\mathscr{L}\{x\} = 8\frac{1}{s} + \frac{1}{24}\frac{4!}{s^5} + \frac{1}{3}\frac{3!}{s^4}$$

and

$$\mathscr{L}\{y\} = \frac{1}{24}\frac{4!}{s^5} - \frac{1}{3}\frac{3!}{s^4}$$

so that

$$x = 8 + \frac{1}{24}t^4 + \frac{1}{3}t^3 \qquad \text{and} \qquad y = \frac{1}{24}t^4 - \frac{1}{3}t^3.$$

12. Taking the Laplace transform of the system gives

$$(s - 4)\,\mathscr{L}\{x\} + 2\mathscr{L}\{y\} = \frac{2e^{-s}}{s}$$

$$-3\,\mathscr{L}\{x\} + (s + 1)\,\mathscr{L}\{y\} = \frac{1}{2} + \frac{e^{-s}}{s}$$

so that

$$\mathscr{L}\{x\} = \frac{-1/2}{(s-1)(s-2)} + e^{-s}\frac{1}{(s-1)(s-2)}$$

$$= \frac{1}{2}\frac{1}{s-1} - \frac{1}{2}\frac{1}{s-2} + e^{-s}\left[-\frac{1}{s-1} + \frac{1}{s-2}\right]$$

and

$$\mathscr{L}\{y\} = \frac{e^{-s}}{s} + \frac{s/4 - 1}{(s-1)(s-2)} + e^{-s}\frac{-s/2 + 2}{(s-1)(s-2)}$$

$$= \frac{3}{4}\frac{1}{s-1} - \frac{1}{2}\frac{1}{s-2} + e^{-s}\left[\frac{1}{s} - \frac{3}{2}\frac{1}{s-1} + \frac{1}{s-2}\right].$$

Then

$$x = \frac{1}{2}e^t - \frac{1}{2}e^{2t} + \left[-e^{t-1} + e^{2(t-1)}\right]\mathscr{U}(t-1)$$

and

$$y = \frac{3}{4}e^t - \frac{1}{2}e^{2t} + \left[1 - \frac{3}{2}e^{t-1} + e^{2(t-1)}\right]\mathscr{U}(t-1).$$

15. (a) By Kirchhoff's first law we have $i_1 = i_2 + i_3$. By Kirchhoff's second law, on each loop we have $E(t) = Ri_1 + L_1 i_2'$ and $E(t) = Ri_1 + L_2 i_3'$ or $L_1 i_2' + Ri_2 + Ri_3 = E(t)$ and $L_2 i_3' + Ri_2 + Ri_3 = E(t)$.

(b) Taking the Laplace transform of the system

$$0.01 i_2' + 5i_2 + 5i_3 = 100$$

$$0.0125 i_3' + 5i_2 + 5i_3 = 100$$

gives

$$(s + 500)\,\mathscr{L}\{i_2\} + 500\mathscr{L}\{i_3\} = \frac{10{,}000}{s}$$

$$400\mathscr{L}\{i_2\} + (s + 400)\,\mathscr{L}\{i_3\} = \frac{8{,}000}{s}$$

so that

$$\mathscr{L}\{i_3\} = \frac{8{,}000}{s^2 + 900s} = \frac{80}{9}\frac{1}{s} - \frac{80}{9}\frac{1}{s + 900}.$$

Then

$$i_3 = \frac{80}{9} - \frac{80}{9}e^{-900t} \qquad \text{and} \qquad i_2 = 20 - 0.0025 i_3' - i_3 = \frac{100}{9} - \frac{100}{9}e^{-900t}.$$

18. Taking the Laplace transform of the system

$$0.5 i_1' + 50 i_2 = 60$$

$$0.005 i_2' + i_2 - i_1 = 0$$

gives

$$s\,\mathscr{L}\{i_1\} + 100\,\mathscr{L}\{i_2\} = \frac{120}{s}$$

$$-200\,\mathscr{L}\{i_1\} + (s + 200)\,\mathscr{L}\{i_2\} = 0$$

so that

$$\mathscr{L}\{i_2\} = \frac{24{,}000}{s(s^2 + 200s + 20{,}000)} = \frac{6}{5}\frac{1}{s} - \frac{6}{5}\frac{s + 100}{(s + 100)^2 + 100^2} - \frac{6}{5}\frac{100}{(s + 100)^2 + 100^2}.$$

Then

$$i_2 = \frac{6}{5} - \frac{6}{5}e^{-100t}\cos 100t - \frac{6}{5}e^{-100t}\sin 100t$$

and

$$i_1 = 0.005 i_2' + i_2 = \frac{6}{5} - \frac{6}{5}e^{-100t}\cos 100t.$$

21. (a) Taking the Laplace transform of the system

$$4\theta_1'' + \theta_2'' + 8\theta_1 = 0$$

$$\theta_1'' + \theta_2'' + 2\theta_2 = 0$$

gives

$$4\left(s^2 + 2\right)\mathscr{L}\{\theta_1\} + s^2 \mathscr{L}\{\theta_2\} = 3s$$

$$s^2 \mathscr{L}\{\theta_1\} + \left(s^2 + 2\right)\mathscr{L}\{\theta_2\} = 0$$

so that

$$\left(3s^2 + 4\right)\left(s^2 + 4\right)\mathscr{L}\{\theta_2\} = -3s^3$$

or

$$\mathscr{L}\{\theta_2\} = \frac{1}{2}\frac{s}{s^2 + 4/3} - \frac{3}{2}\frac{s}{s^2 + 4}.$$

Then

$$\theta_2 = \frac{1}{2}\cos\frac{2}{\sqrt{3}}t - \frac{3}{2}\cos 2t \qquad \text{and} \qquad \theta_1'' = -\theta_2'' - 2\theta_2$$

so that

$$\theta_1 = \frac{1}{4}\cos\frac{2}{\sqrt{3}}t + \frac{3}{4}\cos 2t.$$

(b)

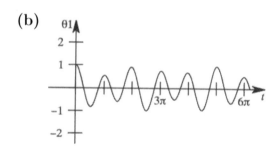 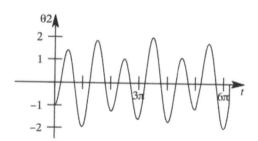

Mass m_2 has extreme displacements of greater magnitude. Mass m_1 first passes through its equilibrium position at about $t = 0.87$, and mass m_2 first passes through its equilibrium position at about $t = 0.66$. The motion of the pendulums is not periodic since $\cos(2t/\sqrt{3})$ has period $\sqrt{3}\,\pi$, $\cos 2t$ has period π, and the ratio of these periods is $\sqrt{3}$, which is not a rational number.

(c) The Lissajous curve is plotted for $0 \leq t \leq 30$.

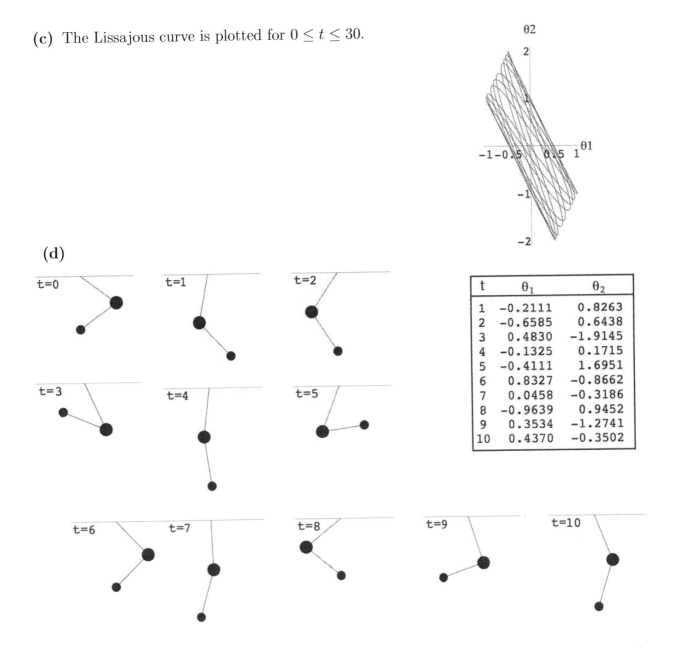

(d)

t	θ_1	θ_2
1	−0.2111	0.8263
2	−0.6585	0.6438
3	0.4830	−1.9145
4	−0.1325	0.1715
5	−0.4111	1.6951
6	0.8327	−0.8662
7	0.0458	−0.3186
8	−0.9639	0.9452
9	0.3534	−1.2741
10	0.4370	−0.3502

(e) Using a CAS to solve $\theta_1(t) = \theta_2(t)$ we see that $\theta_1 = \theta_2$ (so that the double pendulum is straight out) when t is about 0.75 seconds.

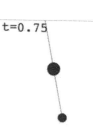

t=0.75

(f) To make a movie of the pendulum it is necessary to locate the mass in the plane as a function of time. Suppose that the upper arm is attached to the origin and that the equilibrium position lies along the negative y-axis. Then mass m_1 is at $(x, (t), y_1(t))$ and mass m_2 is at $(x_2(t), y_2(t))$, where

$$x_1(t) = 16 \sin \theta_1(t) \qquad \text{and} \qquad y_1(t) = -16 \cos \theta_1(t)$$

and

$$x_2(t) = x_1(t) + 16 \sin \theta_2(t) \qquad \text{and} \qquad y_2(t) = y_1(t) - 16 \cos \theta_2(t).$$

A reasonable movie can be constructed by letting t range from 0 to 10 in increments of 0.1 seconds.

7.R   Chapter 7 in Review

3. False; consider $f(t) = t^{-1/2}$.

6. False; consider $f(t) = 1$ and $g(t) = 1$.

9. $\mathscr{L}\{\sin 2t\} = \dfrac{2}{s^2 + 4}$

12. $\mathscr{L}\{\sin 2t\, \mathscr{U}(t - \pi)\} = \mathscr{L}\{\sin 2(t - \pi)\, \mathscr{U}(t - \pi)\} = \dfrac{2}{s^2 + 4} e^{-\pi s}$

15. $\mathscr{L}^{-1}\left\{\dfrac{1}{(s-5)^3}\right\} = \dfrac{1}{2}\dfrac{2}{(s-5)^3} = \dfrac{1}{2}t^2 e^{5t}$

18. $\mathscr{L}^{-1}\left\{\dfrac{1}{s^2} e^{-5s}\right\} = (t - 5)\, \mathscr{U}(t - 5)$

21. $\mathscr{L}\{e^{-5t}\}$ exists for $s > -5$.

24. $\mathscr{L}\left\{\displaystyle\int_0^t e^{a\tau} f(\tau)\, d\tau\right\} = \dfrac{1}{s}\mathscr{L}\{e^{at} f(t)\} = \dfrac{F(s - a)}{s}$, whereas

$$\mathscr{L}\left\{e^{at} \int_0^t f(\tau)\, d\tau\right\} = \mathscr{L}\left\{\int_0^t f(\tau)\, d\tau\right\}\bigg|_{s \to s-a} = \dfrac{F(s)}{s}\bigg|_{s \to s-a} = \dfrac{F(s - a)}{s - a}.$$

27. $f(t - t_0)\,\mathcal{U}(t - t_0)$

30. $f(t) = \sin t\,\mathcal{U}(t - \pi) - \sin t\,\mathcal{U}(t - 3\pi) = -\sin(t - \pi)\,\mathcal{U}(t - \pi) + \sin(t - 3\pi)\,\mathcal{U}(t - 3\pi)$

$$\mathcal{L}\{f(t)\} = -\frac{1}{s^2 + 1}e^{-\pi s} + \frac{1}{s^2 + 1}e^{-3\pi s}$$

$$\mathcal{L}\{e^t f(t)\} = -\frac{1}{(s-1)^2 + 1}e^{-\pi(s-1)} + \frac{1}{(s-1)^2 + 1}e^{-3\pi(s-1)}$$

33. Taking the Laplace transform of the differential equation we obtain

$$\mathcal{L}\{y\} = \frac{5}{(s-1)^2} + \frac{1}{2}\frac{2}{(s-1)^3}$$

so that

$$y = 5te^t + \frac{1}{2}t^2 e^t.$$

36. Taking the Laplace transform of the differential equation we obtain

$$\mathcal{L}\{y\} = \frac{s^3 + 2}{s^3(s - 5)} - \frac{2 + 2s + s^2}{s^3(s - 5)}e^{-s}$$

$$= -\frac{2}{125}\frac{1}{s} - \frac{2}{25}\frac{1}{s^2} - \frac{1}{5}\frac{2}{s^3} + \frac{127}{125}\frac{1}{s-5} - \left[-\frac{37}{125}\frac{1}{s} - \frac{12}{25}\frac{1}{s^2} - \frac{1}{5}\frac{2}{s^3} + \frac{37}{125}\frac{1}{s-5}\right]e^{-s}$$

so that

$$y = -\frac{2}{125} - \frac{2}{25}t - \frac{1}{5}t^2 + \frac{127}{125}e^{5t} - \left[-\frac{37}{125} - \frac{12}{25}(t-1) - \frac{1}{5}(t-1)^2 + \frac{37}{125}e^{5(t-1)}\right]\mathcal{U}(t - 1).$$

39. Taking the Laplace transform of the integral equation we obtain

$$\mathcal{L}\{y\} = \frac{1}{s} + \frac{1}{s^2} + \frac{1}{2}\frac{2}{s^3}$$

so that

$$y(t) = 1 + t + \frac{1}{2}t^2.$$

42. Taking the Laplace transform of the system gives

$$s^2 \mathscr{L}\{x\} + s^2 \mathscr{L}\{y\} = \frac{1}{s-2}$$

$$2s \mathscr{L}\{x\} + s^2 \mathscr{L}\{y\} = -\frac{1}{s-2}$$

so that

$$\mathscr{L}\{x\} = \frac{2}{s(s-2)^2} = \frac{1}{2}\frac{1}{s} - \frac{1}{2}\frac{1}{s-2} + \frac{1}{(s-2)^2}$$

and

$$\mathscr{L}\{y\} = \frac{-s-2}{s^2(s-2)^2} = -\frac{3}{4}\frac{1}{s} - \frac{1}{2}\frac{1}{s^2} + \frac{3}{4}\frac{1}{s-2} - \frac{1}{(s-2)^2}.$$

Then

$$x = \frac{1}{2} - \frac{1}{2}e^{2t} + te^{2t} \quad \text{and} \quad y = -\frac{3}{4} - \frac{1}{2}t + \frac{3}{4}e^{2t} - te^{2t}.$$

45. Taking the Laplace transform of the given differential equation we obtain

$$\mathscr{L}\{y\} = \frac{2w_0}{EIL}\left(\frac{L}{48}\cdot\frac{4!}{s^5} - \frac{1}{120}\cdot\frac{5!}{s^6} + \frac{1}{120}\cdot\frac{5!}{s^6}e^{-sL/2}\right) + \frac{c_1}{2}\cdot\frac{2!}{s^3} + \frac{c_2}{6}\cdot\frac{3!}{s^4}$$

so that

$$y = \frac{2w_0}{EIL}\left[\frac{L}{48}x^4 - \frac{1}{120}x^5 + \frac{1}{120}\left(x - \frac{L}{2}\right)^5 \mathscr{U}\left(x - \frac{L}{2}\right) + \frac{c_1}{2}x^2 + \frac{c_2}{6}x^3\right]$$

where $y''(0) = c_1$ and $y'''(0) = c_2$. Using $y''(L) = 0$ and $y'''(L) = 0$ we find

$$c_1 = w_0 L^2/24EI, \qquad c_2 = -w_0 L/4EI.$$

Hence

$$y = \frac{w_0}{12EIL}\left[-\frac{1}{5}x^5 + \frac{L}{2}x^4 - \frac{L^2}{2}x^3 + \frac{L^3}{4}x^2 + \frac{1}{5}\left(x - \frac{L}{2}\right)^5 \mathscr{U}\left(x - \frac{L}{2}\right)\right].$$

48. (a) We will find the first two times for which $x'(t) = 0$ and then obtain the rest of the times using periodicity of $x'(t)$. The solution of

$$x'' + \omega^2 x = F, \quad x(0) = x_0, \quad x'(0) = 0$$

is

$$x(t) = \left(x_0 - \frac{F}{\omega^2}\right)\cos\omega t + \frac{F}{\omega^2} \quad \text{so} \quad x'(t) = \left(x_0 - \frac{F}{\omega^2}\right)(-\omega\sin\omega t).$$

The latter equation is 0 when $t = \pi/\omega$. The next initial-value problem is then

$$x'' + \omega^2 x = -F, \quad x\left(\frac{\pi}{\omega}\right) = \frac{2F}{\omega^2} - x_0, \quad x'\left(\frac{\pi}{\omega}\right) = 0,$$

where $\pi/\omega = T/2$. From the solution of this problem,

$$x(t) = \left(x_0 - \frac{3F}{\omega^2}\right)\cos\omega t - \frac{F}{\omega^2},$$

we see from

$$x'(t) = \left(x_0 - \frac{3F}{\omega^2}\right)(-\omega\sin\omega t) = 0$$

that $t = 2\pi/\omega = T$. Since $x'(t)$ has period $T = 2\pi/\omega$ we can see that $x'(t) = 0$ at the times

$$\frac{\pi}{\omega} + \frac{2\pi}{\omega}k \quad \text{and} \quad \frac{2\pi}{\omega} + \frac{2\pi}{\omega}k, \quad \text{for } k = 0, 1, 2, \ldots$$

which are

$$0, \frac{\pi}{\omega}, \frac{2\pi}{\omega}, \frac{3\pi}{\omega}, \ldots \quad \text{or} \quad 0, \frac{t}{2}, T, \frac{3T}{2}, \ldots.$$

(b) There is no motion unless the initial displacement is such that the force of the spring is greater than the force due to friction. That is,

$$k|x_0| > f_k \quad \text{or} \quad \frac{k}{m}|x_0| > \frac{f_k}{m} \quad \text{or} \quad \omega^2|x_0| > F.$$

(c) From part **(b)**, an initial displacement $x(0) = x_0$ for which $|x_0| > F/\omega^2$ will result in motion. On the other hand, an initial displacement x_0 for which $|x_0| \leq F/\omega^2$ will be insufficient to overcome the force of friction and the system will be "dead".

(d) The system can be described by the initial-value problem

$$x'' + \omega^2 x = g(t), \quad x(0) = x_0, \quad x'(0) = 0,$$

where g is a version of the meander function shown in Figure 7.4.6 in the text. In this case the amplitude of the function is F instead of 1, and the length of each line segment is $T/2$ rather than a. Then

$$\mathscr{L}\{x''\} + \omega^2\mathscr{L}\{x\} = \mathscr{L}\{g\},$$
$$s^2 X(s) - sx_0 + \omega^2 X(s) = G(s),$$

and

$$X(s) = \frac{sx_0}{s^2 + \omega^2} + \frac{1}{s^2 + \omega^2} G(s).$$

Now, using Problem 49 in Section 7.4 with F instead of 1 and $a = T/2$, we have

$$\mathcal{L}\{g(t)\} = G(s) = \frac{F}{s} \frac{1 - e^{-sT/2}}{1 + e^{-sT/2}}$$

$$= \frac{F}{s}[1 - 2e^{-sT/2} + 2e^{-sT} - 2e^{-3sT/2} + \cdots].$$

Then

$$X(s) = x_0 \frac{s}{s^2 + \omega^2} + \frac{F}{s} \frac{1}{s^2 + \omega^2}[1 - 2e^{-sT/2} + 2e^{-sT} - 2e^{-3sT/2} + \cdots]$$

$$= x_0 \frac{s}{s^2 + \omega^2} + \frac{F}{\omega^2}\left[\frac{1}{s} - \frac{s}{s^2 + \omega^2}\right][1 - 2e^{-sT/2} + 2e^{-sT} - 2e^{-3sT/2} + \cdots]$$

and

$$x(t) = x_0 \cos \omega t + \frac{F}{\omega^2}(1 - \cos \omega t) - \frac{2F}{\omega^2}\left[\mathcal{U}\left(t - \frac{T}{2}\right) - \cos \omega \left(t - \frac{T}{2}\right)\mathcal{U}\left(t - \frac{T}{2}\right)\right]$$

$$+ \frac{2F}{\omega^2}[\mathcal{U}(t - T) - \cos \omega(t - T)\mathcal{U}(t - T)]$$

$$- \frac{2F}{\omega^2}\left[\mathcal{U}\left(t - \frac{3T}{2}\right) - \cos \omega \left(t - \frac{3T}{2}\right)\mathcal{U}\left(t - \frac{3T}{2}\right)\right] + \cdots$$

or

$$x(t) = \begin{cases} x_0 \cos \omega t + \dfrac{F}{\omega^2}(1 - \cos \omega t), & 0 \leq t < T/2 \\[2mm] x_0 \cos \omega t + \dfrac{F}{\omega^2}(1 - \cos \omega t) - \dfrac{2F}{\omega^2}\left[1 - \cos \omega \left(t - \dfrac{T}{2}\right)\right], & T/2 \leq t < T \\[2mm] x_0 \cos \omega t + \dfrac{F}{\omega^2}(1 - \cos \omega t) - \dfrac{2F}{\omega^2}\left[1 - \cos \omega \left(t - \dfrac{T}{2}\right)\right] \\[2mm] \qquad\qquad\qquad\qquad + \dfrac{2F}{\omega^2}[1 - \cos \omega(t - T)], & T \leq t < 3T/2 \\[2mm] \qquad\qquad \vdots \end{cases}$$

(e) The solutions from Problem 28 in Chapter 5 in Review are

$$x(t) = \begin{cases} 4.5 \cos t + 1, & 0 \leq t < \pi \\ 2.5 \cos t - 1, & \pi \leq t < 2\pi. \end{cases}$$

On the interval $[0, 2\pi)$ the solution is

$$x(t) = \begin{cases} x_0 \cos \omega t + \dfrac{F}{\omega^2}(1 - \cos \omega t), & 0 \leq t < T/2 \\[3mm] x_0 \cos \omega t + \dfrac{F}{\omega^2}(1 - \cos \omega t) - \dfrac{2F}{\omega^2}\left[1 - \cos \omega \left(t - \dfrac{T}{2}\right)\right], & T/2 \leq t < T. \end{cases}$$

We let $m = 1$, $k = 1$, $f_k = 1$, and $x_0 = 5.5$. Then, when $T = 2\pi$, the foregoing becomes

$$x(t) = \begin{cases} 5.5\cos t + (1 - \cos t), & 0 \le t < \pi \\ 5.5\cos t + (1 - \cos t) - 2\big[1 - \cos(t - \pi)\big], & \pi \le t < 2\pi. \end{cases}$$

Simplifying this we get

$$x(t) = \begin{cases} 4.5\cos t + 1, & 0 \le t < \pi \\ 2.5\cos t - 1, & \pi \le t < 2\pi. \end{cases}$$

(f) At $t = T/2$, $x(T/2) = -x_0 + 2F/\omega^2$. At $t = T$,

$$x(T) = x_0 - 4F/\omega^2 = (x_0 - 2F/\omega^2) - 2F/\omega^2 = -x(T/2) - 2F/\omega^2.$$

At $t = 3T/2$, $x(3T/2) = -x_0 + 6F/\omega^2 = -x(T) + 2F/\omega^2$. We see in general that each successive oscillation is $2F/\omega^2$ shorter than the preceding one.

(g) The system will stay in motion until the oscillations bring the mass within the "dead zone" at which time the motion ceases.

8 Systems of Linear First-Order Differential Equations

8.1 | Preliminary Theory – Linear Systems

The terminology and concepts listed below provide an outline of the main ideas encountered in this section. These can be useful when preparing for a quiz or test.

Terminology and Concepts

- linear system of first-order DEs

- homogeneous linear system of DEs

- matrix form of a linear system of DEs

- superposition principle

- linear independence of a set of solutions

- Wronskian

- fundamental set of solutions

- general solution of a linear system of DEs

- particular solution of a nonhomogeneous linear system of DEs

- complementary function of a homogeneous linear system of DEs

The basic skills listed below summarize the more mechanical types of problems encountered in the exercise set for this section.

Basic Skills

- express a linear first-order system in both normal form and matrix form

- determine whether a set of vectors forms a fundamental set

3. Let $\mathbf{X} = \begin{pmatrix} x \\ y \\ z \end{pmatrix}$. Then $\mathbf{X}' = \begin{pmatrix} -3 & 4 & -9 \\ 6 & -1 & 0 \\ 10 & 4 & 3 \end{pmatrix} \mathbf{X}$.

6. Let $\mathbf{X} = \begin{pmatrix} x \\ y \\ z \end{pmatrix}$. Then $\mathbf{X}' = \begin{pmatrix} -3 & 4 & 0 \\ 5 & 9 & 0 \\ 0 & 1 & 6 \end{pmatrix} \mathbf{X} + \begin{pmatrix} e^{-t}\sin 2t \\ 4e^{-t}\cos 2t \\ -e^{-t} \end{pmatrix}$.

208

9. $\dfrac{dx}{dt} = x - y + 2z + e^{-t} - 3t; \quad \dfrac{dy}{dt} = 3x - 4y + z + 2e^{-t} + t; \quad \dfrac{dz}{dt} = -2x + 5y + 6z + 2e^{-t} - t$

12. Since

$$\mathbf{X}' = \begin{pmatrix} 5\cos t - 5\sin t \\ 2\cos t - 4\sin t \end{pmatrix} e^t \quad \text{and} \quad \begin{pmatrix} -2 & 5 \\ -2 & 4 \end{pmatrix} \mathbf{X} = \begin{pmatrix} 5\cos t - 5\sin t \\ 2\cos t - 4\sin t \end{pmatrix} e^t$$

we see that

$$\mathbf{X}' = \begin{pmatrix} -2 & 5 \\ -2 & 4 \end{pmatrix} \mathbf{X}.$$

15. Since

$$\mathbf{X}' = \begin{pmatrix} 0 \\ 0 \\ 0 \end{pmatrix} \quad \text{and} \quad \begin{pmatrix} 1 & 2 & 1 \\ 6 & -1 & 0 \\ -1 & -2 & -1 \end{pmatrix} \mathbf{X} = \begin{pmatrix} 0 \\ 0 \\ 0 \end{pmatrix}$$

we see that

$$\mathbf{X}' = \begin{pmatrix} 1 & 2 & 1 \\ 6 & -1 & 0 \\ -1 & -2 & -1 \end{pmatrix} \mathbf{X}.$$

18. Yes, since $W(\mathbf{X}_1, \mathbf{X}_2) = 8e^{2t} \neq 0$ the set $\mathbf{X}_1, \mathbf{X}_2$ is linearly independent on $-\infty < t < \infty$.

21. Since

$$\mathbf{X}'_p = \begin{pmatrix} 2 \\ -1 \end{pmatrix} \quad \text{and} \quad \begin{pmatrix} 1 & 4 \\ 3 & 2 \end{pmatrix} \mathbf{X}_p + \begin{pmatrix} 2 \\ -4 \end{pmatrix} t + \begin{pmatrix} -7 \\ -18 \end{pmatrix} = \begin{pmatrix} 2 \\ -1 \end{pmatrix}$$

we see that

$$\mathbf{X}'_p = \begin{pmatrix} 1 & 4 \\ 3 & 2 \end{pmatrix} \mathbf{X}_p + \begin{pmatrix} 2 \\ -4 \end{pmatrix} t + \begin{pmatrix} -7 \\ -18 \end{pmatrix}.$$

24. Since

$$\mathbf{X}'_p = \begin{pmatrix} 3\cos 3t \\ 0 \\ -3\sin 3t \end{pmatrix} \quad \text{and} \quad \begin{pmatrix} 1 & 2 & 3 \\ -4 & 2 & 0 \\ -6 & 1 & 0 \end{pmatrix} \mathbf{X}_p + \begin{pmatrix} -1 \\ 4 \\ 3 \end{pmatrix} \sin 3t = \begin{pmatrix} 3\cos 3t \\ 0 \\ -3\sin 3t \end{pmatrix}$$

we see that

$$\mathbf{X}'_p = \begin{pmatrix} 1 & 2 & 3 \\ -4 & 2 & 0 \\ -6 & 1 & 0 \end{pmatrix} \mathbf{X}_p + \begin{pmatrix} -1 \\ 4 \\ 3 \end{pmatrix} \sin 3t.$$

8.2 | Homogeneous Linear Systems

The terminology and concepts listed below provide an outline of the main ideas encountered in this section. These can be useful when preparing for a quiz or test.

Terminology and Concepts

- characteristic equation of a matrix
- eigenvalues of a matrix
- eigenvector of a matrix corresponding to an eigenvalue
- general solution of a homogeneous linear system of DEs with constant coefficients
- phase plane
- trajectory
- phase portrait of a linear system of DEs
- repeller
- attractor
- multiplicity of an eigenvalue

The basic skills listed below summarize the more mechanical types of problems encountered in the exercise set for this section.

Basic Skills

- find the general solution of a homogeneous linear system of DEs with constant coefficients when the eigenvalues are real and distinct, real and repeated, or complex

Use of Computers
To find eigenvalues and eigenvectors of a matrix using *Mathematica* or *Maple*, we can input the matrix by rows. For example,

$$m = \begin{pmatrix} -4 & 1 & 1 \\ 1 & 5 & -1 \\ 0 & 1 & -3 \end{pmatrix}$$

is the matrix of the coefficients of the linear system (6) in Example 2 in the text. Using *Mathematica*, we can write **m** as

$$\mathbf{m} = \{\{-4, 1, 1\}, \{1, 5, -1\}, \{0, 1, -3\}\}$$

The commands **Eigenvalues[m]** and **Eigenvectors[m]** given in sequence yield

$$[-4, -3, 5] \quad \text{and} \quad \{\{-10, -1, 1\}, \{1, 0, 1\}, \{1, 8, 1\}\}.$$

Translated this means $\lambda = -4$ is an eigenvalue with corresponding eigenvector

$$\mathbf{K} = \begin{pmatrix} -10 \\ -1 \\ 1 \end{pmatrix},$$

and so on. In *Mathematica* eigenvalues and eigenvectors can be obtained at the same time using the single command **Eigensystem[m]**. In *Maple*, the matrix **m** is written

```
m:=matrix(3,3,[-4,1,1,1,5,-1,0,1,-3])
```

The commands for eigenvalues and eigenvectors are, respectively, **eigenvals(m)** and **eigenvects(m)**.

3. The system is

$$\mathbf{X}' = \begin{pmatrix} -4 & 2 \\ -5/2 & 2 \end{pmatrix} \mathbf{X}$$

and $\det(\mathbf{A} - \lambda\mathbf{I}) = (\lambda - 1)(\lambda + 3) = 0$. For $\lambda_1 = 1$ we obtain

$$\begin{pmatrix} -5 & 2 & | & 0 \\ -5/2 & 1 & | & 0 \end{pmatrix} \implies \begin{pmatrix} -5 & 2 & | & 0 \\ 0 & 0 & | & 0 \end{pmatrix} \quad \text{so that} \quad \mathbf{K}_1 = \begin{pmatrix} 2 \\ 5 \end{pmatrix}.$$

For $\lambda_2 = -3$ we obtain

$$\begin{pmatrix} -1 & 2 & | & 0 \\ -5/2 & 5 & | & 0 \end{pmatrix} \implies \begin{pmatrix} -1 & 2 & | & 0 \\ 0 & 0 & | & 0 \end{pmatrix} \quad \text{so that} \quad \mathbf{K}_2 = \begin{pmatrix} 2 \\ 1 \end{pmatrix}.$$

Then

$$\mathbf{X} = c_1 \begin{pmatrix} 2 \\ 5 \end{pmatrix} e^t + c_2 \begin{pmatrix} 2 \\ 1 \end{pmatrix} e^{-3t}.$$

6. The system is

$$\mathbf{X}' = \begin{pmatrix} -6 & 2 \\ -3 & 1 \end{pmatrix} \mathbf{X}$$

and $\det(\mathbf{A} - \lambda\mathbf{I}) = \lambda(\lambda + 5) = 0$. For $\lambda_1 = 0$ we obtain

$$\begin{pmatrix} -6 & 2 & | & 0 \\ -3 & 1 & | & 0 \end{pmatrix} \implies \begin{pmatrix} 1 & -1/3 & | & 0 \\ 0 & 0 & | & 0 \end{pmatrix} \quad \text{so that} \quad \mathbf{K}_1 = \begin{pmatrix} 1 \\ 3 \end{pmatrix}.$$

For $\lambda_2 = -5$ we obtain

$$\begin{pmatrix} -1 & 2 & | & 0 \\ -3 & 6 & | & 0 \end{pmatrix} \implies \begin{pmatrix} 1 & -2 & | & 0 \\ 0 & 0 & | & 0 \end{pmatrix} \quad \text{so that} \quad \mathbf{K}_2 = \begin{pmatrix} 2 \\ 1 \end{pmatrix}.$$

Then

$$\mathbf{X} = c_1 \begin{pmatrix} 1 \\ 3 \end{pmatrix} + c_2 \begin{pmatrix} 2 \\ 1 \end{pmatrix} e^{-5t}.$$

9. We have $\det(\mathbf{A} - \lambda\mathbf{I}) = -(\lambda + 1)(\lambda - 3)(\lambda + 2) = 0$. For $\lambda_1 = -1$, $\lambda_2 = 3$, and $\lambda_3 = -2$ we obtain

$$\mathbf{K}_1 = \begin{pmatrix} -1 \\ 0 \\ 1 \end{pmatrix}, \quad \mathbf{K}_2 = \begin{pmatrix} 1 \\ 4 \\ 3 \end{pmatrix}, \quad \text{and} \quad \mathbf{K}_3 = \begin{pmatrix} 1 \\ -1 \\ 3 \end{pmatrix},$$

so that

$$\mathbf{X} = c_1 \begin{pmatrix} -1 \\ 0 \\ 1 \end{pmatrix} e^{-t} + c_2 \begin{pmatrix} 1 \\ 4 \\ 3 \end{pmatrix} e^{3t} + c_3 \begin{pmatrix} 1 \\ -1 \\ 3 \end{pmatrix} e^{-2t}.$$

12. We have $\det(\mathbf{A} - \lambda\mathbf{I}) = (\lambda - 3)(\lambda + 5)(6 - \lambda) = 0$. For $\lambda_1 = 3$, $\lambda_2 = -5$, and $\lambda_3 = 6$ we obtain

$$\mathbf{K}_1 = \begin{pmatrix} 1 \\ 1 \\ 0 \end{pmatrix}, \quad \mathbf{K}_2 = \begin{pmatrix} 1 \\ -1 \\ 0 \end{pmatrix}, \quad \text{and} \quad \mathbf{K}_3 = \begin{pmatrix} 2 \\ -2 \\ 11 \end{pmatrix},$$

so that

$$\mathbf{X} = c_1 \begin{pmatrix} 1 \\ 1 \\ 0 \end{pmatrix} e^{3t} + c_2 \begin{pmatrix} 1 \\ -1 \\ 0 \end{pmatrix} e^{-5t} + c_3 \begin{pmatrix} 2 \\ -2 \\ 11 \end{pmatrix} e^{6t}.$$

15. $\mathbf{X} = c_1 \begin{pmatrix} 0.382175 \\ 0.851161 \\ 0.359815 \end{pmatrix} e^{8.58979t} + c_2 \begin{pmatrix} 0.405188 \\ -0.676043 \\ 0.615458 \end{pmatrix} e^{2.25684t} + c_3 \begin{pmatrix} -0.923562 \\ -0.132174 \\ 0.35995 \end{pmatrix} e^{-0.0466321t}$

18. In Problem 2, letting $c_1 = 1$ and $c_2 = 0$ we get $x = -2e^t$, $y = e^t$. Eliminating the parameter we find $y = -\frac{1}{2}x$, $x < 0$. When $c_1 = -1$ and $c_2 = 0$ we find $y = -\frac{1}{2}x$, $x > 0$. Letting $c_1 = 0$ and $c_2 = 1$ we get $x = e^{4t}$, $y = e^{4t}$. Eliminating the parameter we find $y = x$, $x > 0$. When $c_1 = 0$ and $c_2 = -1$ we find $y = x$, $x < 0$.

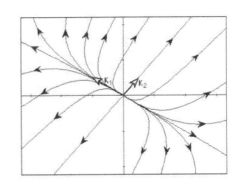

In Problem 4, letting $c_1 = 1$ and $c_2 = 0$ we get $x = -2e^{-7t/2}$, $y = e^{-7t/2}$. Eliminating the parameter we find $y = -\frac{1}{2}x$, $x < 0$. When $c_1 = -1$ and $c_2 = 0$ we find $y = -\frac{1}{2}x$, $x > 0$. Letting $c_1 = 0$ and $c_2 = 1$ we get $x = 4e^{-t}$, $y = 3e^{-t}$. Eliminating the parameter we find $y = \frac{3}{4}x$, $x > 0$. When $c_1 = 0$ and $c_2 = -1$ we find $y = \frac{3}{4}x$, $x < 0$.

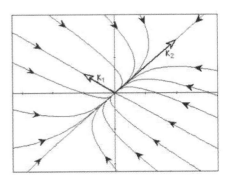

21. We have $\det(\mathbf{A} - \lambda\mathbf{I}) = (\lambda - 2)^2 = 0$. For $\lambda_1 = 2$ we obtain

$$\mathbf{K} = \begin{pmatrix} 1 \\ 1 \end{pmatrix}.$$

A solution of $(\mathbf{A} - \lambda_1\mathbf{I})\mathbf{P} = \mathbf{K}$ is

$$\mathbf{P} = \begin{pmatrix} -1/3 \\ 0 \end{pmatrix}$$

so that

$$\mathbf{X} = c_1 \begin{pmatrix} 1 \\ 1 \end{pmatrix} e^{2t} + c_2 \left[\begin{pmatrix} 1 \\ 1 \end{pmatrix} te^{2t} + \begin{pmatrix} -1/3 \\ 0 \end{pmatrix} e^{2t} \right].$$

24. We have $\det(\mathbf{A} - \lambda\mathbf{I}) = (\lambda - 8)(\lambda + 1)^2 = 0$. For $\lambda_1 = 8$ we obtain

$$\mathbf{K}_1 = \begin{pmatrix} 2 \\ 1 \\ 2 \end{pmatrix}.$$

For $\lambda_2 = -1$ we obtain

$$\mathbf{K}_2 = \begin{pmatrix} 0 \\ -2 \\ 1 \end{pmatrix} \quad \text{and} \quad \mathbf{K}_3 = \begin{pmatrix} 1 \\ -2 \\ 0 \end{pmatrix}.$$

Then

$$\mathbf{X} = c_1 \begin{pmatrix} 2 \\ 1 \\ 2 \end{pmatrix} e^{8t} + c_2 \begin{pmatrix} 0 \\ -2 \\ 1 \end{pmatrix} e^{-t} + c_3 \begin{pmatrix} 1 \\ -2 \\ 0 \end{pmatrix} e^{-t}.$$

27. We have $\det(\mathbf{A} - \lambda\mathbf{I}) = -(\lambda - 1)^3 = 0$. For $\lambda_1 = 1$ we obtain

$$\mathbf{K} = \begin{pmatrix} 0 \\ 1 \\ 1 \end{pmatrix}.$$

Solutions of $(\mathbf{A} - \lambda_1\mathbf{I})\mathbf{P} = \mathbf{K}$ and $(\mathbf{A} - \lambda_1\mathbf{I})\mathbf{Q} = \mathbf{P}$ are

$$\mathbf{P} = \begin{pmatrix} 0 \\ 1 \\ 0 \end{pmatrix} \quad \text{and} \quad \mathbf{Q} = \begin{pmatrix} 1/2 \\ 0 \\ 0 \end{pmatrix}$$

so that

$$\mathbf{X} = c_1 \begin{pmatrix} 0 \\ 1 \\ 1 \end{pmatrix} e^t + c_2 \left[\begin{pmatrix} 0 \\ 1 \\ 1 \end{pmatrix} t e^t + \begin{pmatrix} 0 \\ 1 \\ 0 \end{pmatrix} e^t \right] + c_3 \left[\begin{pmatrix} 0 \\ 1 \\ 1 \end{pmatrix} \frac{t^2}{2} e^t + \begin{pmatrix} 0 \\ 1 \\ 0 \end{pmatrix} t e^t + \begin{pmatrix} 1/2 \\ 0 \\ 0 \end{pmatrix} e^t \right].$$

30. We have $\det(\mathbf{A} - \lambda\mathbf{I}) = -(\lambda + 1)(\lambda - 1)^2 = 0$. For $\lambda_1 = -1$ we obtain

$$\mathbf{K}_1 = \begin{pmatrix} -1 \\ 0 \\ 1 \end{pmatrix}.$$

For $\lambda_2 = 1$ we obtain

$$\mathbf{K}_2 = \begin{pmatrix} 1 \\ 0 \\ 1 \end{pmatrix} \quad \text{and} \quad \mathbf{K}_3 = \begin{pmatrix} 0 \\ 1 \\ 0 \end{pmatrix}$$

so that

$$\mathbf{X} = c_1 \begin{pmatrix} -1 \\ 0 \\ 1 \end{pmatrix} e^{-t} + c_2 \begin{pmatrix} 1 \\ 0 \\ 1 \end{pmatrix} e^t + c_3 \begin{pmatrix} 0 \\ 1 \\ 0 \end{pmatrix} e^t.$$

If

$$\mathbf{X}(0) = \begin{pmatrix} 1 \\ 2 \\ 5 \end{pmatrix}$$

then $c_1 = 2$, $c_2 = 3$, and $c_3 = 2$.

33. We have $\det(\mathbf{A} - \lambda\mathbf{I}) = \lambda^2 - 8\lambda + 17 = 0$. For $\lambda_1 = 4 + i$ we obtain

$$\mathbf{K}_1 = \begin{pmatrix} 2 + i \\ 5 \end{pmatrix}$$

so that

$$\mathbf{X}_1 = \begin{pmatrix} 2 + i \\ 5 \end{pmatrix} e^{(4+i)t} = \begin{pmatrix} 2\cos t - \sin t \\ 5\cos t \end{pmatrix} e^{4t} + i \begin{pmatrix} \cos t + 2\sin t \\ 5\sin t \end{pmatrix} e^{4t}$$

Then

$$\mathbf{X} = c_1 \begin{pmatrix} 2\cos t - \sin t \\ 5\cos t \end{pmatrix} e^{4t} + c_2 \begin{pmatrix} \cos t + 2\sin t \\ 5\sin t \end{pmatrix} e^{4t}.$$

36. We have $\det(\mathbf{A} - \lambda\mathbf{I}) = \lambda^2 - 10\lambda + 34 = 0$. For $\lambda_1 = 5 + 3i$ we obtain

$$\mathbf{K}_1 = \begin{pmatrix} 1 - 3i \\ 2 \end{pmatrix}$$

so that

$$\mathbf{X}_1 = \begin{pmatrix} 1 - 3i \\ 2 \end{pmatrix} e^{(5+3i)t} = \begin{pmatrix} \cos 3t + 3\sin 3t \\ 2\cos 3t \end{pmatrix} e^{5t} + i \begin{pmatrix} \sin 3t - 3\cos 3t \\ 2\sin 3t \end{pmatrix} e^{5t}.$$

Then

$$\mathbf{X} = c_1 \begin{pmatrix} \cos 3t + 3\sin 3t \\ 2\cos 3t \end{pmatrix} e^{5t} + c_2 \begin{pmatrix} \sin 3t - 3\cos 3t \\ 2\sin 3t \end{pmatrix} e^{5t}.$$

39. We have $\det(\mathbf{A} - \lambda\mathbf{I}) = -\lambda\left(\lambda^2 + 1\right) = 0$. For $\lambda_1 = 0$ we obtain

$$\mathbf{K}_1 = \begin{pmatrix} 1 \\ 0 \\ 0 \end{pmatrix}.$$

For $\lambda_2 = i$ we obtain

$$\mathbf{K}_2 = \begin{pmatrix} -i \\ i \\ 1 \end{pmatrix}$$

so that

$$\mathbf{X}_2 = \begin{pmatrix} -i \\ i \\ 1 \end{pmatrix} e^{it} = \begin{pmatrix} \sin t \\ -\sin t \\ \cos t \end{pmatrix} + i \begin{pmatrix} -\cos t \\ \cos t \\ \sin t \end{pmatrix}.$$

Then

$$\mathbf{X} = c_1 \begin{pmatrix} 1 \\ 0 \\ 0 \end{pmatrix} + c_2 \begin{pmatrix} \sin t \\ -\sin t \\ \cos t \end{pmatrix} + c_3 \begin{pmatrix} -\cos t \\ \cos t \\ \sin t \end{pmatrix}.$$

42. We have $\det(\mathbf{A} - \lambda\mathbf{I}) = -(\lambda - 6)(\lambda^2 - 8\lambda + 20) = 0$. For $\lambda_1 = 6$ we obtain

$$\mathbf{K}_1 = \begin{pmatrix} 0 \\ 1 \\ 0 \end{pmatrix}.$$

For $\lambda_2 = 4 + 2i$ we obtain

$$\mathbf{K}_2 = \begin{pmatrix} -i \\ 0 \\ 2 \end{pmatrix}$$

so that

$$\mathbf{X}_2 = \begin{pmatrix} -i \\ 0 \\ 2 \end{pmatrix} e^{(4+2i)t} = \begin{pmatrix} \sin 2t \\ 0 \\ 2\cos 2t \end{pmatrix} e^{4t} + i \begin{pmatrix} -\cos 2t \\ 0 \\ 2\sin 2t \end{pmatrix} e^{4t}.$$

Then

$$\mathbf{X} = c_1 \begin{pmatrix} 0 \\ 1 \\ 0 \end{pmatrix} e^{6t} + c_2 \begin{pmatrix} \sin 2t \\ 0 \\ 2\cos 2t \end{pmatrix} e^{4t} + c_3 \begin{pmatrix} -\cos 2t \\ 0 \\ 2\sin 2t \end{pmatrix} e^{4t}.$$

45. We have $\det(\mathbf{A} - \lambda\mathbf{I}) = (1 - \lambda)(\lambda^2 + 25) = 0$. For $\lambda_1 = 1$ we obtain

$$\mathbf{K}_1 = \begin{pmatrix} 25 \\ -7 \\ 6 \end{pmatrix}.$$

For $\lambda_2 = 5i$ we obtain

$$\mathbf{K}_2 = \begin{pmatrix} 1 + 5i \\ 1 \\ 1 \end{pmatrix}$$

so that

$$\mathbf{X}_2 = \begin{pmatrix} 1 + 5i \\ 1 \\ 1 \end{pmatrix} e^{5it} = \begin{pmatrix} \cos 5t - 5\sin 5t \\ \cos 5t \\ \cos 5t \end{pmatrix} + i \begin{pmatrix} \sin 5t + 5\cos 5t \\ \sin 5t \\ \sin 5t \end{pmatrix}.$$

Then

$$\mathbf{X} = c_1 \begin{pmatrix} 25 \\ -7 \\ 6 \end{pmatrix} e^{t} + c_2 \begin{pmatrix} \cos 5t - 5\sin 5t \\ \cos 5t \\ \cos 5t \end{pmatrix} + c_3 \begin{pmatrix} \sin 5t + 5\cos 5t \\ \sin 5t \\ \sin 5t \end{pmatrix}.$$

If

$$\mathbf{X}(0) = \begin{pmatrix} 4 \\ 6 \\ -7 \end{pmatrix}$$

then $c_1 = c_2 = -1$ and $c_3 = 6$.

48. (a) Letting $x_1 = y_1$, $x_1' = y_2$, $x_2 = y_3$, and $x_2' = y_4$ we have

$$y_2' = x_1'' = -10x_1 + 4x_2 = -10y_1 + 4y_3$$

$$y_4' = x_2'' = 4x_1 - 4x_2 = 4y_1 - 4y_3.$$

The corresponding linear system is

$$y_1' = y_2$$

$$y_2' = -10y_1 + 4y_3$$

$$y_3' = y_4$$

$$y_4' = 4y_1 - 4y_3$$

or

$$\mathbf{Y}' = \begin{pmatrix} 0 & 1 & 0 & 0 \\ -10 & 0 & 4 & 0 \\ 0 & 0 & 0 & 1 \\ 4 & 0 & -4 & 0 \end{pmatrix} \mathbf{Y}.$$

Using a CAS, we find eigenvalues $\pm\sqrt{2}i$ and $\pm 2\sqrt{3}i$ with corresponding eigenvectors

$$\begin{pmatrix} \mp\sqrt{2}i/4 \\ 1/2 \\ \mp\sqrt{2}i/2 \\ 1 \end{pmatrix} = \begin{pmatrix} 0 \\ 1/2 \\ 0 \\ 1 \end{pmatrix} + i \begin{pmatrix} \mp\sqrt{2}/4 \\ 0 \\ \mp\sqrt{2}/2 \\ 0 \end{pmatrix}$$

and

$$\begin{pmatrix} \pm\sqrt{3}i/3 \\ -2 \\ \mp\sqrt{3}i/6 \\ 1 \end{pmatrix} = \begin{pmatrix} 0 \\ -2 \\ 0 \\ 1 \end{pmatrix} + i \begin{pmatrix} \pm\sqrt{3}/3 \\ 0 \\ \mp\sqrt{3}/6 \\ 0 \end{pmatrix}.$$

Thus

$$\mathbf{Y}(t) = c_1 \left[\begin{pmatrix} 0 \\ 1/2 \\ 0 \\ 1 \end{pmatrix} \cos\sqrt{2}t - \begin{pmatrix} -\sqrt{2}/4 \\ 0 \\ -\sqrt{2}/2 \\ 0 \end{pmatrix} \sin\sqrt{2}t \right]$$

$$+ c_2 \left[\begin{pmatrix} -\sqrt{2}/4 \\ 0 \\ -\sqrt{2}/2 \\ 0 \end{pmatrix} \cos\sqrt{2}t + \begin{pmatrix} 0 \\ 1/2 \\ 0 \\ 1 \end{pmatrix} \sin\sqrt{2}t \right]$$

$$+ c_3 \left[\begin{pmatrix} 0 \\ -2 \\ 0 \\ 1 \end{pmatrix} \cos 2\sqrt{3}t - \begin{pmatrix} \sqrt{3}/3 \\ 0 \\ -\sqrt{3}/6 \\ 0 \end{pmatrix} \sin 2\sqrt{3}t \right]$$

$$+ c_4 \left[\begin{pmatrix} \sqrt{3}/3 \\ 0 \\ -\sqrt{3}/6 \\ 0 \end{pmatrix} \cos 2\sqrt{3}t + \begin{pmatrix} 0 \\ -2 \\ 0 \\ 1 \end{pmatrix} \sin 2\sqrt{3}t \right].$$

The initial conditions $y_1(0) = 0$, $y_2(0) = 1$, $y_3(0) = 0$, and $y_4(0) = -1$ imply $c_1 = -\frac{2}{5}$, $c_2 = 0$, $c_3 = -\frac{3}{5}$, and $c_4 = 0$. Thus,

$$x_1(t) = y_1(t) = -\frac{\sqrt{2}}{10} \sin\sqrt{2}t + \frac{\sqrt{3}}{5} \sin 2\sqrt{3}t$$

$$x_2(t) = y_3(t) = -\frac{\sqrt{2}}{5} \sin\sqrt{2}t - \frac{\sqrt{3}}{10} \sin 2\sqrt{3}t.$$

(b) The second-order system is

$$x_1'' = -10x_1 + 4x_2$$

$$x_2'' = 4x_1 - 4x_2$$

or

$$\mathbf{X}'' = \begin{pmatrix} -10 & 4 \\ 4 & -4 \end{pmatrix} \mathbf{X}.$$

We assume solutions of the form $\mathbf{X} = \mathbf{V}\cos\omega t$ and $\mathbf{X} = \mathbf{V}\sin\omega t$. Since the eigenvalues are -2 and -12, $\omega_1 = \sqrt{-(-2)} = \sqrt{2}$ and $\omega_2 = \sqrt{-(-12)} = 2\sqrt{3}$. The corresponding eigenvectors are

$$\mathbf{V}_1 = \begin{pmatrix} 1 \\ 2 \end{pmatrix} \qquad \text{and} \qquad \mathbf{V}_2 = \begin{pmatrix} -2 \\ 1 \end{pmatrix}.$$

Then, the general solution of the system is

$$\mathbf{X} = c_1 \begin{pmatrix} 1 \\ 2 \end{pmatrix} \cos \sqrt{2}t + c_2 \begin{pmatrix} 1 \\ 2 \end{pmatrix} \sin \sqrt{2}t + c_3 \begin{pmatrix} -2 \\ 1 \end{pmatrix} \cos 2\sqrt{3}t + c_4 \begin{pmatrix} -2 \\ 1 \end{pmatrix} \sin 2\sqrt{3}t.$$

The initial conditions

$$\mathbf{X}(0) = \begin{pmatrix} 0 \\ 0 \end{pmatrix} \quad \text{and} \quad \mathbf{X}'(0) = \begin{pmatrix} 1 \\ -1 \end{pmatrix}$$

imply $c_1 = 0$, $c_2 = -\sqrt{2}/10$, $c_3 = 0$, and $c_4 = -\sqrt{3}/10$. Thus

$$x_1(t) = -\frac{\sqrt{2}}{10} \sin \sqrt{2}t + \frac{\sqrt{3}}{5} \sin 2\sqrt{3}t$$

$$x_2(t) = -\frac{\sqrt{2}}{5} \sin \sqrt{2}t - \frac{\sqrt{3}}{10} \sin 2\sqrt{3}t.$$

51. From $x = 2\cos 2t - 2\sin 2t$, $y = -\cos 2t$ we find $x + 2y = -2\sin 2t$. Then

$$(x + 2y)^2 = 4\sin^2 2t = 4(1 - \cos^2 2t) = 4 - 4\cos^2 2t = 4 - 4y^2$$

and

$$x^2 + 4xy + 4y^2 = 4 - 4y^2 \quad \text{or} \quad x^2 + 4xy + 8y^2 = 4.$$

This is a rotated conic section and, from the discriminant $b^2 - 4ac = 16 - 32 < 0$, we see that the curve is an ellipse.

8.3 Nonhomogeneous Linear Systems

The terminology and concepts listed below provide an outline of the main ideas encountered in this section. These can be useful when preparing for a quiz or test.

Terminology and Concepts

- method of undetermined coefficients for finding a particular solution of a nonhomogeneous linear system of DEs

- fundamental matrix of a linear system of DEs

- method of variation of parameters for finding a particular solution of a nonhomogeneous linear system of DEs

The basic skills listed below summarize the more mechanical types of problems encountered in the exercise set for this section.

Basic Skills

- use undetermined coefficients to solve a nonhomogeneous linear system of DEs
- use variation of parameters to solve a nonhomogeneous linear system of DEs

3. Solving

$$\det(\mathbf{A} - \lambda \mathbf{I}) = \begin{vmatrix} 1 - \lambda & 3 \\ 3 & 1 - \lambda \end{vmatrix} = \lambda^2 - 2\lambda - 8 = (\lambda - 4)(\lambda + 2) = 0$$

we obtain eigenvalues $\lambda_1 = -2$ and $\lambda_2 = 4$. Corresponding eigenvectors are

$$\mathbf{K}_1 = \begin{pmatrix} 1 \\ -1 \end{pmatrix} \quad \text{and} \quad \mathbf{K}_2 = \begin{pmatrix} 1 \\ 1 \end{pmatrix}.$$

Thus

$$\mathbf{X}_c = c_1 \begin{pmatrix} 1 \\ -1 \end{pmatrix} e^{-2t} + c_2 \begin{pmatrix} 1 \\ 1 \end{pmatrix} e^{4t}.$$

Substituting

$$\mathbf{X}_p = \begin{pmatrix} a_3 \\ b_3 \end{pmatrix} t^2 + \begin{pmatrix} a_2 \\ b_2 \end{pmatrix} t + \begin{pmatrix} a_1 \\ b_1 \end{pmatrix}$$

into the system yields

$$a_3 + 3b_3 = 2 \qquad a_2 + 3b_2 = 2a_3 \qquad a_1 + 3b_1 = a_2$$

$$3a_3 + b_3 = 0 \qquad 3a_2 + b_2 + 1 = 2b_3 \qquad 3a_1 + b_1 + 5 = b_2$$

from which we obtain $a_3 = -1/4$, $b_3 = 3/4$, $a_2 = 1/4$, $b_2 = -1/4$, $a_1 = -2$, and $b_1 = 3/4$. Then

$$\mathbf{X}(t) = c_1 \begin{pmatrix} 1 \\ -1 \end{pmatrix} e^{-2t} + c_2 \begin{pmatrix} 1 \\ 1 \end{pmatrix} e^{4t} + \begin{pmatrix} -1/4 \\ 3/4 \end{pmatrix} t^2 + \begin{pmatrix} 1/4 \\ -1/4 \end{pmatrix} t + \begin{pmatrix} -2 \\ 3/4 \end{pmatrix}.$$

6. Solving

$$\det(\mathbf{A} - \lambda \mathbf{I}) = \begin{vmatrix} -1 - \lambda & 5 \\ -1 & 1 - \lambda \end{vmatrix} = \lambda^2 + 4 = 0$$

we obtain the eigenvalues $\lambda_1 = 2i$ and $\lambda_2 = -2i$. Corresponding eigenvectors are

$$\mathbf{K}_1 = \begin{pmatrix} 5 \\ 1 + 2i \end{pmatrix} \quad \text{and} \quad \mathbf{K}_2 = \begin{pmatrix} 5 \\ 1 - 2i \end{pmatrix}.$$

Thus

$$\mathbf{X}_c = c_1 \begin{pmatrix} 5\cos 2t \\ \cos 2t - 2\sin 2t \end{pmatrix} + c_2 \begin{pmatrix} 5\sin 2t \\ 2\cos 2t + \sin 2t \end{pmatrix}.$$

Substituting

$$\mathbf{X}_p = \begin{pmatrix} a_2 \\ b_2 \end{pmatrix} \cos t + \begin{pmatrix} a_1 \\ b_1 \end{pmatrix} \sin t$$

into the system yields

$$-a_2 + 5b_2 - a_1 = 0$$

$$-a_2 + b_2 - b_1 - 2 = 0$$

$$-a_1 + 5b_1 + a_2 + 1 = 0$$

$$-a_1 + b_1 + b_2 = 0$$

from which we obtain $a_2 = -3$, $b_2 = -2/3$, $a_1 = -1/3$, and $b_1 = 1/3$. Then

$$\mathbf{X}(t) = c_1 \begin{pmatrix} 5\cos 2t \\ \cos 2t - 2\sin 2t \end{pmatrix} + c_2 \begin{pmatrix} 5\sin 2t \\ 2\cos 2t + \sin 2t \end{pmatrix} + \begin{pmatrix} -3 \\ -2/3 \end{pmatrix} \cos t + \begin{pmatrix} -1/3 \\ 1/3 \end{pmatrix} \sin t.$$

9. Solving

$$\det(\mathbf{A} - \lambda\mathbf{I}) = \begin{vmatrix} -1 - \lambda & -2 \\ 3 & 4 - \lambda \end{vmatrix} = \lambda^2 - 3\lambda + 2 = (\lambda - 1)(\lambda - 2) = 0$$

we obtain the eigenvalues $\lambda_1 = 1$ and $\lambda_2 = 2$. Corresponding eigenvectors are

$$\mathbf{K}_1 = \begin{pmatrix} 1 \\ -1 \end{pmatrix} \quad \text{and} \quad \mathbf{K}_2 = \begin{pmatrix} -4 \\ 6 \end{pmatrix}.$$

Thus

$$\mathbf{X}_c = c_1 \begin{pmatrix} 1 \\ -1 \end{pmatrix} e^t + c_2 \begin{pmatrix} -4 \\ 6 \end{pmatrix} e^{2t}.$$

Substituting

$$\mathbf{X}_p = \begin{pmatrix} a_1 \\ b_1 \end{pmatrix}$$

into the system yields

$$-a_1 - 2b_1 = -3$$

$$3a_1 + 4b_1 = -3$$

from which we obtain $a_1 = -9$ and $b_1 = 6$. Then

$$\mathbf{X}(t) = c_1 \begin{pmatrix} 1 \\ -1 \end{pmatrix} e^t + c_2 \begin{pmatrix} -4 \\ 6 \end{pmatrix} e^{2t} + \begin{pmatrix} -9 \\ 6 \end{pmatrix}.$$

Setting

$$\mathbf{X}(0) = \begin{pmatrix} -4 \\ 5 \end{pmatrix}$$

we obtain

$$c_1 - 4c_2 - 9 = -4$$

$$-c_1 + 6c_2 + 6 = 5.$$

Then $c_1 = 13$ and $c_2 = 2$ so

$$\mathbf{X}(t) = 13 \begin{pmatrix} 1 \\ -1 \end{pmatrix} e^t + 2 \begin{pmatrix} -4 \\ 6 \end{pmatrix} e^{2t} + \begin{pmatrix} -9 \\ 6 \end{pmatrix}.$$

12. From

$$\mathbf{X}' = \begin{pmatrix} 2 & -1 \\ 3 & -2 \end{pmatrix} \mathbf{X} + \begin{pmatrix} 0 \\ 4 \end{pmatrix} t$$

we obtain

$$\mathbf{X}_c = c_1 \begin{pmatrix} 1 \\ 1 \end{pmatrix} e^t + c_2 \begin{pmatrix} 1 \\ 3 \end{pmatrix} e^{-t}.$$

Then

$$\mathbf{\Phi} = \begin{pmatrix} e^t & e^{-t} \\ e^t & 3e^{-t} \end{pmatrix} \quad \text{and} \quad \mathbf{\Phi}^{-1} = \begin{pmatrix} \frac{3}{2}e^{-t} & -\frac{1}{2}e^{-t} \\ -\frac{1}{2}e^t & \frac{1}{2}e^t \end{pmatrix}$$

so that

$$\mathbf{U} = \int \mathbf{\Phi}^{-1}\mathbf{F}\,dt = \int \begin{pmatrix} -2te^{-t} \\ 2te^t \end{pmatrix} dt = \begin{pmatrix} 2te^{-t} + 2e^{-t} \\ 2te^t - 2e^t \end{pmatrix}$$

and

$$\mathbf{X}_p = \mathbf{\Phi}\mathbf{U} = \begin{pmatrix} 4 \\ 8 \end{pmatrix} t + \begin{pmatrix} 0 \\ -4 \end{pmatrix}.$$

15. From

$$\mathbf{X}' = \begin{pmatrix} 0 & 2 \\ -1 & 3 \end{pmatrix} \mathbf{X} + \begin{pmatrix} 1 \\ -1 \end{pmatrix} e^t$$

we obtain

$$\mathbf{X}_c = c_1 \begin{pmatrix} 2 \\ 1 \end{pmatrix} e^t + c_2 \begin{pmatrix} 1 \\ 1 \end{pmatrix} e^{2t}.$$

Then

$$\mathbf{\Phi} = \begin{pmatrix} 2e^t & e^{2t} \\ e^t & e^{2t} \end{pmatrix} \quad \text{and} \quad \mathbf{\Phi}^{-1} = \begin{pmatrix} e^{-t} & -e^{-t} \\ -e^{-2t} & 2e^{-2t} \end{pmatrix}$$

so that

$$\mathbf{U} = \int \mathbf{\Phi}^{-1}\mathbf{F}\,dt = \int \begin{pmatrix} 2 \\ -3e^{-t} \end{pmatrix} dt = \begin{pmatrix} 2t \\ 3e^{-t} \end{pmatrix}$$

and

$$\mathbf{X}_p = \mathbf{\Phi}\mathbf{U} = \begin{pmatrix} 4 \\ 2 \end{pmatrix} te^t + \begin{pmatrix} 3 \\ 3 \end{pmatrix} e^t.$$

18. From

$$\mathbf{X}' = \begin{pmatrix} 1 & 8 \\ 1 & -1 \end{pmatrix} \mathbf{X} + \begin{pmatrix} e^{-t} \\ te^t \end{pmatrix}$$

we obtain

$$\mathbf{X}_c = c_1 \begin{pmatrix} 4 \\ 1 \end{pmatrix} e^{3t} + c_2 \begin{pmatrix} -2 \\ 1 \end{pmatrix} e^{-3t}.$$

Then

$$\boldsymbol{\Phi} = \begin{pmatrix} 4e^{3t} & -2e^{3t} \\ e^{3t} & e^{-3t} \end{pmatrix} \quad \text{and} \quad \boldsymbol{\Phi}^{-1} = \begin{pmatrix} \frac{1}{6}e^{-3t} & \frac{1}{3}e^{-3t} \\ -\frac{1}{6}e^{3t} & \frac{2}{3}e^{3t} \end{pmatrix}$$

so that

$$\mathbf{U} = \int \boldsymbol{\Phi}^{-1}\mathbf{F}\,dt = \int \begin{pmatrix} \frac{1}{6}e^{-4t} + \frac{1}{3}te^{-2t} \\ -\frac{1}{6}e^{2t} + \frac{2}{3}te^{4t} \end{pmatrix} dt = \begin{pmatrix} -\frac{1}{24}e^{-4t} - \frac{1}{6}te^{-2t} - \frac{1}{12}e^{-2t} \\ -\frac{1}{12}e^{2t} + \frac{1}{6}te^{4t} - \frac{1}{24}e^{4t} \end{pmatrix}$$

and

$$\mathbf{X}_p = \boldsymbol{\Phi}\mathbf{U} = \begin{pmatrix} -te^t - \frac{1}{4}e^t \\ -\frac{1}{8}e^{-t} - \frac{1}{8}e^t \end{pmatrix}.$$

21. From

$$\mathbf{X}' = \begin{pmatrix} 0 & -1 \\ 1 & 0 \end{pmatrix} \mathbf{X} + \begin{pmatrix} \sec t \\ 0 \end{pmatrix}$$

we obtain

$$\mathbf{X}_c = c_1 \begin{pmatrix} \cos t \\ \sin t \end{pmatrix} + c_2 \begin{pmatrix} \sin t \\ -\cos t \end{pmatrix}.$$

Then

$$\boldsymbol{\Phi} = \begin{pmatrix} \cos t & \sin t \\ \sin t & -\cos t \end{pmatrix} \quad \text{and} \quad \boldsymbol{\Phi}^{-1} = \begin{pmatrix} \cos t & \sin t \\ \sin t & -\cos t \end{pmatrix}$$

so that

$$\mathbf{U} = \int \boldsymbol{\Phi}^{-1}\mathbf{F}\,dt = \int \begin{pmatrix} 1 \\ \tan t \end{pmatrix} dt = \begin{pmatrix} t \\ -\ln|\cos t| \end{pmatrix}$$

and

$$\mathbf{X}_p = \boldsymbol{\Phi}\mathbf{U} = \begin{pmatrix} t\cos t - \sin t \ln|\cos t| \\ t\sin t + \cos t \ln|\cos t| \end{pmatrix}.$$

24. From

$$\mathbf{X}' = \begin{pmatrix} 2 & -2 \\ 8 & -6 \end{pmatrix} \mathbf{X} + \begin{pmatrix} 1 \\ 3 \end{pmatrix} \frac{1}{t} e^{-2t}$$

we obtain

$$\mathbf{X}_c = c_1 \begin{pmatrix} 1 \\ 2 \end{pmatrix} e^{-2t} + c_2 \left[\begin{pmatrix} 1 \\ 2 \end{pmatrix} t e^{-2t} + \begin{pmatrix} 1/2 \\ 1/2 \end{pmatrix} e^{-2t} \right].$$

Then

$$\mathbf{\Phi} = \begin{pmatrix} 1 & t + \frac{1}{2} \\ 2 & 2t + \frac{1}{2} \end{pmatrix} e^{-2t} \quad \text{and} \quad \mathbf{\Phi}^{-1} = \begin{pmatrix} -4t - 1 & 2t + 1 \\ 4 & -2 \end{pmatrix} e^{2t}$$

so that

$$\mathbf{U} = \int \mathbf{\Phi}^{-1} \mathbf{F}\, dt = \int \begin{pmatrix} 2 + 2/t \\ -2/t \end{pmatrix} dt = \begin{pmatrix} 2t + 2\ln t \\ -2\ln t \end{pmatrix}$$

and

$$\mathbf{X}_p = \mathbf{\Phi}\mathbf{U} = \begin{pmatrix} 2t + \ln t - 2t\ln t \\ 4t + 3\ln t - 4t\ln t \end{pmatrix} e^{-2t}.$$

27. From

$$\mathbf{X}' = \begin{pmatrix} 1 & 2 \\ -1/2 & 1 \end{pmatrix} \mathbf{X} + \begin{pmatrix} \csc t \\ \sec t \end{pmatrix} e^t$$

we obtain

$$\mathbf{X}_c = c_1 \begin{pmatrix} 2\sin t \\ \cos t \end{pmatrix} e^t + c_2 \begin{pmatrix} 2\cos t \\ -\sin t \end{pmatrix} e^t.$$

Then

$$\mathbf{\Phi} = \begin{pmatrix} 2\sin t & 2\cos t \\ \cos t & -\sin t \end{pmatrix} e^t \quad \text{and} \quad \mathbf{\Phi}^{-1} = \begin{pmatrix} \frac{1}{2}\sin t & \cos t \\ \frac{1}{2}\cos t & -\sin t \end{pmatrix} e^{-t}$$

so that

$$\mathbf{U} = \int \mathbf{\Phi}^{-1} \mathbf{F}\, dt = \int \begin{pmatrix} \frac{3}{2} \\ \frac{1}{2}\cot t - \tan t \end{pmatrix} dt = \begin{pmatrix} \frac{3}{2}t \\ \frac{1}{2}\ln|\sin t| + \ln|\cos t| \end{pmatrix}$$

and

$$\mathbf{X}_p = \mathbf{\Phi}\mathbf{U} = \begin{pmatrix} 3\sin t \\ \frac{3}{2}\cos t \end{pmatrix} t e^t + \begin{pmatrix} \cos t \\ -\frac{1}{2}\sin t \end{pmatrix} e^t \ln|\sin t| + \begin{pmatrix} 2\cos t \\ -\sin t \end{pmatrix} e^t \ln|\cos t|.$$

30. From

$$\mathbf{X}' = \begin{pmatrix} 3 & -1 & -1 \\ 1 & 1 & -1 \\ 1 & -1 & 1 \end{pmatrix} \mathbf{X} + \begin{pmatrix} 0 \\ t \\ 2e^t \end{pmatrix}$$

we obtain

$$\mathbf{X}_c = c_1 \begin{pmatrix} 1 \\ 1 \\ 1 \end{pmatrix} e^t + c_2 \begin{pmatrix} 1 \\ 1 \\ 0 \end{pmatrix} e^{2t} + c_3 \begin{pmatrix} 1 \\ 0 \\ 1 \end{pmatrix} e^{2t}.$$

Then

$$\Phi = \begin{pmatrix} e^t & e^{2t} & e^{2t} \\ e^t & e^{2t} & 0 \\ e^t & 0 & e^{2t} \end{pmatrix} \quad \text{and} \quad \Phi^{-1} = \begin{pmatrix} -e^{-t} & e^{-t} & e^{-t} \\ e^{-2t} & 0 & -e^{-2t} \\ e^{-2t} & -e^{-2t} & 0 \end{pmatrix}$$

so that

$$\mathbf{U} = \int \Phi^{-1}\mathbf{F}\, dt = \int \begin{pmatrix} te^{-t} + 2 \\ -2e^{-t} \\ -te^{-2t} \end{pmatrix} dt = \begin{pmatrix} -te^{-t} - e^{-t} + 2t \\ 2e^{-t} \\ \frac{1}{2}te^{-2t} + \frac{1}{4}e^{-2t} \end{pmatrix}$$

and

$$\mathbf{X}_p = \Phi\mathbf{U} = \begin{pmatrix} -1/2 \\ -1 \\ -1/2 \end{pmatrix} t + \begin{pmatrix} -3/4 \\ -1 \\ -3/4 \end{pmatrix} + \begin{pmatrix} 2 \\ 2 \\ 0 \end{pmatrix} e^t + \begin{pmatrix} 2 \\ 2 \\ 2 \end{pmatrix} te^t.$$

33. Let $\mathbf{I} = \begin{pmatrix} i_1 \\ i_2 \end{pmatrix}$ so that

$$\mathbf{I}' = \begin{pmatrix} -11 & 3 \\ 3 & -3 \end{pmatrix} \mathbf{I} + \begin{pmatrix} 100 \sin t \\ 0 \end{pmatrix}$$

and

$$\mathbf{I}_c = c_1 \begin{pmatrix} 1 \\ 3 \end{pmatrix} e^{-2t} + c_2 \begin{pmatrix} 3 \\ -1 \end{pmatrix} e^{-12t}.$$

Then

$$\Phi = \begin{pmatrix} e^{-2t} & 3e^{-12t} \\ 3e^{-2t} & -e^{-12t} \end{pmatrix}, \quad \Phi^{-1} = \begin{pmatrix} \frac{1}{10}e^{2t} & \frac{3}{10}e^{2t} \\ \frac{3}{10}e^{12t} & -\frac{1}{10}e^{12t} \end{pmatrix},$$

$$\mathbf{U} = \int \Phi^{-1}\mathbf{F}\, dt = \int \begin{pmatrix} 10e^{2t} \sin t \\ 30e^{12t} \sin t \end{pmatrix} dt = \begin{pmatrix} 2e^{2t}(2 \sin t - \cos t) \\ \frac{6}{29}e^{12t}(12 \sin t - \cos t) \end{pmatrix},$$

and

$$\mathbf{I}_p = \Phi\mathbf{U} = \begin{pmatrix} \frac{332}{29} \sin t - \frac{76}{29} \cos t \\ \frac{276}{29} \sin t - \frac{168}{29} \cos t \end{pmatrix}$$

so that

$$\mathbf{I} = c_1 \begin{pmatrix} 1 \\ 3 \end{pmatrix} e^{-2t} + c_2 \begin{pmatrix} 3 \\ -1 \end{pmatrix} e^{-12t} + \mathbf{I}_p.$$

If $\mathbf{I}(0) = \begin{pmatrix} 0 \\ 0 \end{pmatrix}$ then $c_1 = 2$ and $c_2 = \frac{6}{29}$.

| 8.4 | Matrix Exponential |

The terminology and concepts listed below provide an outline of the main ideas encountered in this section. These can be useful when preparing for a quiz or test.

Terminology and Concepts

- infinite series representation of the matrix exponential $e^{\mathbf{A}t}$
- the matrix exponential is a fundamental matrix

The basic skills listed below summarize the more mechanical types of problems encountered in the exercise set for this section.

Basic Skills

- compute $e^{\mathbf{A}t}$ for a given matrix $\mathbf{A}$
- use the matrix exponential to solve a nonhomogeneous linear system

Use of Computers   To compute the matrix exponential for a square matrix $\mathbf{A}t$ use

$$\textbf{MatrixExp[At]} \qquad\qquad (\textit{Mathematica})$$

```
with(linalg):                           (Maple)
exponential(A,t);
```

```
expm(At)                                (MATLAB)
```

3. For

$$\mathbf{A} = \begin{pmatrix} 1 & 1 & 1 \\ 1 & 1 & 1 \\ -2 & -2 & -2 \end{pmatrix}$$

we have

$$\mathbf{A}^2 = \begin{pmatrix} 1 & 1 & 1 \\ 1 & 1 & 1 \\ -2 & -2 & -2 \end{pmatrix} \begin{pmatrix} 1 & 1 & 1 \\ 1 & 1 & 1 \\ -2 & -2 & -2 \end{pmatrix} = \begin{pmatrix} 0 & 0 & 0 \\ 0 & 0 & 0 \\ 0 & 0 & 0 \end{pmatrix}.$$

Thus, $\mathbf{A}^3 = \mathbf{A}^4 = \mathbf{A}^5 = \cdots = \mathbf{0}$ and

$$e^{\mathbf{A}t} = \mathbf{I} + \mathbf{A}t = \begin{pmatrix} 1 & 0 & 0 \\ 0 & 1 & 0 \\ 0 & 0 & 1 \end{pmatrix} + \begin{pmatrix} t & t & t \\ t & t & t \\ -2t & -2t & -2t \end{pmatrix} = \begin{pmatrix} t+1 & t & t \\ t & t+1 & t \\ -2t & -2t & -2t+1 \end{pmatrix}.$$

6. Using the result of Problem 2,

$$\mathbf{X} = \begin{pmatrix} \cosh t & \sinh t \\ \sinh t & \cosh t \end{pmatrix} \begin{pmatrix} c_1 \\ c_2 \end{pmatrix} = c_1 \begin{pmatrix} \cosh t \\ \sinh t \end{pmatrix} + c_2 \begin{pmatrix} \sinh t \\ \cosh t \end{pmatrix}.$$

9. To solve

$$\mathbf{X}' = \begin{pmatrix} 1 & 0 \\ 0 & 2 \end{pmatrix} \mathbf{X} + \begin{pmatrix} 3 \\ -1 \end{pmatrix}$$

we identify $t_0 = 0$, $\mathbf{F}(t) = \begin{pmatrix} 3 \\ -1 \end{pmatrix}$, and use the results of Problem 1 and equation (5) in the text.

$$\mathbf{X}(t) = e^{\mathbf{A}t}\mathbf{C} + e^{\mathbf{A}t} \int_{t_0}^{t} e^{-\mathbf{A}s}\mathbf{F}(s)\,ds$$

$$= \begin{pmatrix} e^t & 0 \\ 0 & e^{2t} \end{pmatrix} \begin{pmatrix} c_1 \\ c_2 \end{pmatrix} + \begin{pmatrix} e^t & 0 \\ 0 & e^{2t} \end{pmatrix} \int_0^t \begin{pmatrix} e^{-s} & 0 \\ 0 & e^{-2s} \end{pmatrix} \begin{pmatrix} 3 \\ -1 \end{pmatrix} ds$$

$$= \begin{pmatrix} c_1 e^t \\ c_2 e^{2t} \end{pmatrix} + \begin{pmatrix} e^t & 0 \\ 0 & e^{2t} \end{pmatrix} \int_0^t \begin{pmatrix} 3e^{-s} \\ -e^{-2s} \end{pmatrix} ds$$

$$= \begin{pmatrix} c_1 e^t \\ c_2 e^{2t} \end{pmatrix} + \begin{pmatrix} e^t & 0 \\ 0 & e^{2t} \end{pmatrix} \begin{pmatrix} -3e^{-s} \\ \frac{1}{2}e^{-2s} \end{pmatrix} \Big|_0^t$$

$$= \begin{pmatrix} c_1 e^t \\ c_2 e^{2t} \end{pmatrix} + \begin{pmatrix} e^t & 0 \\ 0 & e^{2t} \end{pmatrix} \begin{pmatrix} -3e^{-t} + 3 \\ \frac{1}{2}e^{-2t} - \frac{1}{2} \end{pmatrix}$$

$$= \begin{pmatrix} c_1 e^t \\ c_2 e^{2t} \end{pmatrix} + \begin{pmatrix} -3 + 3e^t \\ \frac{1}{2} - \frac{1}{2}e^{2t} \end{pmatrix} = c_3 \begin{pmatrix} 1 \\ 0 \end{pmatrix} e^t + c_4 \begin{pmatrix} 0 \\ 1 \end{pmatrix} e^{2t} + \begin{pmatrix} -3 \\ \frac{1}{2} \end{pmatrix}.$$

12. To solve

$$\mathbf{X}' = \begin{pmatrix} 0 & 1 \\ 1 & 0 \end{pmatrix} \mathbf{X} + \begin{pmatrix} \cosh t \\ \sinh t \end{pmatrix}$$

we identify $t_0 = 0$, $\mathbf{F}(t) = \begin{pmatrix} \cosh t \\ \sinh t \end{pmatrix}$, and use the results of Problem 2 and equation (5) in the text.

$$\mathbf{X}(t) = e^{\mathbf{A}t}\mathbf{C} + e^{\mathbf{A}t} \int_{t_0}^{t} e^{-\mathbf{A}s}\mathbf{F}(s)\,ds$$

$$= \begin{pmatrix} \cosh t & \sinh t \\ \sinh t & \cosh t \end{pmatrix} \begin{pmatrix} c_1 \\ c_2 \end{pmatrix} + \begin{pmatrix} \cosh t & \sinh t \\ \sinh t & \cosh t \end{pmatrix} \int_0^t \begin{pmatrix} \cosh s & -\sinh s \\ -\sinh s & \cosh s \end{pmatrix} \begin{pmatrix} \cosh s \\ \sinh s \end{pmatrix} ds$$

$$= \begin{pmatrix} c_1 \cosh t + c_2 \sinh t \\ c_1 \sinh t + c_2 \cosh t \end{pmatrix} + \begin{pmatrix} \cosh t & \sinh t \\ \sinh t & \cosh t \end{pmatrix} \int_0^t \begin{pmatrix} 1 \\ 0 \end{pmatrix} ds$$

$$= \begin{pmatrix} c_1 \cosh t + c_2 \sinh t \\ c_1 \sinh t + c_2 \cosh t \end{pmatrix} + \begin{pmatrix} \cosh t & \sinh t \\ \sinh t & \cosh t \end{pmatrix} \begin{pmatrix} s \\ 0 \end{pmatrix} \Big|_0^t$$

$$= \begin{pmatrix} c_1 \cosh t + c_2 \sinh t \\ c_1 \sinh t + c_2 \cosh t \end{pmatrix} + \begin{pmatrix} \cosh t & \sinh t \\ \sinh t & \cosh t \end{pmatrix} \begin{pmatrix} t \\ 0 \end{pmatrix}$$

$$= \begin{pmatrix} c_1 \cosh t + c_2 \sinh t \\ c_1 \sinh t + c_2 \cosh t \end{pmatrix} + \begin{pmatrix} t \cosh t \\ t \sinh t \end{pmatrix} = c_1 \begin{pmatrix} \cosh t \\ \sinh t \end{pmatrix} + c_2 \begin{pmatrix} \sinh t \\ \cosh t \end{pmatrix} + t \begin{pmatrix} \cosh t \\ \sinh t \end{pmatrix}.$$

15. From $s\mathbf{I} - \mathbf{A} = \begin{pmatrix} s & 4 \\ 4 & s+4 \end{pmatrix}$ we find

$$(s\mathbf{I} - \mathbf{A})^{-1} = \begin{pmatrix} \dfrac{3/2}{s-2} - \dfrac{1/2}{s+2} & \dfrac{3/4}{s-2} - \dfrac{3/4}{s+2} \\[2mm] \dfrac{-1}{s-2} + \dfrac{1}{s+2} & \dfrac{-1/2}{s-2} + \dfrac{3/2}{s+2} \end{pmatrix}$$

and

$$e^{\mathbf{A}t} = \begin{pmatrix} \frac{3}{2} e^{2t} - \frac{1}{2} e^{-2t} & \frac{3}{4} e^{2t} - \frac{3}{4} e^{-2t} \\[2mm] -e^{2t} + e^{-2t} & -\frac{1}{2} e^{2t} + \frac{3}{2} e^{-2t} \end{pmatrix}.$$

The general solution of the system is then

$$\mathbf{X} = e^{\mathbf{A}t}\mathbf{C} = \begin{pmatrix} \frac{3}{2} e^{2t} - \frac{1}{2} e^{-2t} & \frac{3}{4} e^{2t} - \frac{3}{4} e^{-2t} \\[2mm] -e^{2t} + e^{-2t} & -\frac{1}{2} e^{2t} + \frac{3}{2} e^{-2t} \end{pmatrix} \begin{pmatrix} c_1 \\ c_2 \end{pmatrix}$$

$$= c_1 \begin{pmatrix} 3/2 \\ -1 \end{pmatrix} e^{2t} + c_1 \begin{pmatrix} -1/2 \\ 1 \end{pmatrix} e^{-2t} + c_2 \begin{pmatrix} 3/4 \\ -1/2 \end{pmatrix} e^{2t} + c_2 \begin{pmatrix} -3/4 \\ 3/2 \end{pmatrix} e^{-2t}$$

$$= \left(\frac{1}{2} c_1 + \frac{1}{4} c_2 \right) \begin{pmatrix} 3 \\ -2 \end{pmatrix} e^{2t} + \left(-\frac{1}{2} c_1 - \frac{3}{4} c_2 \right) \begin{pmatrix} 1 \\ -2 \end{pmatrix} e^{-2t}$$

$$= c_3 \begin{pmatrix} 3 \\ -2 \end{pmatrix} e^{2t} + c_4 \begin{pmatrix} 1 \\ -2 \end{pmatrix} e^{-2t}.$$

18. From $s\mathbf{I} - \mathbf{A} = \begin{pmatrix} s & -1 \\ 2 & s+2 \end{pmatrix}$ we find

$$(s\mathbf{I} - \mathbf{A})^{-1} = \begin{pmatrix} \dfrac{s+1+1}{(s+1)^2+1} & \dfrac{1}{(s+1)^2+1} \\ \dfrac{-2}{(s+1)^2+1} & \dfrac{s+1-1}{(s+1)^2+1} \end{pmatrix}$$

and

$$e^{\mathbf{A}t} = \begin{pmatrix} e^{-t}\cos t + e^{-t}\sin t & e^{-t}\sin t \\ -2e^{-t}\sin t & e^{-t}\cos t - e^{-t}\sin t \end{pmatrix}.$$

The general solution of the system is then

$$\mathbf{X} = e^{\mathbf{A}t}\mathbf{C} = \begin{pmatrix} e^{-t}\cos t + e^{-t}\sin t & e^{-t}\sin t \\ -2e^{-t}\sin t & e^{-t}\cos t - e^{-t}\sin t \end{pmatrix}\begin{pmatrix} c_1 \\ c_2 \end{pmatrix}$$

$$= c_1\begin{pmatrix} 1 \\ 0 \end{pmatrix}e^{-t}\cos t + c_1\begin{pmatrix} 1 \\ -2 \end{pmatrix}e^{-t}\sin t + c_2\begin{pmatrix} 0 \\ 1 \end{pmatrix}e^{-t}\cos t + c_2\begin{pmatrix} 1 \\ -1 \end{pmatrix}e^{-t}\sin t$$

$$= c_1\begin{pmatrix} \cos t + \sin t \\ -2\sin t \end{pmatrix}e^{-t} + c_2\begin{pmatrix} \sin t \\ \cos t - \sin t \end{pmatrix}e^{-t}.$$

21. From equation (3) in the text

$$e^{t\mathbf{A}} = e^{t\mathbf{PDP}^{-1}} = \mathbf{I} + t(\mathbf{PDP}^{-1}) + \frac{1}{2!}t^2(\mathbf{PDP}^{-1})^2 + \frac{1}{3!}t^3(\mathbf{PDP}^{-1})^3 + \cdots$$

$$= \mathbf{P}\left[\mathbf{I} + t\mathbf{D} + \frac{1}{2!}(t\mathbf{D})^2 + \frac{1}{3!}(t\mathbf{D})^3 + \cdots\right]\mathbf{P}^{-1} = \mathbf{P}e^{t\mathbf{D}}\mathbf{P}^{-1}.$$

24. From Problems 20-22 and equation (1) in the text

$$\mathbf{X} = e^{t\mathbf{A}}\mathbf{C} = \mathbf{P}e^{t\mathbf{D}}\mathbf{P}^{-1}\mathbf{C}$$

$$= \begin{pmatrix} -e^t & e^{3t} \\ e^t & e^{3t} \end{pmatrix}\begin{pmatrix} e^t & 0 \\ 0 & e^{3t} \end{pmatrix}\begin{pmatrix} -\frac{1}{2}e^{-t} & \frac{1}{2}e^{-t} \\ \frac{1}{2}e^{3t} & \frac{1}{2}e^{-3t} \end{pmatrix}\begin{pmatrix} c_1 \\ c_2 \end{pmatrix}$$

$$= \begin{pmatrix} \frac{1}{2}e^t + \frac{1}{2}e^{9t} & -\frac{1}{2}e^t + \frac{1}{2}e^{3t} \\ -\frac{1}{2}e^t + \frac{1}{2}e^{9t} & \frac{1}{2}e^t + \frac{1}{2}e^{3t} \end{pmatrix}\begin{pmatrix} c_1 \\ c_2 \end{pmatrix}.$$

27. (a) The following commands can be used in *Mathematica*:

```
A={{4, 2},{3, 3}};
c={c1, c2};
m=MatrixExp[A t];
sol=Expand[m.c]
Collect[sol, {c1, c2}]//MatrixForm
```

The output gives

$$x(t) = c_1 \left(\frac{2}{5} e^t + \frac{3}{5} e^{6t} \right) + c_2 \left(-\frac{2}{5} e^t + \frac{2}{5} e^{6t} \right)$$

$$y(t) = c_1 \left(-\frac{3}{5} e^t + \frac{3}{5} e^{6t} \right) + c_2 \left(\frac{3}{5} e^t + \frac{2}{5} e^{6t} \right).$$

The eigenvalues are 1 and 6 with corresponding eigenvectors

$$\begin{pmatrix} -2 \\ 3 \end{pmatrix} \quad \text{and} \quad \begin{pmatrix} 1 \\ 1 \end{pmatrix},$$

so the solution of the system is

$$\mathbf{X}(t) = b_1 \begin{pmatrix} -2 \\ 3 \end{pmatrix} e^t + b_2 \begin{pmatrix} 1 \\ 1 \end{pmatrix} e^{6t}$$

or

$$x(t) = -2b_1 e^t + b_2 e^{6t}$$

$$y(t) = 3b_1 e^t + b_2 e^{6t}.$$

If we replace b_1 with $-\frac{1}{5} c_1 + \frac{1}{5} c_2$ and b_2 with $\frac{3}{5} c_1 + \frac{2}{5} c_2$, we obtain the solution found using the matrix exponential.

(b) $x(t) = c_1 e^{-2t} \cos t - (c_1 + c_2) e^{-2t} \sin t$
$y(t) = c_2 e^{-2t} \cos t + (2c_1 + c_2) e^{-2t} \sin t$

8.R Chapter 8 in Review

3. Since

$$\begin{pmatrix} 4 & 6 & 6 \\ 1 & 3 & 2 \\ -1 & -4 & -3 \end{pmatrix} \begin{pmatrix} 3 \\ 1 \\ -1 \end{pmatrix} = \begin{pmatrix} 12 \\ 4 \\ -4 \end{pmatrix} = 4 \begin{pmatrix} 3 \\ 1 \\ -1 \end{pmatrix},$$

we see that $\lambda = 4$ is an eigenvalue with eigenvector $\mathbf{K}_3$. The corresponding solution is $\mathbf{X}_3 = \mathbf{K}_3 e^{4t}$.

6. We have $\det(\mathbf{A} - \lambda\mathbf{I}) = (\lambda + 6)(\lambda + 2) = 0$ so that

$$\mathbf{X} = c_1 \begin{pmatrix} 1 \\ -1 \end{pmatrix} e^{-6t} + c_2 \begin{pmatrix} 1 \\ 1 \end{pmatrix} e^{-2t}.$$

9. We have $\det(\mathbf{A} - \lambda\mathbf{I}) = -(\lambda - 2)(\lambda - 4)(\lambda + 3) = 0$ so that

$$\mathbf{X} = c_1 \begin{pmatrix} -2 \\ 3 \\ 1 \end{pmatrix} e^{2t} + c_2 \begin{pmatrix} 0 \\ 1 \\ 1 \end{pmatrix} e^{4t} + c_3 \begin{pmatrix} 7 \\ 12 \\ -16 \end{pmatrix} e^{-3t}.$$

12. We have

$$\mathbf{X}_c = c_1 \begin{pmatrix} 2\cos t \\ -\sin t \end{pmatrix} e^t + c_2 \begin{pmatrix} 2\sin t \\ \cos t \end{pmatrix} e^t.$$

Then

$$\mathbf{\Phi} = \begin{pmatrix} 2\cos t & 2\sin t \\ -\sin t & \cos t \end{pmatrix} e^t, \quad \mathbf{\Phi}^{-1} = \begin{pmatrix} \frac{1}{2}\cos t & -\sin t \\ \frac{1}{2}\sin t & \cos t \end{pmatrix} e^{-t},$$

and

$$\mathbf{U} = \int \mathbf{\Phi}^{-1}\mathbf{F}\,dt = \int \begin{pmatrix} \cos t - \sec t \\ \sin t \end{pmatrix} dt = \begin{pmatrix} \sin t - \ln|\sec t + \tan t| \\ -\cos t \end{pmatrix},$$

so that

$$\mathbf{X}_p = \mathbf{\Phi}\mathbf{U} = \begin{pmatrix} -2\cos t \ln|\sec t + \tan t| \\ -1 + \sin t \ln|\sec t + \tan t| \end{pmatrix} e^t.$$

15. (a) Letting

$$\mathbf{K} = \begin{pmatrix} k_1 \\ k_2 \\ k_3 \end{pmatrix}$$

we note that $(\mathbf{A} - 2\mathbf{I})\mathbf{K} = \mathbf{0}$ implies that $3k_1 + 3k_2 + 3k_3 = 0$, so $k_1 = -(k_2 + k_3)$. Choosing $k_2 = 0$, $k_3 = 1$ and then $k_2 = 1$, $k_3 = 0$ we get

$$\mathbf{K}_1 = \begin{pmatrix} -1 \\ 0 \\ 1 \end{pmatrix} \quad \text{and} \quad \mathbf{K}_2 = \begin{pmatrix} -1 \\ 1 \\ 0 \end{pmatrix},$$

respectively. Thus,

$$\mathbf{X}_1 = \begin{pmatrix} -1 \\ 0 \\ 1 \end{pmatrix} e^{2t} \quad \text{and} \quad \mathbf{X}_2 = \begin{pmatrix} -1 \\ 1 \\ 0 \end{pmatrix} e^{2t}$$

are two solutions.

(b) From $\det(\mathbf{A} - \lambda\mathbf{I}) = \lambda^2(3 - \lambda) = 0$ we see that $\lambda_1 = 3$, and 0 is an eigenvalue of multiplicity two. Letting

$$\mathbf{K} = \begin{pmatrix} k_1 \\ k_2 \\ k_3 \end{pmatrix},$$

as in part **(a)**, we note that $(\mathbf{A} - 0\mathbf{I})\mathbf{K} = \mathbf{AK} = \mathbf{0}$ implies that $k_1 + k_2 + k_3 = 0$, so $k_1 = -(k_2 + k_3)$. Choosing $k_2 = 0$, $k_3 = 1$, and then $k_2 = 1$, $k_3 = 0$ we get

$$\mathbf{K}_2 = \begin{pmatrix} -1 \\ 0 \\ 1 \end{pmatrix} \quad \text{and} \quad \mathbf{K}_3 = \begin{pmatrix} -1 \\ 1 \\ 0 \end{pmatrix},$$

respectively. Since the eigenvector corresponding to $\lambda_1 = 3$ is

$$\mathbf{K}_1 = \begin{pmatrix} 1 \\ 1 \\ 1 \end{pmatrix},$$

the general solution of the system is

$$\mathbf{X} = c_1 \begin{pmatrix} 1 \\ 1 \\ 1 \end{pmatrix} e^{3t} + c_2 \begin{pmatrix} -1 \\ 0 \\ 1 \end{pmatrix} + c_3 \begin{pmatrix} -1 \\ 1 \\ 0 \end{pmatrix}.$$

9 Numerical Solutions of Ordinary Differential Equations

9.1 Euler Methods and Error Analysis

The terminology and concepts listed below provide an outline of the main ideas encountered in this section. These can be useful when preparing for a quiz or test.

Terminology and Concepts

- error analysis for numerical solutions of ordinary DEs

- round-off error

- local and global truncation errors

- bound for local truncation error

- Euler's method for approximating the solution of a first-order IVP

- improved Euler's method for approximating the solution of a first-order IVP

- a predictor-corrector method

The basic skills listed below summarize the more mechanical types of problems encountered in the exercise set for this section.

Basic Skills

- use the improved Euler's method to approximate the solution of a first-order IVP

- find a bound on the local truncation error when Euler's method or the improved Euler's method is used to approximate the solution of an IVP

233

3. h=0.1   h=0.05

x_n	y_n
0.00	0.0000
0.10	0.1005
0.20	0.2030
0.30	0.3098
0.40	0.4234
0.50	0.5470

x_n	y_n
0.00	0.0000
0.05	0.0501
0.10	0.1004
0.15	0.1512
0.20	0.2028
0.25	0.2554
0.30	0.3095
0.35	0.3652
0.40	0.4230
0.45	0.4832
0.50	0.5465

6. h=0.1   h=0.05

x_n	y_n
0.00	0.0000
0.10	0.0050
0.20	0.0200
0.30	0.0451
0.40	0.0805
0.50	0.1266

x_n	y_n
0.00	0.0000
0.05	0.0013
0.10	0.0050
0.15	0.0113
0.20	0.0200
0.25	0.0313
0.30	0.0451
0.35	0.0615
0.40	0.0805
0.45	0.1022
0.50	0.1266

9. h=0.1   h=0.05

x_n	y_n
1.00	1.0000
1.10	1.0095
1.20	1.0404
1.30	1.0967
1.40	1.1866
1.50	1.3260

x_n	y_n
1.00	1.0000
1.05	1.0024
1.10	1.0100
1.15	1.0228
1.20	1.0414
1.25	1.0663
1.30	1.0984
1.35	1.1389
1.40	1.1895
1.45	1.2526
1.50	1.3315

12. (a)

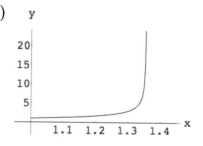

(b)

x_n	Euler	Imp. Euler
1.00	1.0000	1.0000
1.10	1.2000	1.2469
1.20	1.4938	1.6430
1.30	1.9711	2.4042
1.40	2.9060	4.5085

15. (a) Using Euler's method we obtain $y(0.1) \approx y_1 = 0.8$.

(b) Using $y'' = 5e^{-2x}$ we see that the local truncation error is

$$5e^{-2c}\frac{(0.1)^2}{2} = 0.025e^{-2c}.$$

Since e^{-2x} is a decreasing function, $e^{-2c} \leq e^0 = 1$ for $0 \leq c \leq 0.1$. Thus an upper bound for the local truncation error is $0.025(1) = 0.025$.

(c) Since $y(0.1) = 0.8234$, the actual error is $y(0.1) - y_1 = 0.0234$, which is less than 0.025.

(d) Using Euler's method with $h = 0.05$ we obtain $y(0.1) \approx y_2 = 0.8125$.

(e) The error in (d) is $0.8234 - 0.8125 = 0.0109$. With global truncation error $O(h)$, when the step size is halved we expect the error for $h = 0.05$ to be one-half the error when $h = 0.1$. Comparing 0.0109 with 0.0234 we see that this is the case.

18. (a) Using $y''' = -114e^{-3(x-1)}$ we see that the local truncation error is

$$\left| y'''(c) \frac{h^3}{6} \right| = 114e^{-3(x-1)} \frac{h^3}{6} = 19h^3 e^{-3(c-1)}.$$

(b) Since $e^{-3(x-1)}$ is a decreasing function for $1 \le x \le 1.5$, $e^{-3(c-1)} \le e^{-3(1-1)} = 1$ for $1 \le c \le 1.5$ and

$$\left| y'''(c) \frac{h^3}{6} \right| \le 19(0.1)^3 (1) = 0.019.$$

(c) Using the improved Euler's method with $h = 0.1$ we obtain $y(1.5) \approx 2.080108$. With $h = 0.05$ we obtain $y(1.5) \approx 2.059166$.

(d) Since $y(1.5) = 2.053216$, the error for $h = 0.1$ is $E_{0.1} = 0.026892$, while the error for $h = 0.05$ is $E_{0.05} = 0.005950$. With global truncation error $O(h^2)$ we expect $E_{0.1}/E_{0.05} \approx 4$. We actually have $E_{0.1}/E_{0.05} = 4.52$.

21. Because y_{n+1}^* depends on y_n and is used to determine y_{n+1}, all of the y_n^* cannot be computed at one time independently of the corresponding y_n values. For example, the computation of y_4^* involves the value of y_3.

9.2 Runge-Kutta Methods

The terminology and concepts listed below provide an outline of the main ideas encountered in this section. These can be useful when preparing for a quiz or test.

Terminology and Concepts

- Runge-Kutta methods

- fourth-order Runge-Kutta method (RK4 method)

- truncation errors for the RK4 method

The basic skills listed below summarize the more mechanical types of problems encountered in the exercise set for this section.

Basic Skills

- use the RK4 method to approximate the solution of a first-order initial-value problem

3.

x_n	y_n
1.00	5.0000
1.10	3.9724
1.20	3.2284
1.30	2.6945
1.40	2.3163
1.50	2.0533

6.

x_n	y_n
0.00	1.0000
0.10	1.1115
0.20	1.2530
0.30	1.4397
0.40	1.6961
0.50	2.0670

9.

x_n	y_n
0.00	0.5000
0.10	0.5213
0.20	0.5358
0.30	0.5443
0.40	0.5482
0.50	0.5493

12.

x_n	y_n
0.00	0.5000
0.10	0.5250
0.20	0.5498
0.30	0.5744
0.40	0.5987
0.50	0.6225

15. (a)

x_n	$h=0.05$	$h=0.1$
1.00	1.0000	1.0000
1.05	1.1112	
1.10	1.2511	1.2511
1.15	1.4348	
1.20	1.6934	1.6934
1.25	2.1047	
1.30	2.9560	2.9425
1.35	7.8981	
1.40	1.0608×10^{15}	903.0282

(b)

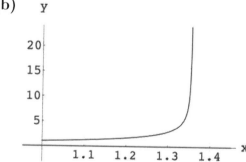

18. (a) Using $y^{(5)} = -1026e^{-3(x-1)}$ we see that the local truncation error is

$$\left| y^{(5)}(c) \frac{h^5}{120} \right| = 8.55 h^5 e^{-3(c-1)}.$$

(b) Since $e^{-3(x-1)}$ is a decreasing function for $1 \le x \le 1.5$, $e^{-3(c-1)} \le e^{-3(1-1)} = 1$ for $1 \le c \le 1.5$ and

$$y^{(5)}(c) \frac{h^5}{120} \le 8.55(0.1)^5(1) = 0.0000855.$$

(c) Using the RK4 method with $h = 0.1$ we obtain $y(1.5) \approx 2.053338827$. With $h = 0.05$ we obtain $y(1.5) \approx 2.053222989$.

21. (a) For $y' + y = 10 \sin 3x$ an integrating factor is e^x so that

$$\frac{d}{dx}[e^x y] = 10 e^x \sin 3x \implies e^x y = e^x \sin 3x - 3 e^x \cos 3x + c$$

$$\implies y = \sin 3x - 3 \cos 3x + ce^{-x}.$$

When $x = 0$, $y = 0$, so $0 = -3 + c$ and $c = 3$. The solution is

$$y = \sin 3x - 3 \cos 3x + 3e^{-x}.$$

Using Newton's method we find that $x = 1.53235$ is the only positive root in $[0, 2]$.

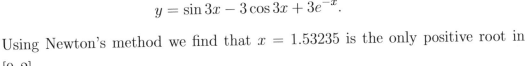

(b) Using the RK4 method with $h = 0.1$ we obtain the table of values shown. These values are used to obtain an interpolating function in *Mathematica*. The graph of the interpolating function is shown. Using *Mathematica*'s root finding capability we see that the only positive root in $[0, 2]$ is $x = 1.53236$.

x_n	y_n	x_n	y_n
0.0	0.0000	1.0	4.2147
0.1	0.1440	1.1	3.8033
0.2	0.5448	1.2	3.1513
0.3	1.1409	1.3	2.3076
0.4	1.8559	1.4	1.3390
0.5	2.6049	1.5	0.3243
0.6	3.3019	1.6	-0.6530
0.7	3.8675	1.7	-1.5117
0.8	4.2356	1.8	-2.1809
0.9	4.3593	1.9	-2.6061
1.0	4.2147	2.0	-2.7539

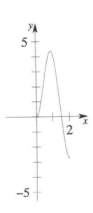

9.3 | Multistep Methods

The terminology and concepts listed below provide an outline of the main ideas encountered in this section. These can be useful when preparing for a quiz or test.

Terminology and Concepts

- predictor-corrector methods
- Adams-Bashforth-Moulton method

The basic skills listed below summarize the more mechanical types of problems encountered in the exercise set for this section.

Basic Skills

- use the Adams-Bashforth-Moulton predictor-corrector method to approximate the solution of a first-order IVP

In the tables in this section "ABM" stands for Adams-Bashforth-Moulton.

3. The first predictor is $y_4^* = 0.73318477$.

x_n	y_n	
0.0	1.00000000	init. cond.
0.2	0.73280000	RK4
0.4	0.64608032	RK4
0.6	0.65851653	RK4
0.8	0.72319464	ABM

6. The first predictor for $h = 0.2$ is $y_4^* = 3.34828434$.

x_n	h=0.2		h=0.1	
0.0	1.00000000	init. cond.	1.00000000	init. cond.
0.1			1.21017082	RK4
0.2	1.44139950	RK4	1.44140511	RK4
0.3			1.69487942	RK4
0.4	1.97190167	RK4	1.97191536	ABM
0.5			2.27400341	ABM
0.6	2.60280694	RK4	2.60283209	ABM
0.7			2.96031780	ABM
0.8	3.34860927	ABM	3.34863769	ABM
0.9			3.77026548	ABM
1.0	4.22797875	ABM	4.22801028	ABM

9.4 | Higher-Order Equations and Systems

The terminology and concepts listed below provide an outline of the main ideas encountered in this section. These can be useful when preparing for a quiz or test.

Terminology and Concepts

- express a second-order IVP as an IVP for a first-order system
- express a higher-order IVP as an IVP for a first-order system
- numerical solution of an IVP for a first-order system using Euler's method and the RK4 method

The basic skills listed below summarize the more mechanical types of problems encountered in the exercise set for this section.

Basic Skills

- solve an IVP for a first-order system consisting of two equations

3. The substitution $y' = u$ leads to the system

$$y' = u, \qquad u' = 4u - 4y.$$

Using formula (4) in the text with x corresponding to t, y corresponding to x, and u corresponding to y, we obtain the table shown.

x_n	h=0.2 y_n	h=0.2 u_n	h=0.1 y_n	h=0.1 u_n
0.0	-2.0000	1.0000	-2.0000	1.0000
0.1			-1.8321	2.4427
0.2	-1.4928	4.4731	-1.4919	4.4753

6. Using $h = 0.1$, the RK4 method for a system, and a numerical solver, we obtain

t_n	h=0.2 i_{1n}	h=0.2 i_{3n}
0.0	0.0000	0.0000
0.1	2.5000	3.7500
0.2	2.8125	5.7813
0.3	2.0703	7.4023
0.4	0.6104	9.1919
0.5	-1.5619	11.4877

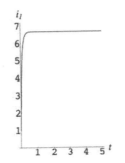
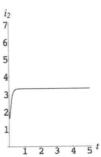

9.

t_n	h=0.2 x_n	h=0.2 y_n	h=0.1 x_n	h=0.1 y_n
0.0	-3.0000	5.0000	-3.0000	5.0000
0.1			-3.4790	4.6707
0.2	-3.9123	4.2857	-3.9123	4.2857

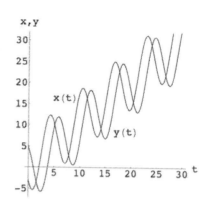

12. Solving for x' and y' we obtain the system

$$x' = \frac{1}{2}y - 3t^2 + 2t - 5$$

$$y' = -\frac{1}{2}y + 3t^2 + 2t + 5.$$

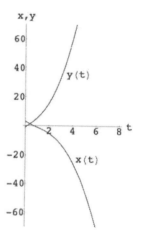

t_n	h=0.2 x_n	h=0.2 y_n	h=0.1 x_n	h=0.1 y_n
0.0	3.0000	-1.0000	3.0000	-1.0000
0.1			2.4727	-0.4527
0.2	1.9867	0.0933	1.9867	0.0933

9.5 | Second-Order Boundary-Value Problems

The terminology and concepts listed below provide an outline of the main ideas encountered in this section. These can be useful when preparing for a quiz or test.

Terminology and Concepts

- difference quotient

- finite difference

- forward difference

- backward difference

- central difference

- finite difference method

The basic skills listed below summarize the more mechanical types of problems encountered in the exercise set for this section.

Basic Skills

- use the finite difference method to approximate the solution of a second-order boundary-value problem

3. We identify $P(x) = 2$, $Q(x) = 1$, $f(x) = 5x$, and $h = (1 - 0)/5 = 0.2$. Then the finite difference equation is

$$1.2y_{i+1} - 1.96y_i + 0.8y_{i-1} = 0.04(5x_i).$$

The solution of the corresponding linear system gives

x	0.0	0.2	0.4	0.6	0.8	1.0
y	0.0000	−0.2259	−0.3356	−0.3308	−0.2167	0.0000

6. We identify $P(x) = 5$, $Q(x) = 0$, $f(x) = 4\sqrt{x}$, and $h = (2 - 1)/6 = 0.1667$. Then the finite difference equation is

$$1.4167y_{i+1} - 2y_i + 0.5833y_{i-1} = 0.2778(4\sqrt{x_i}).$$

The solution of the corresponding linear system gives

x	1.0000	1.1667	1.3333	1.5000	1.6667	1.8333	2.0000
y	1.0000	−0.5918	−1.1626	−1.3070	−1.2704	−1.1541	−1.0000

9. We identify $P(x) = 1 - x$, $Q(x) = x$, $f(x) = x$, and $h = (1 - 0)/10 = 0.1$. Then the finite difference equation is

$$[1 + 0.05(1 - x_i)]y_{i+1} + [-2 + 0.01x_i]y_i + [1 - 0.05(1 - x_i)]y_{i-1} = 0.01x_i.$$

The solution of the corresponding linear system gives

x	0.0	0.1	0.2	0.3	0.4	0.5	0.6
y	0.0000	0.2660	0.5097	0.7357	0.9471	1.1465	1.3353

0.7	0.8	0.9	1.0
1.5149	1.6855	1.8474	2.0000

12. We identify $P(r) = 2/r$, $Q(r) = 0$, $f(r) = 0$, and $h = (4-1)/6 = 0.5$. Then the finite difference equation is

$$\left(1 + \frac{0.5}{r_i}\right) u_{i+1} - 2u_i + \left(1 - \frac{0.5}{r_i}\right) u_{i-1} = 0.$$

The solution of the corresponding linear system gives

r	1.0	1.5	2.0	2.5	3.0	3.5	4.0
u	50.0000	72.2222	83.3333	90.0000	94.4444	97.6190	100.0000

9.R   Chapter 9 in Review

3.

x_n	Euler h=0.1	Euler h=0.05	Imp. Euler h=0.1	Imp. Euler h=0.05	RK4 h=0.1	RK4 h=0.05
0.50	0.5000	0.5000	0.5000	0.5000	0.5000	0.5000
0.55		0.5500		0.5512		0.5512
0.60	0.6000	0.6024	0.6048	0.6049	0.6049	0.6049
0.65		0.6573		0.6609		0.6610
0.70	0.7095	0.7144	0.7191	0.7193	0.7194	0.7194
0.75		0.7739		0.7800		0.7801
0.80	0.8283	0.8356	0.8427	0.8430	0.8431	0.8431
0.85		0.8996		0.9082		0.9083
0.90	0.9559	0.9657	0.9752	0.9755	0.9757	0.9757
0.95		1.0340		1.0451		1.0452
1.00	1.0921	1.1044	1.1163	1.1168	1.1169	1.1169

6. The first predictor is $y_3^* = 1.14822731$.

x_n	y_n	
0.0	2.00000000	init. cond.
0.1	1.65620000	RK4
0.2	1.41097281	RK4
0.3	1.24645047	RK4
0.4	1.14796764	ABM

10 Plane Autonomous Systems

10.1 Autonomous Systems

The terminology and concepts listed below provide an outline of the main ideas encountered in this section. These can be useful when preparing for a quiz or test.

Terminology and Concepts

- plane autonomous system of first-order DEs
- dynamical system of DEs
- state (response) of the system
- vector field
- critical (stationary) point
- equilibrium solution

The basic skills listed below summarize the more mechanical types of problems encountered in the exercise set for this section.

Basic Skills

- express a nonlinear second-order DE as a plane autonomous system
- find the critical points of a plane autonomous system
- find the general solution of a linear system of DEs with constant coefficients
- express a nonlinear plane autonomous system using polar coordinates
- determine if a linear system of DEs has periodic solutions

3. The corresponding plane autonomous system is

$$x' = y, \quad y' = x^2 - y(1 - x^3).$$

If (x, y) is a critical point, $y = 0$ and so $x^2 - y(1 - x^3) = x^2 = 0$. Therefore $(0, 0)$ is the sole critical point.

6. The corresponding plane autonomous system is

$$x' = y, \quad y' = -x + \epsilon x |x|.$$

If (x, y) is a critical point, $y = 0$ and $-x + \epsilon x|x| = x(-1 + \epsilon|x|) = 0$. Hence $x = 0$, $1/\epsilon$, $-1/\epsilon$. The critical points are $(0, 0)$, $(1/\epsilon, 0)$ and $(-1/\epsilon, 0)$.

9. From $x - y = 0$ we have $y = x$. Substituting into $3x^2 - 4y = 0$ we obtain $3x^2 - 4x = x(3x - 4) = 0$. It follows that $(0, 0)$ and $(4/3, 4/3)$ are the critical points of the system.

12. Adding the two equations we obtain $10 - 15y/(y + 5) = 0$. It follows that $y = 10$, and from $-2x + y + 10 = 0$ we can conclude that $x = 10$. Therefore $(10, 10)$ is the sole critical point of the system.

15. From $x(1 - x^2 - 3y^2) = 0$ we have $x = 0$ or $x^2 + 3y^2 = 1$. If $x = 0$, then substituting into $y(3 - x^2 - 3y^2)$ gives $y(3 - 3y^2) = 0$. Therefore $y = 0$, 1, -1. Likewise $x^2 = 1 - 3y^2$ yields $2y = 0$ so that $y = 0$ and $x^2 = 1 - 3(0)^2 = 1$. The critical points of the system are therefore $(0, 0)$, $(0, 1)$, $(0, -1)$, $(1, 0)$, and $(-1, 0)$.

18. (a) From Exercises 8.2, Problem 6, $x = c_1 + 2c_2 e^{-5t}$ and $y = 3c_1 + c_2 e^{-5t}$, which is not periodic.

(b) From $\mathbf{X}(0) = (3, 4)$ it follows that $c_1 = c_2 = 1$. Therefore $x = 1 + 2e^{-5t}$ and $y = 3 + e^{-5t}$ gives $y = \frac{1}{2}(x - 1) + 3$.

(c)

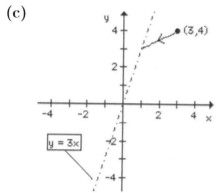

21. (a) From Exercises 8.2, Problem 35, $x = c_1(\sin t - \cos t)e^{4t} + c_2(-\sin t - \cos t)e^{4t}$ and $y = 2c_1(\cos t)e^{4t} + 2c_2(\sin t)e^{4t}$. Because of the presence of e^{4t}, there are no periodic solutions.

(b) From $\mathbf{X}(0) = (-1, 2)$ it follows that $c_1 = 1$ and $c_2 = 0$. Therefore $x = (\sin t - \cos t)e^{4t}$ and $y = 2(\cos t)e^{4t}$.

(c)

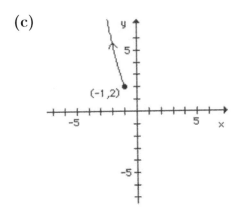

24. Switching to polar coordinates,

$$\frac{dr}{dt} = \frac{1}{r}\left(x\frac{dx}{dt} + y\frac{dy}{dt}\right) = \frac{1}{r}(xy - x^2r^2 - xy + y^2r^2) = r^3$$

$$\frac{d\theta}{dt} = \frac{1}{r^2}\left(-y\frac{dx}{dt} + x\frac{dy}{dt}\right) = \frac{1}{r^2}(-y^2 - xyr^2 - x^2 + xyr^2) = -1.$$

If we use separation of variables, it follows that

$$r = \frac{1}{\sqrt{-2t + c_1}} \quad \text{and} \quad \theta = -t + c_2.$$

Since $\mathbf{X}(0) = (4, 0)$, $r = 4$ and $\theta = 0$ when $t = 0$. It follows that $c_2 = 0$ and $c_1 = \frac{1}{16}$. The final solution can be written as

$$r = \frac{4}{\sqrt{1 - 32t}}, \qquad \theta = -t.$$

Note that $r \to \infty$ as $t \to \left(\frac{1}{32}\right)$. Because $0 \le t \le \frac{1}{32}$, the curve is not a spiral.

27. The system has no critical points, so there are no periodic solutions.

30. The system has no critical points, so there are no periodic solutions.

10.2 | Stability of Linear Systems

The terminology and concepts listed below provide an outline of the main ideas encountered in this section. These can be useful when preparing for a quiz or test.

Terminology and Concepts

- locally stable critical point
- unstable cricital point
- phase plane
- phase portrait

- eigenvalues and eigenvectors of the coefficient matrix of a linear first-order system of DEs
- stable and unstable modes
- saddle point
- degenerate stable and unstable modes
- center
- stable and unstable spiral points

The basic skills listed below summarize the more mechanical types of problems encountered in the exercise set for this section.

Basic Skills

- given the solution of a linear system $\mathbf{X'} = \mathbf{AX}$ where $\mathbf{A}$ is a 2×2 matrix, discuss the nature of the solutions in a neighborhood of $(0,0)$
- classify the critical point $(0,0)$ of a linear system

$$x' = ax + by$$

$$y' = cx + dy$$

3. (a) All solutions are unstable spirals which become unbounded as t increases.

(b)

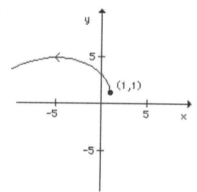

6. (a) All solutions become unbounded and $y = x/2$ serves as the asymptote.

(b)

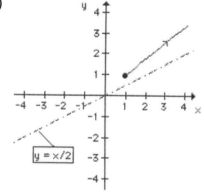

9. Since $\Delta = -41 < 0$, we can conclude from Figure 10.2.12 that $(0,0)$ is a saddle point.

12. Since $\Delta = 1$ and $\tau = -1$, $\tau^2 - 4\Delta = -3$ and so from Figure 10.2.12, $(0,0)$ is a stable spiral point.

15. Since $\Delta = 0.01$ and $\tau = -0.03$, $\tau^2 - 4\Delta < 0$ and so from Figure 10.2.12, $(0,0)$ is a stable spiral point.

18. Note that $\Delta = 1$ and $\tau = \mu$. Therefore we need both $\tau = \mu < 0$ and $\tau^2 - 4\Delta = \mu^2 - 4 < 0$ for $(0,0)$ to be a stable spiral point. These two conditions can be written as $-2 < \mu < 0$.

21. $\mathbf{A}\mathbf{X}_1 + \mathbf{F} = \mathbf{0}$ implies that $\mathbf{A}\mathbf{X}_1 = -\mathbf{F}$ or $\mathbf{X}_1 = -\mathbf{A}^{-1}\mathbf{F}$. Since $\mathbf{X}_p(t) = -\mathbf{A}^{-1}\mathbf{F}$ is a particular solution, it follows from Theorem 8.1.6 that $\mathbf{X}(t) = \mathbf{X}_c(t) + \mathbf{X}_1$ is the general solution to $\mathbf{X}' = \mathbf{A}\mathbf{X} + \mathbf{F}$. If $\tau < 0$ and $\Delta > 0$ then $\mathbf{X}_c(t)$ approaches $(0,0)$ by Theorem 10.1(a). It follows that $\mathbf{X}(t)$ approaches $\mathbf{X}_1$ as $t \to \infty$.

24. **(a)** The critical point is $\mathbf{X}_1 = (-1, -2)$.

(b) From the graph, $\mathbf{X}_1$ appears to be a stable node or a degenerate stable node.

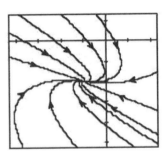

(c) Since $\tau = -16$, $\Delta = 64$, and $\tau^2 - 4\Delta = 0$, $(0,0)$ is a degenerate stable node.

10.3 | Linearization and Local Stability

The terminology and concepts listed below provide an outline of the main ideas encountered in this section. These can be useful when preparing for a quiz or test.

Terminology and Concepts

- stable critical point
- asymptotically stable critical point
- unstable critical point
- linearization of a nonlinear system
- Jacobian matrix
- classification of critical points
- phase-plane method

The basic skills listed below summarize the more mechanical types of problems encountered in the exercise set for this section.

Basic Skills

- find critical points and classify each critical point of a first-order autonomous DE as either asymptotically stable or unstable

- find critical points and classify each critical point of a first-order system of two DEs as a stable node, a stable spiral point, an unstable spiral point, an unstable node, or a saddle point

- find critical points and classify each critical point of a second-order DE as a stable node, a stable spiral point, an unstable spiral point, an unstable node, or a saddle point.

3. The critical points are $x = 0$ and $x = n + 1$. Since $g'(x) = k(n+1) - 2kx$, $g'(0) = k(n+1) > 0$ and $g'(n+1) = -k(n+1) < 0$. Therefore $x = 0$ is unstable while $x = n+1$ is asymptotically stable. See Theorem 10.2.

6. The only critical point is $v = mg/k$. Now $g(v) = g - (k/m)v$ and so $g'(v) = -k/m < 0$. Therefore $v = mg/k$ is an asymptotically stable critical point by Theorem 10.3.1.

9. Critical points occur at $P = a/b$, c but not at $P = 0$. Since $g'(P) = (a - bP) + (P - c)(-b)$,

$$g'(a/b) = (a/b - c)(-b) = -a + bc \quad \text{and} \quad g'(c) = a - bc.$$

Since $a < bc$, $-a + bc > 0$ and $a - bc < 0$. Therefore $P = a/b$ is unstable while $P = c$ is asymptotically stable.

12. Critical points are $(1, 0)$ and $(-1, 0)$, and

$$\mathbf{g}'(\mathbf{X}) = \begin{pmatrix} 2x & -2y \\ 0 & 2 \end{pmatrix}.$$

At $\mathbf{X} = (1, 0)$, $\tau = 4$, $\Delta = 4$, and so $\tau^2 - 4\Delta = 0$. We can conclude that $(1, 0)$ is unstable but we are unable to classify this critical point any further. At $\mathbf{X} = (-1, 0)$, $\Delta = -4 < 0$ and so $(-1, 0)$ is a saddle point.

15. Since $x^2 - y^2 = 0$, $y^2 = x^2$ and so $x^2 - 3x + 2 = (x - 1)(x - 2) = 0$. It follows that the critical points are $(1, 1)$, $(1, -1)$, $(2, 2)$, and $(2, -2)$. We next use the Jacobian

$$\mathbf{g}'(\mathbf{X}) = \begin{pmatrix} -3 & 2y \\ 2x & -2y \end{pmatrix}$$

to classify these four critical points. For $\mathbf{X} = (1,1)$, $\tau = -5$, $\Delta = 2$, and so $\tau^2 - 4\Delta = 17 > 0$. Therefore $(1,1)$ is a stable node. For $\mathbf{X} = (1,-1)$, $\Delta = -2 < 0$ and so $(1,-1)$ is a saddle point. For $\mathbf{X} = (2,2)$, $\Delta = -4 < 0$ and so we have another saddle point. Finally, if $\mathbf{X} = (2,-2)$, $\tau = 1$ $\Delta = 4$, and so $\tau^2 - 4\Delta = -15 < 0$. Therefore $(2,-2)$ is an unstable spiral point.

18. We found that $(0,0)$, $(0,1)$, $(0,-1)$, $(1,0)$ and $(-1,0)$ were the critical points in Exercise 15, Section 10.1. The Jacobian is

$$\mathbf{g}'(\mathbf{X}) = \begin{pmatrix} 1 - 3x^2 - 3y^2 & -6xy \\ -2xy & 3 - x^2 - 9y^2 \end{pmatrix}.$$

For $\mathbf{X} = (0,0)$, $\tau = 4$, $\Delta = 3$ and so $\tau^2 - 4\Delta = 4 > 0$. Therefore $(0,0)$ is an unstable node. Both $(0,1)$ and $(0,-1)$ give $\tau = -8$, $\Delta = 12$, and $\tau^2 - 4\Delta = 16 > 0$. These two critical points are therefore stable nodes. For $\mathbf{X} = (1,0)$ or $(-1,0)$, $\Delta = -4 < 0$ and so saddle points occur.

21. The corresponding plane autonomous system is

$$\theta' = y, \quad y' = \left(\cos\theta - \frac{1}{2}\right)\sin\theta.$$

Since $|\theta| < \pi$, it follows that critical points are $(0,0)$, $(\pi/3, 0)$ and $(-\pi/3, 0)$. The Jacobian matrix is

$$\mathbf{g}'(\mathbf{X}) = \begin{pmatrix} 0 & 1 \\ \cos 2\theta - \frac{1}{2}\cos\theta & 0 \end{pmatrix}$$

and so at $(0,0)$, $\tau = 0$ and $\Delta = -1/2$. Therefore $(0,0)$ is a saddle point. For $\mathbf{X} = (\pm\pi/3, 0)$, $\tau = 0$ and $\Delta = 3/4$. It is not possible to classify either critical point in this borderline case.

24. The corresponding plane autonomous system is

$$x' = y, \quad y' = -\frac{4x}{1+x^2} - 2y$$

and the only critical point is $(0,0)$. Since the Jacobian matrix is

$$\mathbf{g}'(\mathbf{X}) = \begin{pmatrix} 0 & 1 \\ -4\dfrac{1-x^2}{(1+x^2)^2} & -2 \end{pmatrix},$$

$\tau = -2$, $\Delta = 4$, $\tau^2 - 4\Delta = -12$, and so $(0,0)$ is a stable spiral point.

27. The corresponding plane autonomous system is

$$x' = y, \quad y' = -\frac{(\beta + \alpha^2 y^2)x}{1 + \alpha^2 x^2}$$

and the Jacobian matrix is

$$\mathbf{g}'(\mathbf{X}) = \begin{pmatrix} 0 & 1 \\ \dfrac{(\beta + \alpha y^2)(\alpha^2 x^2 - 1)}{(1 + \alpha^2 x^2)^2} & \dfrac{-2\alpha^2 yx}{1 + \alpha^2 x^2} \end{pmatrix}.$$

For $\mathbf{X} = (0,0)$, $\tau = 0$ and $\Delta = \beta$. Since $\beta < 0$, we can conclude that $(0,0)$ is a saddle point.

30. (a) The corresponding plane autonomous system is

$$x' = y, \quad y' = \epsilon\left(y - \frac{1}{3}y^3\right) - x$$

and so the only critical point is $(0,0)$. Since the Jacobian matrix is

$$\mathbf{g}'(\mathbf{X}) = \begin{pmatrix} 0 & 1 \\ -1 & \epsilon(1 - y^2) \end{pmatrix},$$

$\tau = \epsilon$, $\Delta = 1$, and so $\tau^2 - 4\Delta = \epsilon^2 - 4$ at the critical point $(0,0)$.

(b) When $\tau = \epsilon > 0$, $(0,0)$ is an unstable critical point.

(c) When $\epsilon < 0$ and $\tau^2 - 4\Delta = \epsilon^2 - 4 < 0$, $(0,0)$ is a stable spiral point. These two requirements can be written as $-2 < \epsilon < 0$.

(d) When $\epsilon = 0$, $x'' + x = 0$ and so $x = c_1 \cos t + c_2 \sin t$. Therefore all solutions are periodic (with period 2π) and so $(0,0)$ is a center.

33. (a) $x' = 2xy = 0$ implies that either $x = 0$ or $y = 0$. If $x = 0$, then from $1 - x^2 + y^2 = 0$, $y^2 = -1$ and there are no real solutions. If $y = 0$, $1 - x^2 = 0$ and so $(1,0)$ and $(-1,0)$ are critical points. The Jacobian matrix is

$$\mathbf{g}'(\mathbf{X}) = \begin{pmatrix} 2y & 2x \\ -2x & 2y \end{pmatrix}$$

and so $\tau = 0$ and $\Delta = 4$ at either $\mathbf{X} = (1,0)$ or $(-1,0)$. We obtain no information about these critical points in this borderline case.

(b) The differential equation is

$$\frac{dy}{dx} = \frac{y'}{x'} = \frac{1 - x^2 + y^2}{2xy}$$

or

$$2xy\frac{dy}{dx} = 1 - x^2 + y^2.$$

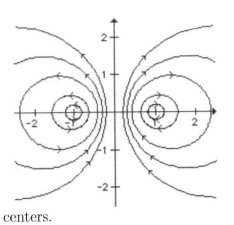

Letting $\mu = y^2/x$, it follows that $d\mu/dx = (1/x^2) - 1$ and so $\mu = -(1/x) - x + 2c$. Therefore $y^2/x = -(1/x) - x + 2c$ which can be put in the form $(x - c)^2 + y^2 = c^2 - 1$. The solution curves are shown and so both $(1, 0)$ and $(-1, 0)$ are centers.

36. The corresponding plane autonomous system is

$$x' = y, \quad y' = \epsilon x^2 - x + 1$$

and so the critical points must satisfy $y = 0$ and

$$x = \frac{1 \pm \sqrt{1 - 4\epsilon}}{2\epsilon}.$$

Therefore we must require that $\epsilon \le \frac{1}{4}$ for real solutions to exist. We will use the Jacobian matrix

$$g'(\mathbf{X}) = \begin{pmatrix} 0 & 1 \\ 2\epsilon x - 1 & 0 \end{pmatrix}$$

to attempt to classify $((1 \pm \sqrt{1 - 4\epsilon})/2\epsilon, 0)$ when $\epsilon \le 1/4$. Note that $\tau = 0$ and $\Delta = \mp\sqrt{1 - 4\epsilon}$. For $\mathbf{X} = ((1 + \sqrt{1 - 4\epsilon})/2\epsilon, 0)$ and $\epsilon < 1/4$, $\Delta < 0$ and so a saddle point occurs. For $\mathbf{X} = ((1 - \sqrt{1 - 4\epsilon})/2\epsilon, 0)$, $\Delta \ge 0$ and we are not able to classify this critical point using linearization.

39. (a) Letting $x = \theta$ and $y = x'$ we obtain the system $x' = y$ and $y' = 1/2 - \sin x$. Since $\sin \pi/6 = \sin 5\pi/6 = 1/2$ we see that $(\pi/6, 0)$ and $(5\pi/6, 0)$ are critical points of the system.

(b) The Jacobian matrix is

$$g'(\mathbf{X}) = \begin{pmatrix} 0 & 1 \\ -\cos x & 0 \end{pmatrix}$$

and so

$$\mathbf{A}_1 = g' = ((\pi/6, 0)) = \begin{pmatrix} 0 & 1 \\ -\sqrt{3}/2 & 0 \end{pmatrix} \quad \text{and} \quad \mathbf{A}_2 = g' = ((5\pi/6, 0)) = \begin{pmatrix} 0 & 1 \\ \sqrt{3}/2 & 0 \end{pmatrix}.$$

Since $\det \mathbf{A}_1 > 0$ and the trace of $\mathbf{A}_1$ is 0, no conclusion can be drawn regarding the critical point $(\pi/6, 0)$. Since $\det \mathbf{A}_2 < 0$, we see that $(5\pi/6, 0)$ is a saddle point.

(c) From the system in part (a) we obtain the first-order differential equation

$$\frac{dy}{dx} = \frac{1/2 - \sin x}{y}.$$

Separating variables and integrating we obtain

and

$$\int y \, dy = \int \left(\frac{1}{2} - \sin x\right) dx$$

$$\frac{1}{2}y^2 = \frac{1}{2}x + \cos x + c_1$$

or

$$y^2 = x + 2\cos x + c_2.$$

For x_0 near $\pi/6$, if $\mathbf{X}(0) = (x_0, 0)$ then $c_2 = -x_0 - 2\cos x_0$ and $y^2 = x + 2\cos x - x_0 - 2\cos x_0$. Thus, there are two values of y for each x in a sufficiently small interval around $\pi/6$. Therefore $(\pi/6, 0)$ is a center.

10.4 Autonomous Systems as Mathematical Models

The terminology and concepts listed below provide an outline of the main ideas encountered in this section. These can be useful when preparing for a quiz or test.

Terminology and Concepts

- nonlinear pendulum

- nonlinear oscillations

- Lotka-Volterra preditor-prey model

- Lotka-Volterra competition model

3. The corresponding plane autonomous system is

$$x' = y, \quad y' = -g\frac{f'(x)}{1 + [f'(x)]^2} - \frac{\beta}{m}y$$

and

$$\frac{\partial}{\partial x}\left(-g\frac{f'(x)}{1 + [f'(x)]^2} - \frac{\beta}{m}y\right) = -g\frac{(1 + [f'(x)]^2)f''(x) - f'(x)2f'(x)f''(x)}{(1 + [f'(x)]^2)^2}.$$

If $\mathbf{X}_1 = (x_1, y_1)$ is a critical point, $y_1 = 0$ and $f'(x_1) = 0$. The Jacobian at this critical point is therefore

$$\mathbf{g}'(\mathbf{X}_1) = \begin{pmatrix} 0 & 1 \\ -gf''(x_1) & -\dfrac{\beta}{m} \end{pmatrix}.$$

6. (a) If $f(x) = \cosh x$, $f'(x) = \sinh x$ and $[f'(x)]^2 + 1 = \sinh^2 x + 1 = \cosh^2 x$. Therefore

$$\frac{dy}{dx} = \frac{y'}{x'} = -g\,\frac{\sinh x}{\cosh^2 x}\,\frac{1}{y}.$$

We can separate variables to show that $y^2 = 2g/\cosh x + c$. But $x(0) = x_0$ and $y(0) = x'(0) = v_0$. Therefore $c = v_0^2 - (2g/\cosh x_0)$ and so

$$y^2 = \frac{2g}{\cosh x} - \frac{2g}{\cosh x_0} + v_0^2.$$

Now

$$\frac{2g}{\cosh x} - \frac{2g}{\cosh x_0} + v_0^2 \geq 0 \quad \text{if and only if} \quad \cosh x \leq \frac{2g\cosh x_0}{2g - v_0^2 \cosh x_0}$$

and the solution to this inequality is an interval $[-a, a]$. Therefore each x in $(-a, a)$ has two corresponding values of y and so the solution is periodic.

(b) Since $z = \cosh x$, the maximum height occurs at the largest value of x on the cycle. From (a), $x_{max} = a$ where $\cosh a = 2g\cosh x_0/(2g - v_0^2 \cosh x_0)$. Therefore

$$z_{max} = \frac{2g\cosh x_0}{2g - v_0^2 \cosh x_0}.$$

9. (a) In the Lotka-Volterra Model the average number of predators is d/c and the average number of prey is a/b. But

$$x' = -ax + bxy - \epsilon_1 x = -(a + \epsilon_1)x + bxy$$

$$y' = -cxy + dy - \epsilon_2 y = -cxy + (d - \epsilon_2)y$$

and so the new critical point in the first quadrant is $(d/c - \epsilon_2/c, a/b + \epsilon_1/b)$.

(b) The average number of predators $d/c - \epsilon_2/c$ has decreased while the average number of prey $a/b + \epsilon_1/b$ has increased. The fishery science model is consistent with Volterra's principle.

12. $\Delta = r_1 r_2$, $\tau = r_1 + r_2$ and $\tau^2 - 4\Delta = (r_1 + r_2)^2 - 4r_1 r_2 = (r_1 - r_2)^2$. Therefore when $r_1 \neq r_2$, $(0, 0)$ is an unstable node.

15. $K_1/\alpha_{12} < K_2 < K_1\alpha_{21}$ and so $\alpha_{12}\alpha_{21} > 1$. Therefore $\Delta = (1 - \alpha_{12}\alpha_{21})\hat{x}\hat{y}\,r_1 r_2/K_1 K_2 < 0$ and so $(\hat{x}, \hat{y})$ is a saddle point.

18. (a) The magnitude of the frictional force between the bead and the wire is $\mu(mg\cos\theta)$ for some $\mu > 0$. The component of this frictional force in the x-direction is

$$(\mu mg\cos\theta)\cos\theta = \mu mg\cos^2\theta.$$

But

$$\cos\theta = \frac{1}{\sqrt{1+[f'(x)]^2}} \quad \text{and so} \quad \mu mg\cos^2\theta = \frac{\mu mg}{1+[f'(x)]^2}.$$

It follows from Newton's Second Law that

$$mx'' = -mg\frac{f'(x)}{1+[f'(x)]^2} - \beta x' + mg\frac{\mu}{1+[f'(x)]^2}$$

and so

$$x'' = g\frac{\mu - f'(x)}{1+[f'(x)]^2} - \frac{\beta}{m}x'.$$

(b) A critical point (x, y) must satisfy $y = 0$ and $f'(x) = \mu$. Therefore critical points occur at $(x_1, 0)$ where $f'(x_1) = \mu$. The Jacobian matrix of the plane autonomous system is

$$\mathbf{g}'(\mathbf{X}) = \begin{pmatrix} 0 & 1 \\ g\dfrac{(1+[f'(x)]^2)(-f''(x)) - (\mu - f'(x))2f'(x)f''(x)}{(1+[f'(x)]^2)^2} & -\dfrac{\beta}{m} \end{pmatrix}$$

and so at a critical point $\mathbf{X}_1$,

$$\mathbf{g}'(\mathbf{X}) = \begin{pmatrix} 0 & 1 \\ \dfrac{-gf''(x_1)}{1+\mu^2} & -\dfrac{\beta}{m} \end{pmatrix}.$$

Therefore $\tau = -\beta/m < 0$ and $\Delta = gf''(x_1)/(1+\mu^2)$. When $f''(x_1) < 0$, $\Delta < 0$ and so a saddle point occurs. When $f''(x_1) > 0$ and

$$\tau^2 - 4\Delta = \frac{\beta^2}{m^2} - 4g\frac{f''(x_1)}{1+\mu^2} < 0,$$

$(x_1, 0)$ is a stable spiral point. This condition can also be written as

$$\beta^2 < 4gm^2\frac{f''(x_1)}{1+\mu^2}.$$

21. The equation

$$x' = \alpha\frac{y}{1+y}x - x = x\left(\frac{\alpha y}{1+y} - 1\right) = 0$$

implies that $x = 0$ or $y = 1/(\alpha - 1)$. When $\alpha > 0$, $\hat{y} = 1/(\alpha - 1) > 0$. If $x = 0$, then from the differential equation for y', $y = \beta$. On the other hand, if $\hat{y} = 1/(\alpha - 1)$, $\hat{y}/(1 + \hat{y}) = 1/\alpha$ and so $\hat{x}/\alpha - 1/(\alpha - 1) + \beta = 0$. It follows that

$$\hat{x} = \alpha \left(\beta - \frac{1}{\alpha - 1} \right) = \frac{\alpha}{\alpha - 1}[(\alpha - 1)\beta - 1]$$

and if $\beta(\alpha - 1) > 1$, $\hat{x} > 0$. Therefore $(\hat{x}, \hat{y})$ is the unique critical point in the first quadrant. The Jacobian matrix is

$$\mathbf{g}'(\mathbf{X}) = \begin{pmatrix} \alpha \dfrac{y}{y + 1} - 1 & \dfrac{\alpha x}{(1 + y)^2} \\[2mm] -\dfrac{y}{1 + y} & \dfrac{-x}{(1 + y)^2} - 1 \end{pmatrix}$$

and for $\mathbf{X} = (\hat{x}, \hat{y})$, the Jacobian can be written in the form

$$\mathbf{g}'((\hat{x}, \hat{y})) = \begin{pmatrix} 0 & \dfrac{(\alpha - 1)^2}{\alpha} \hat{x} \\[2mm] -\dfrac{1}{\alpha} & -\dfrac{(\alpha - 1)^2}{\alpha^2} - 1 \end{pmatrix}.$$

It follows that

$$\tau = -\left[\frac{(\alpha - 1)^2}{\alpha^2} \hat{x} + 1 \right] < 0, \quad \Delta = \frac{(\alpha - 1)^2}{\alpha^2} \hat{x}$$

and so $\tau = -(\Delta + 1)$. Therefore $\tau^2 - 4\Delta = (\Delta + 1)^2 - 4\Delta = (\Delta - 1)^2 > 0$. Therefore $(\hat{x}, \hat{y})$ is a stable node.

10.R   Chapter 10 in Review

3. a center or a saddle point

6. True

9. The system is linear and we identify $\Delta = -\alpha$ and $\tau = \alpha + 1$. Since a critical point will be a center when $\Delta > 0$ and $\tau = 0$ we see that for $\alpha = -1$ critical points will be centers and solutions will be periodic. Note also that when $\alpha = -1$ the system is

$$x' = -x - 2y$$

$$y' = x + y,$$

which does have an isolated critical point at $(0, 0)$.

12. (a) If $\mathbf{X}(0) = \mathbf{X}_0$ lies on the line $y = -2x$, then $\mathbf{X}(t)$ approaches $(0,0)$ along this line. For all other initial conditions, $\mathbf{X}(t)$ approaches $(0,0)$ from the direction determined by the line $y = x$.

(b) If $\mathbf{X}(0) = \mathbf{X}_0$ lies on the line $y = -x$, then $\mathbf{X}(t)$ approaches $(0,0)$ along this line. For all other initial conditions, $\mathbf{X}(t)$ becomes unbounded and $y = 2x$ serves as an asymptote.

15. From $x = r\cos\theta$, $y = r\sin\theta$ we have

$$\frac{dx}{dt} = -r\sin\theta\,\frac{d\theta}{dt} + \frac{dr}{dt}\cos\theta$$

$$\frac{dy}{dt} = r\cos\theta\,\frac{d\theta}{dt} + \frac{dr}{dt}\sin\theta.$$

Then $r' = \alpha r$, $\theta' = 1$ gives

$$\frac{dx}{dt} = -r\sin\theta + \alpha r\cos\theta$$

$$\frac{dy}{dt} = r\cos\theta + \alpha r\sin\theta.$$

We see that $r = 0$, which corresponds to $\mathbf{X} = (0,0)$, is a critical point. Solving $r' = \alpha r$ we have $r = c_1 e^{\alpha t}$. Thus, when $\alpha < 0$, $\lim_{t\to\infty} r(t) = 0$ and $(0,0)$ is a stable critical point. When $\alpha = 0$, $r' = 0$ and $r = c_1$. In this case $(0,0)$ is a center, which is stable. Therefore, $(0,0)$ is a stable critical point for the system when $\alpha \le 0$.

18. Using the phase-plane method we obtain

$$\frac{dy}{dx} = \frac{y'}{x'} = \frac{-2x\sqrt{y^2+1}}{y}.$$

We can separate variables to show that $\sqrt{y^2+1} = -x^2 + c$. But $x(0) = x_0$ and $y(0) = x'(0) = 0$. It follows that $c = 1 + x_0^2$ so that $y^2 = (1 + x_0^2 - x^2)^2 - 1$. Note that $1 + x_0^2 - x^2 > 1$ for $-x_0 < x < x_0$ and $y = 0$ for $x = \pm x_0$. Each x with $-x_0 < x < x_0$ has two corresponding values of y and so the solution $\mathbf{X}(t)$ with $\mathbf{X}(0) = (x_0, 0)$ is periodic.

11 Orthogonal Functions and Fourier Series

11.1 Orthogonal Functions

The terminology and concepts listed below provide an outline of the main ideas encountered in this section. These can be useful when preparing for a quiz or test.

Terminology and Concepts

- inner product of two vectors in 3-space
- inner product of two functions on an interval
- orthogonal functions and orthogonal sets
- norm of a function
- orthonormal set
- orthogonal series expansion of a function
- orthogonality with respect to a weight function

The basic skills listed below summarize the more mechanical types of problems encountered in the exercise set for this section.

Basic Skills

- show that a set of functions is orthogonal on a given interval
- show that a set of functions is orthogonal with respect to a given weight function on an interval

3. $\displaystyle\int_0^2 e^x(xe^{-x} - e^{-x})dx = \int_0^2 (x-1)dx = \left(\frac{1}{2}x^2 - x\right)\Big|_0^2 = 0$

6. $\displaystyle\int_{\pi/4}^{5\pi/4} e^x \sin x\, dx = \left(\frac{1}{2}e^x \sin x - \frac{1}{2}e^x \cos x\right)\Big|_{\pi/4}^{5\pi/4} = 0$

9. For $m \neq n$

$$\int_0^\pi \sin nx \sin mx\, dx = \frac{1}{2}\int_0^\pi \Big(\cos(n-m)x - \cos(n+m)x\Big)\, dx$$

257

$$= \frac{1}{2(n-m)} \sin(n-m)x \Big|_0^\pi - \frac{1}{2(n+m)} \sin(n+m)x \Big|_0^\pi$$

$$= 0.$$

For $m = n$

$$\int_0^\pi \sin^2 nx \, dx = \int_0^\pi \left(\frac{1}{2} - \frac{1}{2} \cos 2nx \right) dx = \frac{1}{2}x \Big|_0^\pi - \frac{1}{4n} \sin 2nx \Big|_0^\pi = \frac{\pi}{2}$$

so that

$$\| \sin nx \| = \sqrt{\frac{\pi}{2}}.$$

12. For $m \neq n$, we use Problems 11 and 10:

$$\int_{-p}^p \cos \frac{n\pi}{p}x \cos \frac{m\pi}{p}x \, dx = 2 \int_0^p \cos \frac{n\pi}{p}x \cos \frac{m\pi}{p}x \, dx = 0$$

$$\int_{-p}^p \sin \frac{n\pi}{p}x \sin \frac{m\pi}{p}x \, dx = 2 \int_0^p \sin \frac{n\pi}{p}x \sin \frac{m\pi}{p}x \, dx = 0.$$

Also

$$\int_{-p}^p \sin \frac{n\pi}{p}x \cos \frac{m\pi}{p}x \, dx = \frac{1}{2} \int_{-p}^p \left(\sin \frac{(n-m)\pi}{p}x + \sin \frac{(n+m)\pi}{p}x \right) dx = 0,$$

$$\int_{-p}^p 1 \cdot \cos \frac{n\pi}{p}x \, dx = \frac{p}{n\pi} \sin \frac{n\pi}{p}x \Big|_{-p}^p = 0,$$

$$\int_{-p}^p 1 \cdot \sin \frac{n\pi}{p}x \, dx = -\frac{p}{n\pi} \cos \frac{n\pi}{p}x \Big|_{-p}^p = 0,$$

and

$$\int_{-p}^p \sin \frac{n\pi}{p}x \cos \frac{n\pi}{p}x \, dx = \int_{-p}^p \frac{1}{2} \sin \frac{2n\pi}{p}x \, dx = -\frac{p}{4n\pi} \cos \frac{2n\pi}{p}x \Big|_{-p}^p = 0.$$

For $m = n$

$$\int_{-p}^p \cos^2 \frac{n\pi}{p}x \, dx = \int_{-p}^p \left(\frac{1}{2} + \frac{1}{2} \cos \frac{2n\pi}{p}x \right) dx = p,$$

$$\int_{-p}^p \sin^2 \frac{n\pi}{p}x \, dx = \int_{-p}^p \left(\frac{1}{2} - \frac{1}{2} \cos \frac{2n\pi}{p}x \right) dx = p,$$

and

$$\int_{-p}^{p} 1^2 dx = 2p$$

so that

$$\|1\| = \sqrt{2p}, \quad \left\|\cos\frac{n\pi}{p}x\right\| = \sqrt{p}, \quad \text{and} \quad \left\|\sin\frac{n\pi}{p}x\right\| = \sqrt{p}.$$

15. By orthogonality $\int_a^b \phi_0(x)\phi_n(x)dx = 0$ for $n = 1, 2, 3, \ldots$; that is, $\int_a^b \phi_n(x)dx = 0$ for $n = 1, 2, 3, \ldots$.

18. Setting

$$0 = \int_{-2}^{2} f_3(x)f_1(x)\,dx = \int_{-2}^{2}\left(x^2 + c_1 x^3 + c_2 x^4\right)dx = \frac{16}{3} + \frac{64}{5}c_2$$

and

$$0 = \int_{-2}^{2} f_3(x)f_2(x)\,dx = \int_{-2}^{2}\left(x^3 + c_1 x^4 + c_2 x^5\right)dx = \frac{64}{5}c_1$$

we obtain $c_1 = 0$ and $c_2 = -5/12$.

21. **(a)** The fundamental period is $2\pi/2\pi = 1$.

(b) The fundamental period is $2\pi/(4/L) = \frac{1}{2}\pi L$.

(c) The fundamental period of $\sin x + \sin 2x$ is 2π.

(d) The fundamental period of $\sin 2x + \cos 4x$ is $2\pi/2 = \pi$.

(e) The fundamental period of $\sin 3x + \cos 4x$ is 2π since the smallest integer multiples of $2\pi/3$ and $2\pi/4 = \pi/2$ that are equal are 3 and 4, respectively.

(f) The fundamental period of $f(x)$ is $2\pi/(\pi/p) = 2p$.

11.2 Fourier Series

The terminology and concepts listed below provide an outline of the main ideas encountered in this section. These can be useful when preparing for a quiz or test.

Terminology and Concepts

- Fourier series of a function
- Fourier coefficients of a function

- conditions under which a Fourier series converges
- convergence at a point of finite discontinuity
- fundamental period of a Fourier series
- periodic extension of a function
- sequence of partial sums of a Fourier series

The basic skills listed below summarize the more mechanical types of problems encountered in the exercise set for this section.

Basic Skills

- find the Fourier series of a function on a given interval

If x_0 is a point of discontinuity of $f(x)$ on an interval, then the Fourier series of $f(x)$ converges to the midpoint of $f(x-)$ and $f(x+)$.

3. $a_0 = \int_{-1}^{1} f(x)\,dx = \int_{-1}^{0} 1\,dx + \int_{0}^{1} x\,dx = \dfrac{3}{2}$

$a_n = \int_{-1}^{1} f(x)\cos n\pi x\,dx = \int_{-1}^{0} \cos n\pi x\,dx + \int_{0}^{1} x\cos n\pi x\,dx = \dfrac{1}{n^2\pi^2}[(-1)^n - 1]$

$b_n = \int_{-1}^{1} f(x)\sin n\pi x\,dx = \int_{-1}^{0} \sin n\pi x\,dx + \int_{0}^{1} x\sin n\pi x\,dx = -\dfrac{1}{n\pi}$

$f(x) = \dfrac{3}{4} + \displaystyle\sum_{n=1}^{\infty} \left[\dfrac{(-1)^n - 1}{n^2\pi^2}\cos n\pi x - \dfrac{1}{n\pi}\sin n\pi x\right]$

$f(x)$ is discontinuous at $x = 0$ and converges to $\frac{1}{2}$ there.

6. $a_0 = \dfrac{1}{\pi}\int_{-\pi}^{\pi} f(x)\,dx = \dfrac{1}{\pi}\int_{-\pi}^{0} \pi^2\,dx + \dfrac{1}{\pi}\int_{0}^{\pi} \left(\pi^2 - x^2\right)\,dx = \dfrac{5}{3}\pi^2$

$a_n = \dfrac{1}{\pi}\int_{-\pi}^{\pi} f(x)\cos nx\,dx = \dfrac{1}{\pi}\int_{-\pi}^{0} \pi^2\cos nx\,dx + \dfrac{1}{\pi}\int_{0}^{\pi} \left(\pi^2 - x^2\right)\cos nx\,dx$

$= \dfrac{1}{\pi}\left(\dfrac{\pi^2 - x^2}{n}\sin nx\,\Big|_{0}^{\pi} + \dfrac{2}{n}\int_{0}^{\pi} x\sin nx\,dx\right) = \dfrac{2}{n^2}(-1)^{n+1}$

$$b_n = \frac{1}{\pi}\int_{-\pi}^{\pi} f(x)\sin nx\, dx = \frac{1}{\pi}\int_{-\pi}^{0} \pi^2 \sin nx\, dx + \frac{1}{\pi}\int_{0}^{\pi} \left(\pi^2 - x^2\right)\sin nx\, dx$$

$$= \frac{\pi}{n}[(-1)^n - 1] + \frac{1}{\pi}\left(\frac{x^2 - \pi^2}{n}\cos nx\,\Big|_0^\pi - \frac{2}{n}\int_0^\pi x\cos nx\, dx\right) = \frac{\pi}{n}(-1)^n + \frac{2}{n^3\pi}[1 - (-1)^n]$$

$$f(x) = \frac{5\pi^2}{6} + \sum_{n=1}^{\infty}\left[\frac{2}{n^2}(-1)^{n+1}\cos nx + \left(\frac{\pi}{n}(-1)^n + \frac{2[1 - (-1)^n]}{n^3\pi}\right)\sin nx\right]$$

$f(x)$ is continuous on the interval.

9. $\displaystyle a_0 = \frac{1}{\pi}\int_{-\pi}^{\pi} f(x)\, dx = \frac{1}{\pi}\int_0^\pi \sin x\, dx = \frac{2}{\pi}$

$$a_n = \frac{1}{\pi}\int_{-\pi}^{\pi} f(x)\cos nx\, dx = \frac{1}{\pi}\int_0^\pi \sin x\,\cos nx\, dx = \frac{1}{2\pi}\int_0^\pi\left(\sin(1 + n)x + \sin(1 - n)x\right)dx$$

$$= \frac{1 + (-1)^n}{\pi(1 - n^2)}\quad \text{for } n = 2, 3, 4, \ldots$$

$$a_1 = \frac{1}{2\pi}\int_0^\pi \sin 2x\, dx = 0$$

$$b_n = \frac{1}{\pi}\int_{-\pi}^{\pi} f(x)\sin nx\, dx = \frac{1}{\pi}\int_0^\pi \sin x\,\sin nx\, dx$$

$$= \frac{1}{2\pi}\int_0^\pi\left(\cos(1 - n)x - \cos(1 + n)x\right)dx = 0\quad \text{for } n = 2, 3, 4, \ldots$$

$$b_1 = \frac{1}{2\pi}\int_0^\pi (1 - \cos 2x)\, dx = \frac{1}{2}$$

$$f(x) = \frac{1}{\pi} + \frac{1}{2}\sin x + \sum_{n=2}^{\infty} \frac{1 + (-1)^n}{\pi(1 - n^2)}\cos nx$$

$f(x)$ is continuous on the interval.

12. $\displaystyle a_0 = \frac{1}{2}\int_{-2}^{2} f(x)\, dx = \frac{1}{2}\left(\int_0^1 x\, dx + \int_1^2 1\, dx\right) = \frac{3}{4}$

$$a_n = \frac{1}{2}\int_{-2}^{2} f(x)\cos\frac{n\pi}{2}x\, dx = \frac{1}{2}\left(\int_0^1 x\cos\frac{n\pi}{2}x\, dx + \int_1^2 \cos\frac{n\pi}{2}x\, dx\right) = \frac{2}{n^2\pi^2}\left(\cos\frac{n\pi}{2} - 1\right)$$

$$b_n = \frac{1}{2} \int_{-2}^{2} f(x) \sin \frac{n\pi}{2} x \, dx = \frac{1}{2} \left(\int_{0}^{1} x \sin \frac{n\pi}{2} x \, dx + \int_{1}^{2} \sin \frac{n\pi}{2} x \, dx \right)$$

$$= \frac{2}{n^2 \pi^2} \left(\sin \frac{n\pi}{2} + \frac{n\pi}{2} (-1)^{n+1} \right)$$

$$f(x) = \frac{3}{8} + \sum_{n=1}^{\infty} \left[\frac{2}{n^2 \pi^2} \left(\cos \frac{n\pi}{2} - 1 \right) \cos \frac{n\pi}{2} x + \frac{2}{n^2 \pi^2} \left(\sin \frac{n\pi}{2} + \frac{n\pi}{2} (-1)^{n+1} \right) \sin \frac{n\pi}{2} x \right]$$

$f(x)$ is continuous on the interval.

15. $a_0 = \frac{1}{\pi} \int_{-\pi}^{\pi} f(x) \, dx = \frac{1}{\pi} \int_{-\pi}^{\pi} e^x \, dx = \frac{1}{\pi} (e^\pi - e^{-\pi})$

$$a_n = \frac{1}{\pi} \int_{-\pi}^{\pi} f(x) \cos nx \, dx = \frac{(-1)^n (e^\pi - e^{-\pi})}{\pi (1 + n^2)}$$

$$b_n = \frac{1}{\pi} \int_{-\pi}^{\pi} f(x) \sin nx \, dx = \frac{1}{\pi} \int_{-\pi}^{\pi} e^x \sin nx \, dx = \frac{(-1)^n n (e^{-\pi} - e^\pi)}{\pi (1 + n^2)}$$

$$f(x) = \frac{e^\pi - e^{-\pi}}{2\pi} + \sum_{n=1}^{\infty} \left[\frac{(-1)^n (e^\pi - e^{-\pi})}{\pi (1 + n^2)} \cos nx + \frac{(-1)^n n (e^{-\pi} - e^\pi)}{\pi (1 + n^2)} \sin nx \right]$$

$f(x)$ is continuous on the interval.

18.

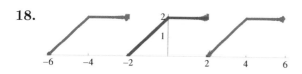

21. The function in Problem 7 is continuous at $x = \pi/2$ so

$$\frac{3\pi}{2} = f\left(\frac{\pi}{2} \right) = \pi + \sum_{n=1}^{\infty} \frac{2}{n} (-1)^{n+1} \sin \frac{n\pi}{2} = \pi + 2 \left(1 - \frac{1}{3} + \frac{1}{5} - \frac{1}{7} + \cdots \right)$$

and

$$\frac{\pi}{4} = 1 - \frac{1}{3} + \frac{1}{5} - \frac{1}{7} + \cdots .$$

24. Identifying $f(x) = e^{-x}$ and $p = \pi$, we have

$$c_n = \frac{1}{2\pi} \int_{-\pi}^{\pi} e^{-x} e^{-inx} dx = \frac{1}{2\pi} \int_{-\pi}^{\pi} e^{-(in+1)x} dx$$

$$= -\frac{1}{2(in+1)\pi} e^{-(in+1)x} \bigg|_{-\pi}^{\pi}$$

$$= -\frac{1}{2(in+1)\pi} \left[e^{-(in+1)\pi} - e^{(in+1)\pi} \right]$$

$$= \frac{e^{(in+1)\pi} - e^{-(in+1)\pi}}{2(in+1)\pi}$$

$$= \frac{e^{\pi}(\cos n\pi + i \sin n\pi) - e^{-\pi}(\cos n\pi - i \sin n\pi)}{2(in+1)\pi}$$

$$= \frac{(e^{\pi} - e^{-\pi})\cos n\pi}{2(in+1)\pi} = \frac{(e^{\pi} - e^{-\pi})(-1)^n}{2(in+1)\pi}.$$

Thus

$$f(x) = \sum_{n=-\infty}^{\infty} (-1)^n \frac{e^{\pi} - e^{-\pi}}{2(in+1)\pi} e^{inx}.$$

11.3 Fourier Cosine and Sine Series

The terminology and concepts listed below provide an outline of the main ideas encountered in this section. These can be useful when preparing for a quiz or test.

Terminology and Concepts

- properties of even and odd functions
- cosine series of an even function
- sine series of an odd function
- Gibbs phenomenon
- half-range expansions

The basic skills listed below summarize the more mechanical types of problems encountered in the exercise set for this section.

Basic Skills

- determine if a given function is even or odd
- expand a given function in an appropriate cosine or sine series

- find half-range cosine and sine expansions of a given function

- expand a function defined on an interval of the form (a, b) in a Fourier series

- use a Fourier series to find a particular solution of an ordinary DE with a periodic driving force

3. Since $f(-x) = (-x)^2 - x = x^2 - x$, $f(x)$ is neither even nor odd.

6. Since $f(-x) = e^{-x} - e^x = -f(x)$, $f(x)$ is an odd function.

9. Since $f(x)$ is not defined for $x < 0$, it is neither even nor odd.

12. Since $f(x)$ is an even function, we expand in a cosine series:

$$a_0 = \int_1^2 1\, dx = 1$$

$$a_n = \int_1^2 \cos \frac{n\pi}{2} x\, dx = -\frac{2}{n\pi} \sin \frac{n\pi}{2}.$$

Thus

$$f(x) = \frac{1}{2} + \sum_{n=1}^{\infty} \frac{-2}{n\pi} \sin \frac{n\pi}{2} \cos \frac{n\pi}{2} x.$$

15. Since $f(x)$ is an even function, we expand in a cosine series:

$$a_0 = 2\int_0^1 x^2\, dx = \frac{2}{3}$$

$$a_n = 2\int_0^1 x^2 \cos n\pi x\, dx = 2\left(\frac{x^2}{n\pi} \sin n\pi x \Big|_0^1 - \frac{2}{n\pi}\int_0^1 x \sin n\pi x\, dx\right) = \frac{4}{n^2\pi^2}(-1)^n.$$

Thus

$$f(x) = \frac{1}{3} + \sum_{n=1}^{\infty} \frac{4}{n^2\pi^2}(-1)^n \cos n\pi x.$$

18. Since $f(x)$ is an odd function, we expand in a sine series:

$$b_n = \frac{2}{\pi}\int_0^\pi x^3 \sin nx\, dx = \frac{2}{\pi}\left(-\frac{x^3}{n} \cos nx \Big|_0^\pi + \frac{3}{n}\int_0^\pi x^2 \cos nx\, dx\right)$$

$$= \frac{2\pi^2}{n}(-1)^{n+1} - \frac{12}{n^2\pi}\int_0^\pi x \sin nx\, dx$$

$$= \frac{2\pi^2}{n}(-1)^{n+1} - \frac{12}{n^2\pi}\left(-\frac{x}{n} \cos nx \Big|_0^\pi + \frac{1}{n}\int_0^\pi \cos nx\, dx\right) = \frac{2\pi^2}{n}(-1)^{n+1} + \frac{12}{n^3}(-1)^n.$$

Thus

$$f(x) = \sum_{n=1}^{\infty} \left(\frac{2\pi^2}{n}(-1)^{n+1} + \frac{12}{n^3}(-1)^n \right) \sin nx.$$

21. Since $f(x)$ is an even function, we expand in a cosine series:

$$a_0 = \int_0^1 x\, dx + \int_1^2 1\, dx = \frac{3}{2}$$

$$a_n = \int_0^1 x \cos \frac{n\pi}{2} x\, dx + \int_1^2 \cos \frac{n\pi}{2} x\, dx = \frac{4}{n^2\pi^2} \left(\cos \frac{n\pi}{2} - 1 \right).$$

Thus

$$f(x) = \frac{3}{4} + \sum_{n=1}^{\infty} \frac{4}{n^2\pi^2} \left(\cos \frac{n\pi}{2} - 1 \right) \cos \frac{n\pi}{2} x.$$

24. Since $f(x)$ is an even function, we expand in a cosine series. [See the solution of Problem 10 in Exercise 11.2 for the computation of the integrals.]

$$a_0 = \frac{2}{\pi/2} \int_0^{\pi/2} \cos x\, dx = \frac{4}{\pi}$$

$$a_n = \frac{2}{\pi/2} \int_0^{\pi/2} \cos x \cos \frac{n\pi}{\pi/2} x\, dx = \frac{4(-1)^{n+1}}{\pi(4n^2 - 1)}$$

Thus

$$f(x) = \frac{2}{\pi} + \sum_{n=1}^{\infty} \frac{4(-1)^{n+1}}{\pi(4n^2 - 1)} \cos 2nx.$$

27. $a_0 = \frac{4}{\pi} \int_0^{\pi/2} \cos x\, dx = \frac{4}{\pi}$

$$a_n = \frac{4}{\pi} \int_0^{\pi/2} \cos x \cos 2nx\, dx = \frac{2}{\pi} \int_0^{\pi/2} [\cos(2n+1)x + \cos(2n-1)x]\, dx = \frac{4(-1)^n}{\pi(1 - 4n^2)}$$

$$b_n = \frac{4}{\pi} \int_0^{\pi/2} \cos x \sin 2nx\, dx = \frac{2}{\pi} \int_0^{\pi/2} [\sin(2n+1)x + \sin(2n-1)x]\, dx = \frac{8n}{\pi(4n^2 - 1)}$$

$$f(x) = \frac{2}{\pi} + \sum_{n=1}^{\infty} \frac{4(-1)^n}{\pi(1 - 4n^2)} \cos 2nx$$

$$f(x) = \sum_{n=1}^{\infty} \frac{8n}{\pi(4n^2 - 1)} \sin 2nx$$

30. $a_0 = \frac{1}{\pi} \int_{\pi}^{2\pi} (x - \pi) \, dx = \frac{\pi}{2}$

$a_n = \frac{1}{\pi} \int_{\pi}^{2\pi} (x - \pi) \cos \frac{n}{2} x \, dx = \frac{4}{n^2 \pi} \left((-1)^n - \cos \frac{n\pi}{2} \right)$

$b_n = \frac{1}{\pi} \int_{\pi}^{2\pi} (x - \pi) \sin \frac{n}{2} x \, dx = \frac{2}{n} (-1)^{n+1} - \frac{4}{n^2 \pi} \sin \frac{n\pi}{2}$

$f(x) = \frac{\pi}{4} + \sum_{n=1}^{\infty} \frac{4}{n^2 \pi} \left((-1)^n - \cos \frac{n\pi}{2} \right) \cos \frac{n}{2} x$

$f(x) = \sum_{n=1}^{\infty} \left(\frac{2}{n} (-1)^{n+1} - \frac{4}{n^2 \pi} \sin \frac{n\pi}{2} \right) \sin \frac{n}{2} x$

33. $a_0 = 2 \int_0^1 (x^2 + x) \, dx = \frac{5}{3}$

$a_n = 2 \int_0^1 (x^2 + x) \cos n\pi x \, dx = \frac{2(x^2 + x)}{n\pi} \sin n\pi x \bigg|_0^1 - \frac{2}{n\pi} \int_0^1 (2x+1) \sin n\pi x \, dx = \frac{2}{n^2 \pi^2} [3(-1)^n - 1]$

$b_n = 2 \int_0^1 (x^2 + x) \sin n\pi x \, dx = -\frac{2(x^2 + x)}{n\pi} \cos n\pi x \bigg|_0^1 + \frac{2}{n\pi} \int_0^1 (2x + 1) \cos n\pi x \, dx$

$\qquad = \frac{4}{n\pi} (-1)^{n+1} + \frac{4}{n^3 \pi^3} [(-1)^n - 1]$

$f(x) = \frac{5}{6} + \sum_{n=1}^{\infty} \frac{2}{n^2 \pi^2} [3(-1)^n - 1] \cos n\pi x$

$f(x) = \sum_{n=1}^{\infty} \left(\frac{4}{n\pi} (-1)^{n+1} + \frac{4}{n^3 \pi^3} [(-1)^n - 1] \right) \sin n\pi x$

36. $a_0 = \frac{2}{\pi} \int_0^{\pi} x \, dx = \pi$

$a_n = \frac{2}{\pi} \int_0^{\pi} x \cos 2nx \, dx = 0$

$b_n = \frac{2}{\pi} \int_0^{\pi} x \sin 2nx \, dx = -\frac{1}{n}$

$f(x) = \frac{\pi}{2} - \sum_{n=1}^{\infty} \frac{1}{n} \sin 2nx$

39. We have

$$b_n = \frac{2}{\pi} \int_0^\pi 5 \sin nt \, dt = \frac{10}{n\pi}[1 - (-1)^n]$$

so that

$$f(t) = \sum_{n=1}^\infty \frac{10[1 - (-1)^n]}{n\pi} \sin nt.$$

Substituting the assumption $x_p(t) = \sum_{n=1}^\infty B_n \sin nt$ into the differential equation then gives

$$x_p'' + 10x_p = \sum_{n=1}^\infty B_n(10 - n^2) \sin nt = \sum_{n=1}^\infty \frac{10[1 - (-1)^n]}{n\pi} \sin nt$$

and so $B_n = 10[1 - (-1)^n]/n\pi(10 - n^2)$. Thus

$$x_p(t) = \frac{10}{\pi} \sum_{n=1}^\infty \frac{1 - (-1)^n}{n(10 - n^2)} \sin nt.$$

42. We have

$$a_0 = \frac{2}{1/2} \int_0^{1/2} t \, dt = \frac{1}{2}$$

$$a_n = \frac{2}{1/2} \int_0^{1/2} t \cos 2n\pi t \, dt = \frac{1}{n^2\pi^2}[(-1)^n - 1]$$

so that

$$f(t) = \frac{1}{4} + \sum_{n=1}^\infty \frac{(-1)^n - 1}{n^2\pi^2} \cos 2n\pi t.$$

Substituting the assumption

$$x_p(t) = \frac{A_0}{2} + \sum_{n=1}^\infty A_n \cos 2n\pi t$$

into the differential equation then gives

$$\frac{1}{4} x_p'' + 12x_p = 6A_0 + \sum_{n=1}^\infty A_n(12 - n^2\pi^2) \cos 2n\pi t = \frac{1}{4} + \sum_{n=1}^\infty \frac{(-1)^n - 1}{n^2\pi^2} \cos 2n\pi t$$

and $A_0 = 1/24$, $A_n = [(-1)^n - 1]/n^2\pi^2(12 - n^2\pi^2)$. Thus

$$x_p(t) = \frac{1}{48} + \frac{1}{\pi^2} \sum_{n=1}^\infty \frac{(-1)^n - 1}{n^2(12 - n^2\pi^2)} \cos 2n\pi t.$$

45. (a) We have

$$b_n = \frac{2}{L} \int_0^L \frac{w_0 x}{L} \sin \frac{n\pi}{L} x \, dx = \frac{2w_0}{n\pi} (-1)^{n+1}$$

so that

$$w(x) = \sum_{n=1}^{\infty} \frac{2w_0}{n\pi} (-1)^{n+1} \sin \frac{n\pi}{L} x.$$

(b) If we assume $y_p(x) = \sum_{n=1}^{\infty} B_n \sin(n\pi x/L)$ then

$$y_p^{(4)} = \sum_{n=1}^{\infty} \frac{n^4 \pi^4}{L^4} B_n \sin \frac{n\pi}{L} x$$

and so the differential equation $EIy_p^{(4)} = w(x)$ gives

$$B_n = \frac{2w_0 (-1)^{n+1} L^4}{EI n^5 \pi^5}.$$

Thus

$$y_p(x) = \frac{2w_0 L^4}{EI \pi^5} \sum_{n=1}^{\infty} \frac{(-1)^{n+1}}{n^5} \sin \frac{n\pi}{L} x.$$

48. (a) If f and g are even and $h(x) = f(x)g(x)$ then

$$h(-x) = f(-x)g(-x) = f(x)g(x) = h(x)$$

and h is even.

(c) If f is even and g is odd and $h(x) = f(x)g(x)$ then

$$h(-x) = f(-x)g(-x) = f(x)[-g(x)] = -h(x)$$

and h is odd.

(d) Let $h(x) = f(x) \pm g(x)$ where f and g are even. Then

$$h(-x) = f(-x) \pm g(-x) = f(x) \pm g(x) = h(x),$$

and so h is an even function.

(f) If f is even then

$$\int_{-a}^{a} f(x) \, dx = -\int_{a}^{0} f(-u) \, du + \int_{0}^{a} f(x) \, dx = \int_{0}^{a} f(u) \, du + \int_{0}^{a} f(x) \, dx = 2 \int_{0}^{a} f(x) \, dx.$$

(g) If f is odd then

$$\int_{-a}^{a} f(x) \, dx = -\int_{-a}^{0} f(-x) \, dx + \int_{0}^{a} f(x) \, dx = \int_{a}^{0} f(u) \, du + \int_{0}^{a} f(x) \, dx$$

$$= -\int_{0}^{a} f(u) \, du + \int_{0}^{a} f(x) \, dx = 0.$$

51. The graph is obtained by summing the series from $n = 1$ to 20. It appears that

$$f(x) = \begin{cases} x, & 0 < x < \pi \\ -\pi, & \pi < x < 2\pi. \end{cases}$$

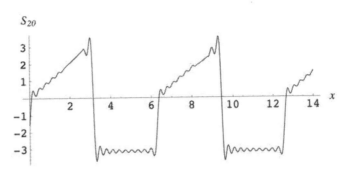

11.4 | Sturm-Liouville Problem

The terminology and concepts listed below provide an outline of the main ideas encountered in this section. These can be useful when preparing for a quiz or test.

Terminology and Concepts

- eigenvalues and eigenfunctions of a BVP
- regular Sturm-Liouville problem
- homogeneous boundary conditions
- separated boundary conditions
- singular Sturm-Liouville problem
- periodic boundary-value problem
- self-adjoint form of a linear second-order DE

The basic skills listed below summarize the more mechanical types of problems encountered in the exercise set for this section.

Basic Skills

- for a given BVP find the equation that defines the eigenvalues and find the corresponding eigenfunctions
- put a second-order DE into self-adjoint form and give an orthogonality relation

3. For $\lambda = 0$ the solution of $y'' = 0$ is $y = c_1 x + c_2$. The condition $y'(0) = 0$ implies $c_1 = 0$, so $\lambda = 0$ is an eigenvalue with corresponding eigenfunction 1.

For $\lambda = -\alpha^2 < 0$ we have $y = c_1 \cosh \alpha x + c_2 \sinh \alpha x$ and $y' = c_1 \alpha \sinh \alpha x + c_2 \alpha \cosh \alpha x$. The condition $y'(0) = 0$ implies $c_2 = 0$ and so $y = c_1 \cosh \alpha x$. Now the condition $y'(L) = 0$ implies $c_1 = 0$. Thus $y = 0$ and there are no negative eigenvalues.

For $\lambda = \alpha^2 > 0$ we have $y = c_1 \cos \alpha x + c_2 \sin \alpha x$ and $y' = -c_1 \alpha \sin \alpha x + c_2 \alpha \cos \alpha x$. The condition $y'(0) = 0$ implies $c_2 = 0$ and so $y = c_1 \cos \alpha x$. Now the condition $y'(L) = 0$ implies $-c_1 \alpha \sin \alpha L = 0$. For $c_1 \neq 0$ this condition will hold when $\alpha L = n\pi$ or $\lambda = \alpha^2 = n^2 \pi^2 / L^2$, where $n = 1, 2, 3, \ldots$. These are the positive eigenvalues with corresponding eigenfunctions $\cos(n\pi x/L)$, $n = 1, 2, 3, \ldots$.

6. The eigenfunctions are $\sin \alpha_n x$ where $\tan \alpha_n = -\alpha_n$. Thus

$$\| \sin \alpha_n x \|^2 = \int_0^1 \sin^2 \alpha_n x \, dx = \frac{1}{2} \int_0^1 (1 - \cos 2\alpha_n x) \, dx$$

$$= \frac{1}{2} \left(x - \frac{1}{2\alpha_n} \sin 2\alpha_n x \right) \Big|_0^1 = \frac{1}{2} \left(1 - \frac{1}{2\alpha_n} \sin 2\alpha_n \right)$$

$$= \frac{1}{2} \left[1 - \frac{1}{2\alpha_n} \left(2 \sin \alpha_n \cos \alpha_n \right) \right]$$

$$= \frac{1}{2} \left[1 - \frac{1}{\alpha_n} \tan \alpha_n \cos \alpha_n \cos \alpha_n \right]$$

$$= \frac{1}{2} \left[1 - \frac{1}{\alpha_n} \left(-\alpha_n \cos^2 \alpha_n \right) \right] = \frac{1}{2} \left(1 + \cos^2 \alpha_n \right).$$

9. To obtain the self-adjoint form we note that an integrating factor is $(1/x)e^{\int (1-x)dx/x} = e^{-x}$. Thus, the differential equation is

$$x e^{-x} y'' + (1 - x) e^{-x} y' + n e^{-x} y = 0$$

and the self-adjoint form is

$$\frac{d}{dx} \left[x e^{-x} y' \right] + n e^{-x} y = 0.$$

Identifying the weight function $p(x) = e^{-x}$ and noting that since $r(x) = x e^{-x}$, $r(0) = 0$ and $\lim_{x \to \infty} r(x) = 0$, we have the orthogonality relation

$$\int_0^\infty e^{-x} L_m(x) L_n(x) \, dx = 0, \quad m \neq n.$$

12. (a) Letting $\lambda = \alpha^2$ the differential equation becomes $x^2 y'' + x y' + (\alpha^2 x^2 - 1) y = 0$. This is the parametric Bessel equation with $\nu = 1$. The general solution is

$$y = c_1 J_1(\alpha x) + c_2 Y_1(\alpha x).$$

Since Y is unbounded at 0 we must have $c_2 = 0$, so that $y = c_1 J_1(\alpha x)$. The condition $J_1(3\alpha) = 0$ defines the eigenvalues $\lambda_n = \alpha_n^2$ for $n = 1, 2, 3, \ldots$. The corresponding eigenfunctions are $J_1(\alpha_n x)$.

(b) Using a CAS or Table 6.1 in Section 6.3 of the text to solve $J_1(3\alpha) = 0$ we find $3\alpha_1 = 3.8317$, $3\alpha_2 = 7.0156$, $3\alpha_3 = 10.1735$, and $3\alpha_4 = 13.3237$. The corresponding eigenvalues are $\lambda_1 = \alpha_1^2 = 1.6313$, $\lambda_2 = \alpha_2^2 = 5.4687$, $\lambda_3 = \alpha_3^2 = 11.4999$, and $\lambda_4 = \alpha_4^2 = 19.7245$.

15. (a) An orthogonality relation is

$$\int_0^1 (x_m \cos x_m x - \sin x_m x)(x_n \cos x_n x - \sin x_n x)\, dx = 0$$

where $x_m \neq x_n$ are positive solutions of $\tan x = x$.

(b) Referring to Problem 2 we use a CAS to compute

$$\int_0^1 (4.4934 \cos 4.4934 x - \sin 4.4934 x)(7.7253 \cos 7.7253 x - \sin 7.7253 x)\, dx = -2.5650 \times 10^{-4} \approx 0.$$

11.5 Bessel and Legendre Series

The terminology and concepts listed below provide an outline of the main ideas encountered in this section. These can be useful when preparing for a quiz or test.

Terminology and Concepts

- Bessel functions
- Fourier-Bessel series
- differential recurrence relations
- convergence of a Fourier-Bessel series
- Fourier-Legendre series
- convergence of a Fourier-Legendre series

The basic skills listed below summarize the more mechanical types of problems encountered in the exercise set for this section.

Basic Skills

- expand a function in a Fourier-Bessel series using Bessel functions of the same order as in the boundary condition
- expand a function in a Fourier-Legendre series

3. The boundary condition indicates that we use (15) and (16) in the text. With $b = 2$ we obtain

$$c_i = \frac{2}{4J_1^2(2\alpha_i)} \int_0^2 xJ_0(\alpha_i x)\,dx$$

$$\boxed{t = \alpha_i x \qquad dt = \alpha_i\,dx}$$

$$= \frac{1}{2J_1^2(2\alpha_i)} \cdot \frac{1}{\alpha_i^2} \int_0^{2\alpha_i} tJ_0(t)\,dt$$

$$= \frac{1}{2\alpha_i^2 J_1^2(2\alpha_i)} \int_0^{2\alpha_i} \frac{d}{dt}[tJ_1(t)]\,dt \qquad \text{[From (5) in the text]}$$

$$= \frac{1}{2\alpha_i^2 J_1^2(2\alpha_i)} tJ_1(t)\Big|_0^{2\alpha_i}$$

$$= \frac{1}{\alpha_i J_1(2\alpha_i)}.$$

Thus

$$f(x) = \sum_{i=1}^{\infty} \frac{1}{\alpha_i J_1(2\alpha_i)} J_0(\alpha_i x).$$

6. Writing the boundary condition in the form

$$2J_0(2\alpha) + 2\alpha J_0'(2\alpha) = 0$$

we identify $b = 2$ and $h = 2$. Using (17) and (18) in the text we obtain

$$c_i = \frac{2\alpha_i^2}{(4\alpha_i^2 + 4)J_0^2(2\alpha_i)} \int_0^2 xJ_0(\alpha_i x)\,dx$$

$$\boxed{t = \alpha_i x \qquad dt = \alpha_i\,dx}$$

$$= \frac{\alpha_i^2}{2(\alpha_i^2 + 1)J_0^2(2\alpha_i)} \cdot \frac{1}{\alpha_i^2} \int_0^{2\alpha_i} tJ_0(t)\,dt$$

$$= \frac{1}{2(\alpha_i^2 + 1)J_0^2(2\alpha_i)} \int_0^{2\alpha_i} \frac{d}{dt}[tJ_1(t)]\,dt \qquad \text{[From (5) in the text]}$$

$$= \frac{1}{2(\alpha_i^2 + 1)J_0^2(2\alpha_i)} tJ_1(t)\Big|_0^{2\alpha_i}$$

$$= \frac{\alpha_i J_1(2\alpha_i)}{(\alpha_i^2 + 1)J_0^2(2\alpha_i)}.$$

Thus

$$f(x) = \sum_{i=1}^{\infty} \frac{\alpha_i J_1(2\alpha_i)}{(\alpha_i^2 + 1)J_0^2(2\alpha_i)} J_0(\alpha_i x).$$

9. The boundary condition indicates that we use (19) and (20) in the text. With $b = 3$ we obtain

$$c_1 = \frac{2}{9} \int_0^3 x x^2 \, dx = \frac{2}{9} \frac{x^4}{4} \Big|_0^3 = \frac{9}{2},$$

$$c_i = \frac{2}{9 J_0^2(3\alpha_i)} \int_0^3 x J_0(\alpha_i x) x^2 \, dx$$

$$\boxed{t = \alpha_i x \qquad dt = \alpha_i \, dx}$$

$$= \frac{2}{9 J_0^2(3\alpha_i)} \cdot \frac{1}{\alpha_i^4} \int_0^{3\alpha_i} t^3 J_0(t) \, dt$$

$$= \frac{2}{9\alpha_i^4 J_0^2(3\alpha_i)} \int_0^{3\alpha_i} t^2 \frac{d}{dt} [t J_1(t)] \, dt$$

$$\boxed{\begin{array}{ll} u = t^2 & dv = \frac{d}{dt}[t J_1(t)]\, dt \\ du = 2t\, dt & v = t J_1(t) \end{array}}$$

$$= \frac{2}{9\alpha_i^4 J_0^2(3\alpha_i)} \left(t^3 J_1(t) \Big|_0^{3\alpha_i} - 2 \int_0^{3\alpha_i} t^2 J_1(t) \, dt \right).$$

With $n = 0$ in equation (6) in the text we have $J_0'(x) = -J_1(x)$, so the boundary condition $J_0'(3\alpha_i) = 0$ implies $J_1(3\alpha_i) = 0$. Then

$$c_i = \frac{2}{9\alpha_i^4 J_0^2(3\alpha_i)} \left(-2 \int_0^{3\alpha_i} \frac{d}{dt} [t^2 J_2(t)] \, dt \right) = \frac{2}{9\alpha_i^4 J_0^2(3\alpha_i)} \left(-2t^2 J_2(t) \Big|_0^{3\alpha_i} \right)$$

$$= \frac{2}{9\alpha_i^4 J_0^2(3\alpha_i)} \left[-18\alpha_i^2 J_2(3\alpha_i) \right] = \frac{-4 J_2(3\alpha_i)}{\alpha_i^2 J_0^2(3\alpha_i)}.$$

Thus

$$f(x) = \frac{9}{2} - 4 \sum_{i=1}^{\infty} \frac{J_2(3\alpha_i)}{\alpha_i^2 J_0^2(3\alpha_i)} J_0(\alpha_i x).$$

12. (a) From Problem 7, the coefficients of the Fourier-Bessel series are

$$c_i = \frac{20\alpha_i J_2(4\alpha_i)}{(2\alpha_i^2 + 1)J_1^2(4\alpha_i)}.$$

Using a CAS we find $c_1 = 26.7896$, $c_2 = -12.4624$, $c_3 = 7.1404$, $c_4 = -4.68705$, and $c_5 = 3.35619$.

(b)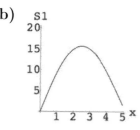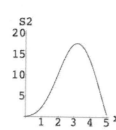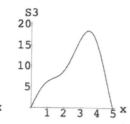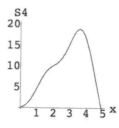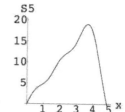

(c)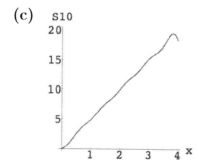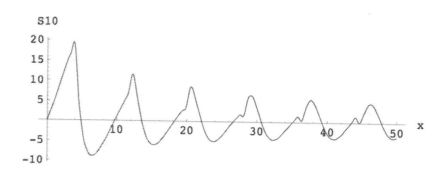

15. We compute

$$c_0 = \frac{1}{2}\int_0^1 xP_0(x)\,dx = \frac{1}{2}\int_0^1 x\,dx = \frac{1}{4}$$

$$c_1 = \frac{3}{2}\int_0^1 xP_1(x)\,dx = \frac{3}{2}\int_0^1 x^2\,dx = \frac{1}{2}$$

$$c_2 = \frac{5}{2}\int_0^1 xP_2(x)\,dx = \frac{5}{2}\int_0^1 \frac{1}{2}(3x^3 - x)\,dx = \frac{5}{16}$$

$$c_3 = \frac{7}{2}\int_0^1 xP_3(x)\,dx = \frac{7}{2}\int_0^1 \frac{1}{2}(5x^4 - 3x^2)\,dx = 0$$

$$c_4 = \frac{9}{2}\int_0^1 xP_4(x)\,dx = \frac{9}{2}\int_0^1 \frac{1}{8}(35x^5 - 30x^3 + 3x)\,dx = -\frac{3}{32}$$

$$c_5 = \frac{11}{2} \int_0^1 x P_5(x)\, dx = \frac{11}{2} \int_0^1 \frac{1}{8}(63x^6 - 70x^4 + 15x^2)\, dx = 0$$

$$c_6 = \frac{13}{2} \int_0^1 x P_6(x)\, dx = \frac{13}{2} \int_0^1 \frac{1}{16}(231x^7 - 315x^5 + 105x^3 - 5x)\, dx = \frac{13}{256}.$$

Thus

$$f(x) = \frac{1}{4} P_0(x) + \frac{1}{2} P_1(x) + \frac{5}{16} P_2(x) - \frac{3}{32} P_4(x) + \frac{13}{256} P_6(x) + \cdots.$$

The figure above is the graph of $S_5(x) = \frac{1}{4} P_0(x) + \frac{1}{2} P_1(x) + \frac{5}{16} P_2(x) - \frac{3}{32} P_4(x) + \frac{13}{256} P_6(x)$.

18. From Problem 17 we have

$$P_2(\cos\theta) = \frac{1}{4}(3\cos 2\theta + 1) \qquad \text{or} \qquad \cos 2\theta = \frac{4}{3} P_2(\cos\theta) - \frac{1}{3}.$$

Then, using $P_0(\cos\theta) = 1$,

$$F(\theta) = 1 - \cos 2\theta = 1 - \left[\frac{4}{3} P_2(\cos\theta) - \frac{1}{3}\right]$$

$$= \frac{4}{3} - \frac{4}{3} P_2(\cos\theta) = \frac{4}{3} P_0(\cos\theta) - \frac{4}{3} P_2(\cos\theta).$$

21. From (26) in Problem 19 in the text we find

$$c_0 = \int_0^1 x P_0(x)\, dx = \int_0^1 x\, dx = \frac{1}{2},$$

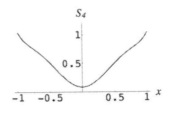

$$c_2 = 5 \int_0^1 x P_2(x)\, dx = 5 \int_0^1 \frac{1}{2}(3x^3 - x)\, dx = \frac{5}{8},$$

$$c_4 = 9 \int_0^1 x P_4(x)\, dx = 9 \int_0^1 \frac{1}{8}(35x^5 - 30x^3 + 3x)\, dx = -\frac{3}{16},$$

and

$$c_6 = 13 \int_0^1 x P_6(x)\, dx = 13 \int_0^1 \frac{1}{16}(231x^7 - 315x^5 + 105x^3 - 5x)\, dx = \frac{13}{128}.$$

Hence, from (25) in the text,

$$f(x) = \frac{1}{2} P_0(x) + \frac{5}{8} P_2(x) - \frac{3}{16} P_4(x) + \frac{13}{128} P_6 + \cdots.$$

On the interval $-1 < x < 1$ this series represents the function $f(x) = |x|$.

24. For $f(x) = x^2$ we have

$$c_0 = \frac{1}{2} \int_{-1}^{1} x^2 \cdot 1 \, dx = \frac{1}{3}$$

$$c_1 = \frac{3}{2} \int_{-1}^{1} x^2 \cdot x \, dx = 0$$

$$c_2 = \frac{5}{2} \int_{-1}^{1} x^2 \cdot \frac{1}{2}(3x^2 - 1) dx = \frac{2}{3},$$

so

$$x^2 = \frac{1}{3} P_0(x) + \frac{2}{3} P_2(x).$$

For $f(x) = x^3$ we have

$$c_0 = \frac{1}{2} \int_{-1}^{1} x^3 \cdot 1 \, dx = 0$$

$$c_1 = \frac{3}{2} \int_{-1}^{1} x^3 \cdot x \, dx = \frac{3}{5}$$

$$c_2 = \frac{5}{2} \int_{-1}^{1} x^3 \cdot \frac{1}{2}(3x^2 - 1) \, dx = 0$$

$$c_3 = \frac{7}{2} \int_{-1}^{1} x^3 \cdot \frac{1}{2}(5x^3 - 3x) \, dx = \frac{2}{5},$$

so

$$x^3 = \frac{3}{5} P_1(x) + \frac{2}{5} P_3(x).$$

11.R Chapter 11 in Review

3. Cosine, since f is even

6. Periodically extending the function we see that at $x = -1$ the function converges to $\frac{1}{2}(-1 + 0) = -\frac{1}{2}$; at $x = 0$ it converges to $\frac{1}{2}(0 + 1) = \frac{1}{2}$, and at $x = 1$ it converges to $\frac{1}{2}(-1 + 0) = -\frac{1}{2}$.

9. Since the coefficient of y in the differential equation is n^2, the weight function is the integrating factor

$$\frac{1}{a(x)} e^{\int (b/a) dx} = \frac{1}{1 - x^2} e^{\int -\frac{x}{1 - x^2} \, dx} = \frac{1}{1 - x^2} e^{\frac{1}{2} \ln(1 - x^2)} = \frac{\sqrt{1 - x^2}}{1 - x^2} = \frac{1}{\sqrt{1 - x^2}}$$

on the interval $[-1, 1]$. The orthogonality relation is

$$\int_{-1}^{1} \frac{1}{\sqrt{1 - x^2}} T_m(x) T_n(x) \, dx = 0, \quad m \neq n.$$

12. (a) For $m \neq n$

$$\int_0^L \sin \frac{(2n+1)\pi}{2L} x \, \sin \frac{(2m+1)\pi}{2L} x \, dx = \frac{1}{2} \int_0^L \left(\cos \frac{n-m}{L} \pi x - \cos \frac{n+m+1}{L} \pi x \right) dx = 0.$$

(b) From

$$\int_0^L \sin^2 \frac{(2n+1)\pi}{2L} x \, dx = \int_0^L \left(\frac{1}{2} - \frac{1}{2} \cos \frac{(2n+1)\pi}{L} x \right) dx = \frac{L}{2}$$

we see that

$$\left\| \sin \frac{(2n+1)\pi}{2L} x \right\| = \sqrt{\frac{L}{2}}.$$

15. (a) Since

$$a_0 = 2 \int_0^1 e^{-x} dx = 2(1 - e^{-1})$$

and

$$a_n = 2 \int_{-1}^1 e^{-x} \cos n\pi x \, dx = \frac{2}{1 + n^2 \pi^2} [1 - (-1)^n e^{-1}]$$

for $n = 1, 2, 3, \ldots$ we have

$$f(x) = 1 - e^{-1} + 2 \sum_{n=1}^{\infty} \frac{1 - (-1)^n e^{-1}}{1 + n^2 \pi^2} \cos n\pi x.$$

(b) Since

$$b_n = 2 \int_0^1 e^{-x} \sin n\pi x \, dx = \frac{2n\pi}{1 + n^2 \pi^2} [1 - (-1)^n e^{-1}]$$

for $n = 1, 2, 3, \ldots$ we have

$$f(x) = \sum_{n=1}^{\infty} \frac{2n\pi}{1 + n^2 \pi^2} [1 - (-1)^n e^{-1}] \sin n\pi x.$$

18. Expanding in a full Fourier series we have

$$a_0 = \frac{1}{2} \left(\int_0^2 x \, dx + \int_2^4 2 \, dx \right) = 3$$

$$a_n = \frac{1}{2} \left(\int_0^2 x \cos \frac{n\pi x}{2} \, dx + \int_2^4 2 \cos \frac{n\pi x}{2} \, dx \right) = 2 \frac{(-1)^n - 1}{n^2 \pi^2}$$

$$b_n = \frac{1}{2} \left(\int_0^2 x \sin \frac{n\pi x}{2} \, dx + \int_2^4 2 \sin \frac{n\pi x}{2} \, dx \right) = 4 \frac{-1}{n\pi}$$

so

$$f(x) = \frac{3}{2} + 2 \sum_{n=1}^{\infty} \left(\frac{(-1)^n - 1}{n^2 \pi^2} \cos \frac{n\pi x}{2} - \frac{2}{n\pi} \sin \frac{n\pi x}{2} \right).$$

21. The boundary condition indicates that we use (15) and (16) of Section 11.5 in the text. With $b = 4$ we obtain

$$c_i = \frac{2}{16J_1^2(4\alpha_i)} \int_0^4 x J_0(\alpha_i x) f(x)\, dx$$

$$= \frac{1}{8J_1^2(4\alpha_i)} \int_0^2 x J_0(\alpha_i x)\, dx \qquad \boxed{t = \alpha_i x \qquad dt = \alpha_i\, dx}$$

$$= \frac{1}{8J_1^2(4\alpha_i)} \cdot \frac{1}{\alpha_i^2} \int_0^{2\alpha_i} t J_0(t)\, dt$$

$$= \frac{1}{8J_1^2(4\alpha_i)} \int_0^{2\alpha_i} \frac{d}{dt}[t J_1(t)]\, dt \qquad \text{[From (5) in 11.5 in the text]}$$

$$= \frac{1}{8J_1^2(4\alpha_i)} t J_1(t) \Big|_0^{2\alpha_i} = \frac{J_1(2\alpha_i)}{4\alpha_i J_1^2(4\alpha_i)}.$$

Thus

$$f(x) = \frac{1}{4} \sum_{i=1}^{\infty} \frac{J_1(2\alpha_i)}{\alpha_i J_1^2(4\alpha_i)} J_0(\alpha_i x).$$

24. Identifying

$$f_e(x) = \frac{f(x) + f(-x)}{2} = \frac{e^x + e^{-x}}{2} \quad \text{and} \quad f_o(x) = \frac{f(x) - f(-x)}{2} = \frac{e^x - e^{-x}}{2},$$

we have

$$e^x = \frac{e^x + e^{-x}}{2} + \frac{e^x - e^{-x}}{2}.$$

12 Boundary-Value Problems in Rectangular Coordinates

12.1 Separable Partial Differential Equations

The terminology and concepts listed below provide an outline of the main ideas encountered in this section. These can be useful when preparing for a quiz or test.

Terminology and Concepts

- general form of a linear second-order PDE
- method of separation of variables for finding a solution of a PDE
- Superposition Principle for solutions of a homogeneous linear PDE
- hyperbolic PDE
- parabolic PDE
- elliptic PDE

The basic skills listed below summarize the more mechanical types of problems encountered in the exercise set for this section.

Basic Skills

- use separation of variables to find a product solution for a PDE
- classify a PDE as hyperbolic, parabolic, or elliptic

3. Substituting $u(x,y) = X(x)Y(y)$ into the partial differential equation yields $X'Y + XY' = XY$. Separating variables and using the separation constant $-\lambda$ we obtain

$$\frac{X'}{X} = \frac{Y - Y'}{Y} = -\lambda.$$

Then

$$X' + \lambda X = 0 \qquad \text{and} \qquad Y' - (1 + \lambda)Y = 0$$

so that

$$X = c_1 e^{-\lambda x} \qquad \text{and} \qquad Y = c_2 e^{(1+\lambda)y}.$$

A particular product solution of the partial differential equation is

$$u = XY = c_3 e^{y + \lambda(y-x)}.$$

6. Substituting $u(x, y) = X(x)Y(y)$ into the partial differential equation yields $yX'Y + xXY' = 0$. Separating variables and using the separation constant $-\lambda$ we obtain

$$\frac{X'}{xX} = -\frac{Y'}{yY} = -\lambda.$$

When $\lambda \neq 0$

$$X' + \lambda x X = 0 \quad \text{and} \quad Y' - \lambda y Y = 0$$

so that

$$X = c_1 e^{\lambda x^2/2} \quad \text{and} \quad Y = c_2 e^{-\lambda y^2/2}.$$

A particular product solution of the partial differential equation is

$$u = XY = c_3 e^{\lambda(x^2 - y^2)/2}.$$

When $\lambda = 0$ the differential equations become $X' = 0$ and $Y' = 0$, so in this case $X = c_4$, $Y = c_5$, and $u = XY = c_6$.

9. Substituting $u(x, t) = X(x)T(t)$ into the partial differential equation yields $kX''T - XT = XT'$. Separating variables and using the separation constant $-\lambda$ we obtain

$$\frac{kX'' - X}{X} = \frac{T'}{T} = -\lambda.$$

Then

$$X'' + \frac{\lambda - 1}{k} X = 0 \quad \text{and} \quad T' + \lambda T = 0.$$

The second differential equation implies $T(t) = c_1 e^{-\lambda t}$. For the first differential equation we consider three cases:

I. If $(\lambda - 1)/k = 0$ then $\lambda = 1$, $X'' = 0$, and $X(x) = c_2 x + c_3$, so

$$u = XT = e^{-t}(A_1 x + A_2).$$

II. If $(\lambda - 1)/k = -\alpha^2 < 0$, then $\lambda = 1 - k\alpha^2$, $X'' - \alpha^2 X = 0$, and $X(x) = c_4 \cosh \alpha x + c_5 \sinh \alpha x$, so

$$u = XT = (A_3 \cosh \alpha x + A_4 \sinh \alpha x)e^{-(1 - k\alpha^2)t}.$$

III. If $(\lambda - 1)/k = \alpha^2 > 0$, then $\lambda = 1 + \lambda\alpha^2$, $X'' + \alpha^2 X = 0$, and $X(x) = c_6 \cos \alpha x + c_7 \sin \alpha x$, so

$$u = XT = (A_5 \cos \alpha x + A_6 \sin \alpha x)e^{-(1 + \lambda\alpha^2)t}.$$

12. Substituting $u(x,t) = X(x)T(t)$ into the partial differential equation yields $a^2 X''T = XT'' + 2kXT'$. Separating variables and using the separation constant $-\lambda$ we obtain

$$\frac{X''}{X} = \frac{T'' + 2kT'}{a^2 T} = -\lambda.$$

Then

$$X'' + \lambda X = 0 \quad \text{and} \quad T'' + 2kT' + a^2\lambda T = 0.$$

We consider three cases:

I. If $\lambda = 0$ then $X'' = 0$ and $X(x) = c_1 x + c_2$. Also, $T'' + 2kT' = 0$ and $T(t) = c_3 + c_4 e^{-2kt}$, so

$$u = XT = (c_1 x + c_2)(c_3 + c_4 e^{-2kt}).$$

II. If $\lambda = -\alpha^2 < 0$, then $X'' - \alpha^2 X = 0$, and $X(x) = c_5 \cosh \alpha x + c_6 \sinh \alpha x$. The auxiliary equation of $T'' + 2kT' - \alpha^2 a^2 T = 0$ is $m^2 + 2km - \alpha^2 a^2 = 0$. Solving for m we obtain $m = -k \pm \sqrt{k^2 + \alpha^2 a^2}$, so $T(t) = c_7 e^{(-k+\sqrt{k^2+\alpha^2 a^2})t} + c_8 e^{(-k-\sqrt{k^2+\alpha^2 a^2})t}$. Then

$$u = XT = (c_5 \cosh \alpha x + c_6 \sinh \alpha x)(c_7 e^{(-k+\sqrt{k^2+\alpha^2 a^2})t} + c_8 e^{(-k-\sqrt{k^2+\alpha^2 a^2})t}).$$

III. If $\lambda = \alpha^2 > 0$, then $X'' + \alpha^2 X = 0$, and $X(x) = c_9 \cos \alpha x + c_{10} \sin \alpha x$. The auxiliary equation of $T'' + 2kT' + \alpha^2 a^2 T = 0$ is $m^2 + 2km + \alpha^2 a^2 = 0$. Solving for m we obtain $m = -k \pm \sqrt{k^2 - \alpha^2 a^2}$. We consider three possibilities for the discriminant $k^2 - \alpha^2 a^2$:

(i) If $k^2 - \alpha^2 a^2 = 0$ then $T(t) = c_{11} e^{-kt} + c_{12} t e^{-kt}$ and

$$u = XT = (c_9 \cos \alpha x + c_{10} \sin \alpha x)(c_{11} e^{-kt} + c_{12} t e^{-kt}).$$

From $k^2 - \alpha^2 a^2 = 0$ we have $\alpha = k/a$ so the solution can be written

$$u = XT = (c_9 \cos kx/a + c_{10} \sin kx/a)(c_{11} e^{-kt} + c_{12} t e^{-kt}).$$

(ii) If $k^2 - \alpha^2 a^2 < 0$ then $T(t) = e^{-kt}\left(c_{13} \cos \sqrt{\alpha^2 a^2 - k^2}\, t + c_{14} \sin \sqrt{\alpha^2 a^2 - k^2}\, t\right)$ and

$$u = XT = (c_9 \cos \alpha x + c_{10} \sin \alpha x)e^{-kt}\left(c_{13} \cos \sqrt{\alpha^2 a^2 - k^2}\, t + c_{14} \sin \sqrt{\alpha^2 a^2 - k^2}\, t\right).$$

(iii) If $k^2 - \alpha^2 a^2 > 0$ then $T(t) = c_{15} e^{(-k+\sqrt{k^2-\alpha^2 a^2})t} + c_{16} e^{(-k-\sqrt{k^2-\alpha^2 a^2})t}$ and

$$u = XT = (c_9 \cos \alpha x + c_{10} \sin \alpha x)\left(c_{15} e^{(-k+\sqrt{k^2-\alpha^2 a^2})t} + c_{16} e^{(-k-\sqrt{k^2-\alpha^2 a^2})t}\right).$$

15. Substituting $u(x,y) = X(x)Y(y)$ into the partial differential equation yields $X''Y + XY'' = XY$. Separating variables and using the separation constant $-\lambda$ we obtain

$$\frac{X''}{X} = \frac{Y - Y''}{Y} = -\lambda.$$

Then

$$X'' + \lambda X = 0 \qquad \text{and} \qquad Y'' - (1 + \lambda)Y = 0.$$

We consider three cases:

I. If $\lambda = 0$ then $X'' = 0$ and $X(x) = c_1 x + c_2$. Also $Y'' - Y = 0$ and $Y(y) = c_3 \cosh y + c_4 \sinh y$ so

$$u = XY = (c_1 x + c_2)(c_3 \cosh y + c_4 \sinh y).$$

II. If $\lambda = -\alpha^2 < 0$ then $X'' - \alpha^2 X = 0$ and $Y'' + (\alpha^2 - 1)Y = 0$. The solution of the first differential equation is $X(x) = c_5 \cosh \alpha x + c_6 \sinh \alpha x$. The solution of the second differential equation depends on the nature of $\alpha^2 - 1$. We consider three cases:

(i) If $\alpha^2 - 1 = 0$, or $\alpha^2 = 1$, then $Y(y) = c_7 y + c_8$ nad

$$u = XY = (c_5 \cosh x + c_6 \sinh x)(c_7 y + c_8).$$

(ii) If $\alpha^2 - 1 < 0$, or $0 < \alpha^2 < 1$, then $Y(y) = c_9 \cosh \sqrt{1 - \alpha^2}\, y + c_{10} \sinh \sqrt{1 - \alpha^2}\, y$ and

$$u = XY = (c_5 \cosh \alpha x + c_6 \sinh \alpha x)\left(c_9 \cosh \sqrt{1 - \alpha^2}\, y + c_{10} \sinh \sqrt{1 - \alpha^2}\, y\right).$$

(iii) If $\alpha^2 - 1 > 0$, or $\alpha^2 > 1$, then $Y(y) = c_{11} \cos \sqrt{\alpha^2 - 1}\, y + c_{12} \sin \sqrt{\alpha^2 - 1}\, y$ and

$$u = XY = (c_5 \cosh \alpha x + c_6 \sinh \alpha x)\left(c_{11} \cos \sqrt{\alpha^2 - 1}\, y + c_{12} \sin \sqrt{\alpha^2 - 1}\, y\right).$$

III. If $\lambda = \alpha^2 > 0$, then $X'' + \alpha^2 X = 0$ and $X(x) = c_{13} \cos \alpha x + c_{14} \sin \alpha x$. Also,
$Y'' - (1 + \alpha^2)Y = 0$ and $Y(y) = c_{15} \cosh \sqrt{1 + \alpha^2}\, y + c_{16} \sinh \sqrt{1 + \alpha^2}\, y$ so

$$u = XY = (c_{13} \cos \alpha x + c_{14} \sin \alpha x)\left(c_{15} \cosh \sqrt{1 + \alpha^2}\, y + c_{16} \sinh \sqrt{1 + \alpha^2}\, y\right).$$

18. Identifying $A = 3$, $B = 5$, and $C = 1$, we compute $B^2 - 4AC = 13 > 0$. The equation is hyperbolic.

21. Identifying $A = 1$, $B = -9$, and $C = 0$, we compute $B^2 - 4AC = 81 > 0$. The equation is hyperbolic.

24. Identifying $A = 1$, $B = 0$, and $C = 1$, we compute $B^2 - 4AC = -4 < 0$. The equation is elliptic.

27. Substituting $u(r, t) = R(r)T(t)$ into the partial differential equation yields

$$k\left(R''T + \frac{1}{r}R'T\right) = RT'.$$

Separating variables and using the separation constant $-\lambda$ we obtain

$$\frac{rR'' + R'}{rR} = \frac{T'}{kT} = -\lambda.$$

Then

$$rR'' + R' + \lambda rR = 0 \quad \text{and} \quad T' + \lambda kT = 0.$$

Letting $\lambda = \alpha^2$ and writing the first equation as $r^2R'' + rR' = \alpha^2 r^2 R = 0$ we see that it is a parametric Bessel equation of order 0. As discussed in Chapter 6 of the text, it has solution $R(r) = c_1 J_0(\alpha r) + c_2 Y_0(\alpha r)$. Since a solution of $T' + \alpha^2 kT$ is $T(t) = e^{-k\alpha^2 t}$, we see that a solution of the partial differential equation is

$$u = RT = e^{-k\alpha^2 t}[c_1 J_0(\alpha r) + c_2 Y_0(\alpha r)].$$

30. We identify $A = xy + 1$, $B = x + 2y$, and $C = 1$. Then $B^2 - 4AC = x^2 + 4y^2 - 4$. The equation $x^2 + 4y^2 = 4$ defines an ellipse. The partial differential equation is hyperbolic outside the ellipse, parabolic on the ellipse, and elliptic inside the ellipse.

12.2 Classical PDEs and Boundary-Value Problems

The terminology and concepts listed below provide an outline of the main ideas encountered in this section. These can be useful when preparing for a quiz or test.

Terminology and Concepts

- one-dimensional heat equation
- one-dimensional wave equation
- two-dimensional form of Laplace's equation
- Laplacian
- difference between initial conditions and boundary conditions
- Dirichlet condition
- Newmann condition
- Robin condition
- boundary-value problem

The basic skills listed below summarize the more mechanical types of problems encountered in the exercise set for this section.

Basic Skills

- set up a boundary-value problem involving the heat equation, the wave equation, or Laplace's equation given a set of initial and boundary conditions

3. $k\dfrac{\partial^2 u}{\partial x^2} = \dfrac{\partial u}{\partial t}$, $0 < x < L,\ t > 0$

$u(0,t) = 100,\quad \left.\dfrac{\partial u}{\partial x}\right|_{x=L} = -hu(L,t),\quad t > 0$

$u(x,0) = f(x),\quad 0 < x < L$

6. $k\dfrac{\partial^2 u}{\partial x^2} + h(u - 50) = \partial t$, $0 < x < L,\ t > 0$

$\left.\dfrac{\partial u}{\partial x}\right|_{x=0} = 0,\quad \left.\dfrac{\partial u}{\partial x}\right|_{x=L} = 0,\quad t > 0$

$u(x,0) = 100,\quad 0 < x < L$

9. $a^2\dfrac{\partial^2 u}{\partial x^2} - 2\beta\dfrac{\partial u}{\partial t} = \dfrac{\partial^2 u}{\partial t^2}$, $0 < x < L,\ t > 0$

$u(0,t) = 0,\quad u(L,t) = \sin \pi t,\quad t > 0$

$u(x,0) = f(x),\quad \left.\dfrac{\partial u}{\partial t}\right|_{t=0} = 0,\quad 0 < x < L$

12. $\dfrac{\partial^2 u}{\partial x^2} + \dfrac{\partial^2 u}{\partial y^2} = 0$, $0 < x < \pi,\ y > 0$

$u(0,y) = e^{-y},\quad u(\pi,y) = \begin{cases} 100, & 0 < y \le 1 \\ 0, & y > 1 \end{cases}$

$u(x,0) = f(x),\quad 0 < x < \pi$

12.3 | Heat Equation

The terminology and concepts listed below provide an outline of the main ideas encountered in this section. These can be useful when preparing for a quiz or test.

Terminology and Concepts

- solution of the heat equation

The basic skills listed below summarize the more mechanical types of problems encountered in the exercise set for this section.

Basic Skills

- solve the heat equation subject to a given set of initial and boundary conditions

3. Using $u = XT$ and $-\lambda$ as a separation constant we obtain

$$X'' + \lambda X = 0,$$
$$X'(0) = 0,$$
$$X'(L) = 0$$

and

$$T' + k\lambda T = 0.$$

This leads to

$$X = c_1 \cos \frac{n\pi}{L} x \qquad \text{and} \qquad T = c_2 e^{-kn^2\pi^2 t/L^2}$$

for $n = 0, 1, 2, \ldots$ ($\lambda = 0$ is an eigenvalue in this case) so that

$$u = \sum_{n=0}^{\infty} A_n e^{-kn^2\pi^2 t/L^2} \cos \frac{n\pi}{L} x.$$

Imposing

$$u(x,0) = f(x) = A_0 + \sum_{n=1}^{\infty} A_n \cos \frac{n\pi}{L} x$$

gives

$$u(x,t) = \frac{1}{L} \int_0^L f(x)\,dx + \frac{2}{L} \sum_{n=1}^{\infty} \left(\int_0^L f(x) \cos \frac{n\pi}{L} x\,dx \right) e^{-kn^2\pi^2 t/L^2} \cos \frac{n\pi}{L} x.$$

6. In Problem 5 we instead find that $X(0) = 0$ and $X(L) = 0$ so that

$$X = c_1 \sin \frac{n\pi}{L} x$$

and

$$u = \frac{2e^{-ht}}{L} \sum_{n=1}^{\infty} \left(\int_0^L f(x) \sin \frac{n\pi}{L} x\,dx \right) e^{-kn^2\pi^2 t/L^2} \sin \frac{n\pi}{L} x.$$

The terminology and concepts listed below provide an outline of the main ideas encountered in this section. These can be useful when preparing for a quiz or test.

Terminology and Concepts

- solution of the wave equation
- standing waves

The basic skills listed below summarize the more mechanical types of problems encountered in the exercise set for this section.

Basic Skills

- solve the wave equation subject to a given set of inital and boundary conditions

3. Using $u = XT$ and $-\lambda$ as a separation constant we obtain

$$X'' + \lambda X = 0,$$

$$X(0) = 0,$$

and

$$X(L) = 0,$$

$$T'' + \lambda a^2 T = 0,$$

$$T'(0) = 0.$$

Solving the differential equations we get

$$X = c_1 \sin \frac{n\pi}{L} x + c_2 \cos \frac{n\pi}{L} x \quad \text{and} \quad T = c_3 \cos \frac{n\pi a}{L} t + c_4 \sin \frac{n\pi a}{L} t$$

for $n = 1, 2, 3, \ldots$. The boundary and initial conditions give

Imposing

$$u = \sum_{n=1}^{\infty} A_n \cos \frac{n\pi a}{L} t \sin \frac{n\pi}{L} x.$$

$$u(x, 0) = \sum_{n=1}^{\infty} A_n \sin \frac{n\pi}{L} x$$

gives

$$A_n = \frac{2}{L} \left(\int_0^{L/3} \frac{3}{L} x \sin \frac{n\pi}{L} x\, dx + \int_{L/3}^{2L/3} \sin \frac{n\pi}{L} x\, dx + \int_{2L/3}^{L} \left(3 - \frac{3}{L} x \right) \sin \frac{n\pi}{L} x\, dx \right)$$

so that

$$A_1 = \frac{6\sqrt{3}}{\pi^2},$$

$$A_2 = A_3 = A_4 = 0,$$

$$A_5 = -\frac{6\sqrt{3}}{5^2\pi^2},$$

$$A_6 = 0,$$

$$A_7 = \frac{6\sqrt{3}}{7^2\pi^2}$$

$$\vdots$$

and

$$u(x,t) = \frac{6\sqrt{3}}{\pi^2}\left(\cos\frac{\pi a}{L}t\,\sin\frac{\pi}{L}x - \frac{1}{5^2}\cos\frac{5\pi a}{L}t\,\sin\frac{5\pi}{L}x + \frac{1}{7^2}\cos\frac{7\pi a}{L}t\,\sin\frac{7\pi}{L}x - \cdots\right).$$

6. Using $u = XT$ and $-\lambda$ as a separation constant we obtain

$$X'' + \lambda X = 0,$$

$$X(0) = 0,$$

$$X(1) = 0,$$

and

$$T'' + \lambda a^2 T = 0,$$

$$T'(0) = 0.$$

Solving the differential equations we get

$$X = c_1\sin n\pi x + c_2\cos n\pi x \qquad \text{and} \qquad T = c_3\cos n\pi at + c_4\sin n\pi at$$

for $n = 1, 2, 3, \ldots$. The boundary and initial conditions give

$$u = \sum_{n=1}^{\infty} A_n\cos nt\,\sin nx.$$

Imposing

$$u(x,0) = 0.01\sin 3\pi x = \sum_{n=1}^{\infty} A_n\sin n\pi x$$

gives $A_3 = 0.01$, and $A_n = 0$ for $n = 1, 2, 4, 5, 6, \ldots$ so that

$$u(x,t) = 0.01\sin 3\pi x\,\cos 3\pi at.$$

9. Using $u = XT$ and $-\lambda$ as a separation constant we obtain

$$X'' + \lambda X = 0,$$

$$X(0) = 0,$$

$$X(\pi) = 0,$$

and

$$T'' + 2\beta T' + \lambda T = 0,$$

$$T'(0) = 0.$$

Solving the differential equations we get

$$X = c_1 \sin nx + c_2 \cos nx \quad \text{and} \quad T = e^{-\beta t}\left(c_3 \cos \sqrt{n^2 - \beta^2}\, t + c_4 \sin \sqrt{n^2 - \beta^2}\, t\right)$$

The boundary conditions on X imply $c_2 = 0$ so

$$X = c_1 \sin nx \quad \text{and} \quad T = e^{-\beta t}\left(c_3 \cos \sqrt{n^2 - \beta^2}\, t + c_4 \sin \sqrt{n^2 - \beta^2}\, t\right)$$

and

$$u = \sum_{n=1}^{\infty} e^{-\beta t}\left(A_n \cos \sqrt{n^2 - \beta^2}\, t + B_n \sin \sqrt{n^2 - \beta^2}\, t\right) \sin nx.$$

Imposing

$$u(x,0) = f(x) = \sum_{n=1}^{\infty} A_n \sin nx$$

and

$$u_t(x,0) = 0 = \sum_{n=1}^{\infty} \left(B_n \sqrt{n^2 - \beta^2} - \beta A_n \right) \sin nx$$

gives

$$u(x,t) = e^{-\beta t} \sum_{n=1}^{\infty} A_n \left(\cos \sqrt{n^2 - \beta^2}\, t + \frac{\beta}{\sqrt{n^2 - \beta^2}} \sin \sqrt{n^2 - \beta^2}\, t \right) \sin nx,$$

where

$$A_n = \frac{2}{\pi} \int_0^{\pi} f(x) \sin nx\, dx.$$

12. (a) Write the differential equation in X from Problem 11 as $X^{(4)} - \alpha^4 X = 0$ where the eigenvalues are $\lambda = \alpha^2$. Then

$$X = c_1 \cosh \alpha x + c_2 \sinh \alpha x + c_3 \cos \alpha x + c_4 \sin \alpha x$$

and using $X(0) = 0$ and $X'(0) = 0$ we find $c_3 = -c_1$ and $c_4 = -c_2$. The conditions $X(L) = 0$ and $X'(L) = 0$ yield the system of equations

$$c_1(\cosh \alpha L - \cos \alpha L) + c_2(\sinh \alpha L - \sin \alpha L) = 0$$
$$c_1(\alpha \sinh \alpha L + \alpha \sin \alpha L) + c_2(\alpha \cosh \alpha L - \alpha \cos \alpha L) = 0.$$

In order for this system to have nontrivial solutions the determinant of the coefficients must be zero. That is,

$$\alpha(\cosh\alpha L - \cos\alpha L)^2 - \alpha(\sinh^2\alpha L - \sin^2\alpha L) = 0.$$

Since $\alpha = 0$ leads to $X = 0$, $\lambda = \alpha^2 = 0^2 = 0$ is not an eigenvalue. Then, dividing the above equation by α, we have

$$(\cosh\alpha L - \cos\alpha L)^2 - (\sinh^2\alpha L - \sin^2\alpha L)$$

$$= \cosh^2\alpha L - 2\cosh\alpha L\cos\alpha L + \cos^2\alpha L - \sinh^2\alpha L + \sin^2\alpha L$$

$$= -2\cosh\alpha L\cos\alpha L + 2 = 0$$

or $\cosh\alpha L\cos\alpha L = 1$. Letting $x = \alpha L$ we see that the eigenvalues are $\lambda_n = \alpha_n^2 = x_n^2/L^2$ where x_n, $n = 1, 2, 3, \dots$, are the positive roots of the equation $\cosh x\cos x = 1$.

(b) The equation $\cosh x\cos x = 1$ is the same as $\cos x = \operatorname{sech} x$. The figure indicates that the equation has an infinite number of roots.

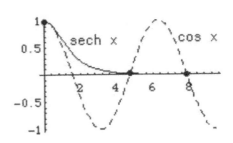

(c) Using a CAS we find the first four positive roots of $\cosh x\cos x = 1$ to be $x_1 = 4.7300$, $x_2 = 7.8532$, $x_3 = 10.9956$, and $x_4 = 14.1372$. Thus the first four eigenvalues are $\lambda_1 = x_1^2/L = 22.3733/L$, $\lambda_2 = x_2^2/L = 61.6728/L$, $\lambda_3 = x_3^2/L = 120.9034/L$, and $\lambda_4 = x_4^2/L = 199.8594/L$.

15. $u(x,t) = \dfrac{1}{2}[\sin(x+at) + \sin(x-at)] + \dfrac{1}{2a}\displaystyle\int_{x-at}^{x+at} ds$

$\qquad = \dfrac{1}{2}[\sin x\cos at + \cos x\sin at + \sin x\cos at - \cos x\sin at] + \dfrac{1}{2a}s\,\Big|_{x-at}^{x+at} = \sin x\cos at + t$

18. $u(x,t) = \dfrac{1}{2}\left[e^{-(x+at)^2} + e^{-(x-at)^2}\right]$

$\qquad = \dfrac{1}{2}\left[e^{-(x^2+2axt+a^2t^2)} + e^{-(x^2-2axt+a^2t^2)}\right]$

$\qquad = e^{-(x^2+a^2t^2)}\left[\dfrac{e^{-2axt} + e^{2axt}}{2}\right] = e^{-(x^2+a^2t^2)}\cosh 2axt$

21. (a) With $a = 1$, d'Alembert's solution is

$$u(x,t) = \frac{1}{2} \int_{x-t}^{x+t} g(s)\, ds \qquad \text{where} \qquad g(s) = \begin{cases} 1, & |s| \leq 0.1 \\ 0, & |s| > 0.1 \end{cases}.$$

Sample plots are shown below.

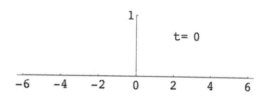

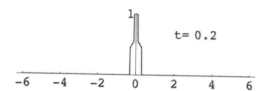

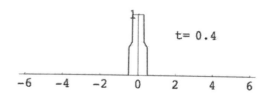

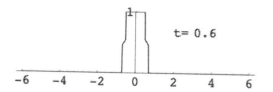

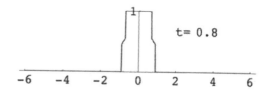

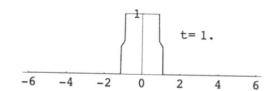

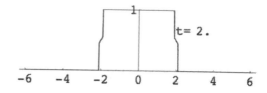

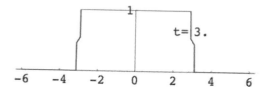

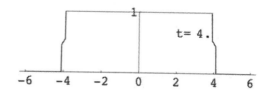

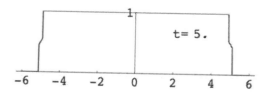

(b) Some frames of the movie are shown in part (a), The string has a roughly rectangular shape with the base on the x-axis increasing in length.

12.5 Laplace's Equation

The terminology and concepts listed below provide an outline of the main ideas encountered in this section. These can be useful when preparing for a quiz or test.

Terminology and Concepts

- solution of Laplace's equation
- Dirichlet problem

The basic skills listed below summarize the more mechanical types of problems encountered in the exercise set for this section.

Basic Skills

- solve Laplace's equation subject to a given set of boundary conditions

3. Using $u = XY$ and $-\lambda$ as a separation constant we obtain

$$X'' + \lambda X = 0,$$

$$X(0) = 0,$$

$$X(a) = 0,$$

and

$$Y'' - \lambda Y = 0,$$

$$Y(b) = 0.$$

With $\lambda = \alpha^2 > 0$ the solutions of the differential equations are

$$X = c_1 \cos \alpha x + c_2 \sin \alpha x \quad \text{and} \quad Y = c_3 \cosh \alpha y + c_4 \sinh \alpha y$$

The boundary and initial conditions imply

$$X = c_2 \sin \frac{n\pi}{a} x \quad \text{and} \quad Y = c_2 \cosh \frac{n\pi}{a} y - c_2 \frac{\cosh \frac{n\pi b}{a}}{\sinh \frac{n\pi b}{a}} \sinh \frac{n\pi}{a} y$$

for $n = 1, 2, 3, \ldots$ so that

$$u = \sum_{n=1}^{\infty} A_n \left(\cosh \frac{n\pi}{a} y - \frac{\cosh \frac{n\pi b}{a}}{\sinh \frac{n\pi b}{a}} \sinh \frac{n\pi}{a} y \right) \sin \frac{n\pi}{a} x.$$

Imposing

$$u(x,0) = f(x) = \sum_{n=1}^{\infty} A_n \sin \frac{n\pi}{a} x$$

gives

$$A_n = \frac{2}{a} \int_0^a f(x) \sin \frac{n\pi}{a} x \, dx$$

so that

$$u(x,y) = \frac{2}{a} \sum_{n=1}^{\infty} \left(\int_0^a f(x) \sin \frac{n\pi}{a} x \, dx \right) \left(\cosh \frac{n\pi}{a} y - \frac{\cosh \frac{n\pi b}{a}}{\sinh \frac{n\pi b}{a}} \sinh \frac{n\pi}{a} y \right) \sin \frac{n\pi}{a} x.$$

6. Using $u = XY$ and $-\lambda$ as a separation constant we obtain

$$X'' + \lambda X = 0,$$

$$X'(1) = 0$$

and

$$Y'' - \lambda Y = 0,$$

$$Y'(0) = 0,$$

$$Y'(\pi) = 0.$$

With $\lambda = \alpha^2 < 0$ the solutions of the differential equations are

$$X = c_1 \cosh \alpha x + c_2 \sinh \alpha x \qquad \text{and} \qquad Y = c_3 \cos \alpha y + c_4 \sin \alpha y$$

The boundary and initial conditions imply

$$X = c_1 \cosh nx - c_1 \frac{\sinh n}{\cosh n} \sinh nx \qquad \text{and} \qquad Y = c_3 \cos ny$$

for $n = 1, 2, 3, \ldots$. Since $\lambda = 0$ is an eigenvalue for both differential equations with corresponding eigenfunctions 1 and 1 we have

$$u = A_0 + \sum_{n=1}^{\infty} A_n \left(\cosh nx - \frac{\sinh n}{\cosh n} \sinh nx \right) \cos ny.$$

Imposing

$$u(0,y) = g(y) = A_0 + \sum_{n=1}^{\infty} A_n \cos ny$$

gives

$$A_0 = \frac{1}{\pi} \int_0^\pi g(y) \, dy \qquad \text{and} \qquad A_n = \frac{2}{\pi} \int_0^\pi g(y) \cos ny \, dy$$

for $n = 1, 2, 3, \ldots$ so that

$$u(x, y) = \frac{1}{\pi}\int_0^\pi g(y)\,dy + \sum_{n=1}^\infty \left(\frac{2}{\pi}\int_0^\pi g(y)\cos ny\,dy\right)\left(\cosh nx - \frac{\sinh n}{\cosh n}\sinh nx\right)\cos ny.$$

9. This boundary-value problem has the form of Problem 1 in this section, with $a = b = 1$, $f(x) = 100$, and $g(x) = 200$. The solution, then, is

$$u(x, y) = \sum_{n=1}^\infty (A_n \cosh n\pi y + B_n \sinh n\pi y)\sin n\pi x,$$

where

$$A_n = 2\int_0^1 100 \sin n\pi x\,dx = 200\left(\frac{1-(-1)^n}{n\pi}\right)$$

and

$$B_n = \frac{1}{\sinh n\pi}\left[2\int_0^1 200 \sin n\pi x\,dx - A_n \cosh n\pi\right]$$

$$= \frac{1}{\sinh n\pi}\left[400\left(\frac{1-(-1)^n}{n\pi}\right) - 200\left(\frac{1-(-1)^n}{n\pi}\right)\cosh n\pi\right]$$

$$= 200\left[\frac{1-(-1)^n}{n\pi}\right][2\operatorname{csch} n\pi - \coth n\pi].$$

12. Using $u = XY$ and $-\lambda$ as a separation constant we obtain

$$X'' + \lambda X = 0,$$

$$X'(0) = 0,$$

$$X'(\pi) = 0,$$

and

$$Y'' - \lambda Y = 0.$$

With $\lambda = \alpha^2 > 0$ the solutions of the differential equations are

$$X = c_1 \cos \alpha x + c_2 \sin \alpha x \quad \text{and} \quad Y = c_3 e^{\alpha y} + c_4 e^{-\alpha y}$$

The boundary conditions at $x = 0$ and $x = \pi$ imply $c_2 = 0$ so $X = c_1 \cos nx$ for $n = 1, 2, 3, \ldots$. Now the boundedness of u as $y \to \infty$ implies $c_3 = 0$, so $Y = c_4 e^{-ny}$. In this problem $\lambda = 0$ is also an eigenvalue with corresponding eigenfunction 1 so that

$$u = A_0 + \sum_{n=1}^\infty A_n e^{-ny} \cos nx.$$

Imposing

$$u(x,0) = f(x) = A_0 + \sum_{n=1}^{\infty} A_n \cos nx$$

gives

$$A_0 = \frac{1}{\pi} \int_0^{\pi} f(x)\,dx \qquad \text{and} \qquad A_n = \frac{2}{\pi} \int_0^{\pi} f(x) \cos nx\,dx$$

so that

$$u(x,y) = \frac{1}{\pi} \int_0^{\pi} f(x)\,dx + \sum_{n=1}^{\infty} \left(\frac{2}{\pi} \int_0^{\pi} f(x) \cos nx\,dx \right) e^{-ny} \cos nx.$$

15. Referring to the discussion in this section of the text we identify $a = b = \pi$, $f(x) = 0$, $g(x) = 1$, $F(y) = 1$, and $G(y) = 1$. Then $A_n = 0$ and

$$u_1(x,y) = \sum_{n=1}^{\infty} B_n \sinh ny \sin nx$$

where

$$B_n = \frac{2}{\pi \sinh n\pi} \int_0^{\pi} \sin nx\,dx = \frac{2[1 - (-1)^n]}{n\pi \sinh n\pi}.$$

Next

$$u_2(x,y) = \sum_{n=1}^{\infty} \left(A_n \cosh nx + B_n \sinh nx \right) \sin ny$$

where

$$A_n = \frac{2}{\pi} \int_0^{\pi} \sin ny\,dy = \frac{2[1 - (-1)^n]}{n\pi}$$

and

$$B_n = \frac{1}{\sinh n\pi} \left(\frac{2}{\pi} \int_0^{\pi} \sin ny\,dy - A_n \cosh n\pi \right)$$

$$= \frac{1}{\sinh n\pi} \left(\frac{2[1 - (-1)^n]}{n\pi} - \frac{2[1 - (-1)^n]}{n\pi} \cosh n\pi \right)$$

$$= \frac{2[1 - (-1)^n]}{n\pi \sinh n\pi} (1 - \cosh n\pi).$$

Now

$$A_n \cosh nx + B_n \sinh nx = \frac{2[1 - (-1)^n]}{n\pi} \left[\cosh nx + \frac{\sinh nx}{\sinh n\pi}(1 - \cosh n\pi) \right]$$

$$= \frac{2[1 - (-1)^n]}{n\pi \sinh n\pi} [\cosh nx \sinh n\pi + \sinh nx - \sinh nx \cosh n\pi]$$

$$= \frac{2[1 - (-1)^n]}{n\pi \sinh n\pi} [\sinh nx + \sinh n(\pi - x)]$$

and

$$u(x,y) = u_1 + u_2 = \frac{2}{\pi}\sum_{n=1}^{\infty}\frac{1-(-1)^n}{n\sinh n\pi}\sinh ny \sin nx$$

$$+ \frac{2}{\pi}\sum_{n=1}^{\infty}\frac{[1-(-1)^n][\sinh nx + \sinh n(\pi - x)]}{n\sinh n\pi}\sin ny.$$

18. From the figure showing the boundary conditions we see that the maximum value of the temperature is 1 at $(1, 2)$ and $(2, 1)$.

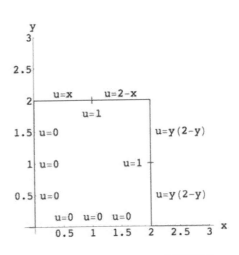

12.6 | Nonhomogeneous Boundary-Value Problems

The terminology and concepts listed below provide an outline of the main ideas encountered in this section. These can be useful when preparing for a quiz or test.

Terminology and Concepts

- nonhomogeneous BVP
- steady-state solution
- transient solution

The basic skills listed below summarize the more mechanical types of problems encountered in the exercise set for this section.

Basic Skills

- solve a nonhomogeneous BVP

3. If we let $u(x,t) = v(x,t) + \psi(x)$, then we obtain as in Example 1 in the text

$$k\psi'' + r = 0$$

or

$$\psi(x) = -\frac{r}{2k}x^2 + c_1 x + c_2.$$

The boundary conditions become

$$u(0,t) = v(0,t) + \psi(0) = u_0$$

$$u(1,t) = v(1,t) + \psi(1) = u_0.$$

Letting $\psi(0) = \psi(1) = u_0$ we obtain homogeneous boundary conditions in v:

$$v(0,t) = 0 \quad \text{and} \quad v(1,t) = 0.$$

Now $\psi(0) = \psi(1) = u_0$ implies $c_2 = u_0$ and $c_1 = r/2k$. Thus

$$\psi(x) = -\frac{r}{2k}x^2 + \frac{r}{2k}x + u_0 = u_0 - \frac{r}{2k}x(x-1).$$

To determine $v(x,t)$ we solve

$$k\frac{\partial^2 v}{\partial x^2} = \frac{\partial v}{dt}, \quad 0 < x < 1, \ t > 0$$

$$v(0,t) = 0, \quad v(1,t) = 0,$$

$$v(x,0) = \frac{r}{2k}x(x-1) - u_0.$$

Separating variables, we find

$$v(x,t) = \sum_{n=1}^{\infty} A_n e^{-kn^2\pi^2 t} \sin n\pi x,$$

where

$$A_n = 2\int_0^1 \left[\frac{r}{2k}x(x-1) - u_0\right]\sin n\pi x\, dx = 2\left[\frac{u_0}{n\pi} + \frac{r}{kn^3\pi^3}\right][(-1)^n - 1]. \tag{1}$$

Hence, a solution of the original problem is

$$u(x,t) = \psi(x) + v(x,t)$$

$$= u_0 - \frac{r}{2k}x(x-1) + \sum_{n=1}^{\infty} A_n e^{-kn^2\pi^2 t} \sin n\pi x,$$

where A_n is defined in (1).

6. Substituting $u(x,t) = v(x,t) + \psi(x)$ into the partial differential equation gives

$$k\frac{\partial^2 v}{\partial x^2} + k\psi'' - hv - h\psi = \frac{\partial v}{\partial t}.$$

This equation will be homogeneous provided ψ satisfies

$$k\psi'' - h\psi = 0.$$

Since k and h are positive, the general solution of this latter linear second-order equation is

$$\psi(x) = c_1 \cosh\sqrt{\frac{h}{k}}\,x + c_2 \sinh\sqrt{\frac{h}{k}}\,x.$$

From $\psi(0) = 0$ and $\psi(\pi) = u_0$ we find $c_1 = 0$ and $c_2 = u_0/\sinh\sqrt{h/k}\,\pi$. Hence

$$\psi(x) = u_0\frac{\sinh\sqrt{h/k}\,x}{\sinh\sqrt{h/k}\,\pi}.$$

Now the new problem is

$$k\frac{\partial^2 v}{\partial x^2} - hv = \frac{\partial v}{\partial t}, \quad 0 < x < \pi, \ t > 0$$

$$v(0,t) = 0, \quad v(\pi,t) = 0, \quad t > 0$$

$$v(x,0) = -\psi(x), \quad 0 < x < \pi.$$

If we let $v = XT$ then

$$\frac{X''}{X} = \frac{T' + hT}{kT} = -\lambda.$$

With $\lambda = \alpha^2 > 0$, the separated differential equations

$$X'' + \alpha^2 X = 0 \quad \text{and} \quad T' + \left(h + k\alpha^2\right)T = 0.$$

have the respective solutions

$$X(x) = c_3 \cos\alpha x + c_4 \sin\alpha x$$

$$T(t) = c_5 e^{-\left(h + k\alpha^2\right)t}.$$

From $X(0) = 0$ we get $c_3 = 0$ and from $X(\pi) = 0$ we find $\alpha = n$ for $n = 1, 2, 3, \ldots$. Consequently, it follows that

$$v(x,t) = \sum_{n=1}^{\infty} A_n e^{-\left(h + kn^2\right)t}\sin nx$$

where

$$A_n = -\frac{2}{\pi} \int_0^\pi \psi(x) \sin nx \, dx.$$

Hence a solution of the original problem is

$$u(x,t) = u_0 \frac{\sinh \sqrt{h/k}\, x}{\sinh \sqrt{h/k}\, \pi} + e^{-ht} \sum_{n=1}^\infty A_n e^{-kn^2 t} \sin nx$$

where

$$A_n = -\frac{2}{\pi} \int_0^\pi u_0 \frac{\sinh \sqrt{h/k}\, x}{\sinh \sqrt{h/k}\, \pi} \sin nx \, dx.$$

Using the exponential definition of the hyperbolic sine and integration by parts we find

$$A_n = \frac{2u_0 nk(-1)^n}{\pi\,(h + kn^2)}.$$

9. Substituting $u(x,t) = v(x,t) + \psi(x)$ into the partial differential equation gives

$$a^2 \frac{\partial^2 v}{\partial x^2} + a^2 \psi'' + Ax = \frac{\partial^2 v}{\partial t^2}.$$

This equation will be homogeneous provided ψ satisfies

$$a^2 \psi'' + Ax = 0.$$

The solution of this differential equation is

$$\psi(x) = -\frac{A}{6a^2} x^3 + c_1 x + c_2.$$

From $\psi(0) = 0$ we obtain $c_2 = 0$, and from $\psi(1) = 0$ we obtain $c_1 = A/6a^2$. Hence

$$\psi(x) = \frac{A}{6a^2}(x - x^3).$$

Now the new problem is

$$a^2 \frac{\partial^2 v}{\partial x^2} = \frac{\partial^2 v}{\partial t^2}$$

$$v(0,t) = 0, \quad v(1,t) = 0, \quad t > 0,$$

$$v(x,0) = -\psi(x), \quad v_t(x,0) = 0, \quad 0 < x < 1.$$

Identifying this as the wave equation solved in Section 12.4 in the text with $L = 1$, $f(x) = -\psi(x)$, and $g(x) = 0$ we obtain

$$v(x,t) = \sum_{n=1}^{\infty} A_n \cos n\pi at \sin n\pi x$$

where

$$A_n = 2 \int_0^1 [-\psi(x)] \sin n\pi x \, dx = \frac{A}{3a^2} \int_0^1 (x^3 - x) \sin n\pi x \, dx = \frac{2A(-1)^n}{a^2 \pi^3 n^3}.$$

Thus

$$u(x,t) = \frac{A}{6a^2}(x - x^3) + \frac{2A}{a^2 \pi^3} \sum_{n=1}^{\infty} \frac{(-1)^n}{n^3} \cos n\pi at \sin n\pi x.$$

12. Substituting $u(x,y) = v(x,y) + \psi(x)$ into Poisson's equation we obtain

$$\frac{\partial^2 v}{\partial x^2} + \psi''(x) + h + \frac{\partial^2 v}{\partial y^2} = 0.$$

The equation will be homogeneous provided ψ satisfies $\psi''(x) + h = 0$ or $\psi(x) = -\frac{h}{2}x^2 + c_1 x + c_2$. From $\psi(0) = 0$ we obtain $c_2 = 0$. From $\psi(\pi) = 1$ we obtain

$$c_1 = \frac{1}{\pi} + \frac{h\pi}{2}.$$

Then

$$\psi(x) = \left(\frac{1}{\pi} + \frac{h\pi}{2}\right) x - \frac{h}{2}x^2.$$

The new boundary-value problem is

$$\frac{\partial^2 v}{\partial x^2} + \frac{\partial^2 v}{\partial y^2} = 0$$

$$v(0,y) = 0, \quad v(\pi, y) = 0,$$

$$v(x,0) = -\psi(x), \quad 0 < x < \pi.$$

This is Problem 11 in Section 12.5. The solution is

$$v(x,y) = \sum_{n=1}^{\infty} A_n e^{-ny} \sin nx$$

where

$$A_n = \frac{2}{\pi} \int_0^\pi [-\psi(x) \sin nx] \, dx$$

$$= \frac{2(-1)^n}{m}\left(\frac{1}{\pi} + \frac{h\pi}{2}\right) - h(-1)^n\left(\frac{\pi}{n} + \frac{2}{n^2}\right).$$

Thus

$$u(x,y) = v(x,y) + \psi(x) = \left(\frac{1}{\pi} + \frac{h\pi}{2}\right)x - \frac{h}{2}x^2 + \sum_{n=1}^\infty A_n e^{-ny}\sin nx.$$

15. With $k = 1$ and $L = 1$ in Method 2 the eigenfunctions of $X'' + \lambda X = 0$, $X(0) = 0$, $X(1) = 0$ are $\sin n\pi x$, $n = 1, 2, 3, \ldots$. Assuming that $u(x,t) = \sum_{n=1}^\infty u_n(t)\sin n\pi x$, the formal partial derivatives of u are

$$\frac{\partial^2 u}{\partial x^2} = \sum_{n=1}^\infty u_n(t)(-n^2\pi^2)\sin n\pi x \qquad \text{and} \qquad \frac{\partial u}{\partial t} = \sum_{n=1}^\infty u_n'(t)\sin n\pi x.$$

Assuming that $-1 + x - x\cos t = \sum_{n=1}^\infty F_n(t)\sin n\pi x$ we have

$$F_n(t) = \frac{2}{1}\int_0^1 (-1 + x - x\cos t)\sin n\pi x \, dx = \frac{2[-1 + (-1)^n\cos t]}{n\pi}.$$

Then

$$-1 + x - x\cos t = \frac{2}{\pi}\sum_{n=1}^\infty \frac{-1 + (-1)^n\cos t}{n}\sin n\pi x$$

and

$$u_t - u_{xx} = \sum_{n=1}^\infty [u_n'(t) + n^2\pi^2 u_n(t)]\sin n\pi x$$

$$= -1 + x - x\cos t = \frac{2}{\pi}\sum_{n=1}^\infty \frac{-1 + (-1)^n\cos t}{n}\sin n\pi x.$$

Equating coefficients we obtain

$$u_n'(t) + n^2\pi^2 u_n(t) = \frac{2[-1 + (-1)^n\cos t]}{n\pi}.$$

This is a linear first-order differential equation whose solution is

$$u_n(t) = \frac{2}{n\pi}\left[-\frac{1}{n^2\pi^2} + (-1)^n\frac{n^2\pi^2\cos t + \sin t}{n^4\pi^4 + 1}\right] + C_n e^{-n^2\pi^2 t}.$$

Thus

$$u(x,t) = \sum_{n=1}^{\infty} \frac{2}{n\pi} \left[-\frac{1}{n^2\pi^2} + (-1)^n \frac{n^2\pi^2 \cos t + \sin t}{n^4\pi^4 + 1} \right] \sin n\pi x + \sum_{n=1}^{\infty} C_n e^{-n^2\pi^2 t} \sin n\pi x$$

and $u(x,0) = x(1-x)$ implies

$$\sum_{n=1}^{\infty} \frac{2}{n\pi} \left[-\frac{1}{n^2\pi^2} + (-1)^n \frac{n^2\pi^2}{n^4\pi^4 + 1} + C_n \right] \sin n\pi x = x(1-x).$$

Hence

$$\frac{2}{n\pi} \left[-\frac{1}{n^2\pi^2} + (-1)^n \frac{n^2\pi^2}{n^4\pi^4 + 1} + C_n \right] = \frac{2}{1} \int_0^1 x(1-x) \sin n\pi x \, dx = 2 \left[\frac{1 - (-1)^n}{n^3\pi^3} \right]$$

and

$$C_n = \frac{4 - 2(-1)^n}{n^3\pi^3} - (-1)^n \frac{2n\pi}{n^4\pi^4 + 1}.$$

Therefore

$$u(x,t) = \sum_{n=1}^{\infty} \frac{2}{n\pi} \left[-\frac{1}{n^2\pi^2} + (-1)^n \frac{n^2\pi^2 \cos t + \sin t}{n^4\pi^4 + 1} \right] \sin n\pi x$$

$$+ \sum_{n=1}^{\infty} \left[\frac{4 - 2(-1)^n}{n^3\pi^3} - (-1)^n \frac{2n\pi}{n^4\pi^4 + 1} \right] e^{-n^2\pi^2 t} \sin n\pi x.$$

18. If this problem appears in your text, then it should be deleted since it cannot be solved by methods discussed in the text.

12.7 | Orthogonal Series Expansions

The terminology and concepts listed below provide an outline of the main ideas encountered in this section. These can be useful when preparing for a quiz or test.

Terminology and Concepts

- orthogonal series expansion

The basic skills listed below summarize the more mechanical types of problems encountered in the exercise set for this section.

Basic Skills

- solve a BVP involving orthogonal series expansions or generalized Fourier series

3. Separating variables in Laplace's equation gives

$$X'' + \alpha^2 X = 0$$

$$Y'' - \alpha^2 Y = 0$$

and

$$X(x) = c_1 \cos \alpha x + c_2 \sin \alpha x$$

$$Y(y) = c_3 \cosh \alpha y + c_4 \sinh \alpha y.$$

From $u(0, y) = 0$ we obtain $X(0) = 0$ and $c_1 = 0$. From $u_x(a, y) = -hu(a, y)$ we obtain $X'(a) = -hX(a)$ and

$$\alpha \cos \alpha a = -h \sin \alpha a \qquad \text{or} \qquad \tan \alpha a = -\frac{\alpha}{h}.$$

Let α_n, where $n = 1, 2, 3, \ldots$, be the consecutive positive roots of this equation. From $u(x, 0) = 0$ we obtain $Y(0) = 0$ and $c_3 = 0$. Thus

$$u(x, y) = \sum_{n=1}^{\infty} A_n \sinh \alpha_n y \sin \alpha_n x.$$

Now

$$f(x) = \sum_{n=1}^{\infty} A_n \sinh \alpha_n b \sin \alpha_n x$$

and

$$A_n \sinh \alpha_n b = \frac{\int_0^a f(x) \sin \alpha_n x \, dx}{\int_0^a \sin^2 \alpha_n x \, dx}.$$

Since

$$\int_0^a \sin^2 \alpha_n x \, dx = \frac{1}{2} \left[a - \frac{1}{2\alpha_n} \sin 2\alpha_n a \right] = \frac{1}{2} \left[a - \frac{1}{\alpha_n} \sin \alpha_n a \cos \alpha_n a \right]$$

$$= \frac{1}{2} \left[a - \frac{1}{h\alpha_n} (h \sin \alpha_n a) \cos \alpha_n a \right]$$

$$= \frac{1}{2} \left[a - \frac{1}{h\alpha_n} (-\alpha_n \cos \alpha_n a) \cos \alpha_n a \right] = \frac{1}{2h} \left[ah + \cos^2 \alpha_n a \right],$$

we have

$$A_n = \frac{2h}{\sinh \alpha_n b [ah + \cos^2 \alpha_n a]} \int_0^a f(x) \sin \alpha_n x \, dx.$$

6. Substituting $u(x,t) = v(x,t) + \psi(x)$ into the partial differential equation gives

$$a^2 \frac{\partial^2 v}{\partial x^2} + \psi''(x) = \frac{\partial^2 v}{\partial t^2}.$$

This equation will be homogeneous if $\psi''(x) = 0$ or $\psi(x) = c_1 x + c_2$. The boundary condition $u(0,t) = 0$ implies $\psi(0) = 0$ which implies $c_2 = 0$. Thus $\psi(x) = c_1 x$. Using the second boundary condition, we obtain

$$E\left(\frac{\partial v}{\partial x} + \psi'\right)\Big|_{x=L} = F_0,$$

which will be homogeneous when

$$E\psi'(L) = F_0.$$

Since $\psi'(x) = c_1$ we conclude that $c_1 = F_0/E$ and

$$\psi(x) = \frac{F_0}{E}x.$$

The new boundary-value problem is

$$a^2 \frac{\partial^2 v}{\partial x^2} = \frac{\partial^2 v}{\partial t^2}, \quad 0 < x < L, \quad t > 0$$

$$v(0,t) = 0, \quad \frac{\partial v}{\partial x}\Big|_{x=L} = 0, \quad t > 0,$$

$$v(x,0) = -\frac{F_0}{E}x, \quad \frac{\partial v}{\partial t}\Big|_{t=0} = 0, \quad 0 < x < L.$$

Referring to Example 2 in the text we see that

$$v(x,t) = \sum_{n=1}^{\infty} A_n \cos a\left(\frac{2n-1}{2L}\right)\pi t \, \sin\left(\frac{2n-1}{2L}\right)\pi x$$

where

$$-\frac{F_0}{E}x = \sum_{n=1}^{\infty} A_n \sin\left(\frac{2n-1}{2L}\right)\pi x$$

and

$$A_n = \frac{-F_0 \int_0^L x \sin\left(\frac{2n-1}{2L}\right)\pi x \, dx}{E \int_0^L \sin^2\left(\frac{2n-1}{2L}\right)\pi x \, dx} = \frac{8F_0 L(-1)^n}{E\pi^2(2n-1)^2}.$$

Thus

$$u(x,t) = v(x,t) + \psi(x)$$

$$= \frac{F_0}{E}x + \frac{8F_0 L}{E\pi^2}\sum_{n=1}^{\infty} \frac{(-1)^n}{(2n-1)^2}\cos a\left(\frac{2n-1}{2L}\right)\pi t \, \sin\left(\frac{2n-1}{2L}\right)\pi x.$$

9. The eigenfunctions of the associated homogeneous boundary-value problem are $\sin \alpha_n x$, $n = 1, 2,$ 3, ... , where the α_n are the consecutive positive roots of $\tan \alpha = -\alpha$. We assume that

$$u(x,t) = \sum_{n=1}^{\infty} u_n(t) \sin \alpha_n x \qquad \text{and} \qquad xe^{-2t} = \sum_{n=1}^{\infty} F_n(t) \sin \alpha_n x.$$

Then

$$F_n(t) = \frac{e^{-2t} \int_0^1 x \sin \alpha_n x \, dx}{\int_0^1 \sin^2 \alpha_n x \, dx}.$$

Since $\alpha_n \cos \alpha_n = -\sin \alpha_n$ and

$$\int_0^1 \sin^2 \alpha_n x \, dx = \frac{1}{2} \left[1 - \frac{1}{2\alpha_n} \sin 2\alpha_n \right],$$

we have

$$e^{-2t} \int_0^1 x \sin \alpha_n x \, dx = e^{-2t} \left(\frac{\sin \alpha_n - \alpha_n \cos \alpha_n}{\alpha_n^2} \right) = \frac{2 \sin \alpha_n}{\alpha_n^2} e^{-2t}$$

$$\int_0^1 \sin^2 \alpha_n x \, dx = \frac{1}{2}[1 + \cos^2 \alpha_n]$$

and so

$$F_n(t) = \frac{4 \sin \alpha_n}{\alpha_n^2 (1 + \cos^2 \alpha_n)} e^{-2t}.$$

Substituting the assumptions into $u_t - k u_{xx} = x e^{-2t}$ and equating coefficients leads to the linear first-order differential equation

$$u_n'(t) + k \alpha_n^2 u(t) = \frac{4 \sin \alpha_n}{\alpha_n^2 (1 + \cos^2 \alpha_n)} e^{-2t}$$

whose solution is

$$u_n(t) = \frac{4 \sin \alpha_n}{\alpha_n^2 (1 + \cos^2 \alpha_n)(k\alpha_n^2 - 2)} e^{-2t} + C_n e^{-k\alpha_n^2 t}.$$

From

$$u(x,t) = \sum_{n=1}^{\infty} \left[\frac{4 \sin \alpha_n}{\alpha_n^2 (1 + \cos^2 \alpha_n)(k\alpha_n^2 - 2)} e^{-2t} + C_n e^{-k\alpha_n^2 t} \right] \sin \alpha_n x$$

and the initial condition $u(x,0) = 0$ we see

$$C_n = -\frac{4 \sin \alpha_n}{\alpha_n^2 (1 + \cos^2 \alpha_n)(k\alpha_n^2 - 2)}.$$

The formal solution of the original problem is then

$$u(x,t) = \sum_{n=1}^{\infty} \frac{4 \sin \alpha_n}{\alpha_n^2 (1 + \cos^2 \alpha_n)(k\alpha_n^2 - 2)} \left(e^{-2t} - e^{-k\alpha_n^2 t} \right) \sin \alpha_n x.$$

12.8 Higher-Dimensional Problems

The terminology and concepts listed below provide an outline of the main ideas encountered in this section. These can be useful when preparing for a quiz or test.

Terminology and Concepts

- two-dimensional heat equation
- two-dimensional wave equation
- double sine series
- double cosine series

The basic skills listed below summarize the more mechanical types of problems encountered in the exercise set for this section.

Basic Skills

- solve a two-dimensional heat equation with specified boundary and initial conditions
- solve a two-dimensional wave equation with specified boundary and initial conditions

3. The conditions $X(0) = 0$ and $Y(0) = 0$ give $c_1 = 0$ and $c_3 = 0$. The conditions $X(\pi) = 0$ and $Y(\pi) = 0$ yield two sets of eigenvalues:

$$\alpha = m, \ m = 1, 2, 3, \ldots \qquad \text{and} \qquad \beta = n, \ n = 1, 2, 3, \ldots .$$

A product solution of the partial differential equation that satisfies the boundary conditions is

$$u_{mn}(x, y, t) = \left(A_{mn} \cos a\sqrt{m^2 + n^2}\, t + B_{mn} \sin a\sqrt{m^2 + n^2}\, t \right) \sin mx \sin ny.$$

To satisfy the initial conditions we use the Superposition Principle:

$$u(x, y, t) = \sum_{m=1}^{\infty} \sum_{n=1}^{\infty} \left(A_{mn} \cos a\sqrt{m^2 + n^2}\, t + B_{mn} \sin a\sqrt{m^2 + n^2}\, t \right) \sin mx \sin ny.$$

The initial condition $u_t(x, y, 0) = 0$ implies $B_{mn} = 0$ and

$$u(x, y, t) = \sum_{m=1}^{\infty} \sum_{n=1}^{\infty} A_{mn} \cos a\sqrt{m^2 + n^2}\, t \sin mx \sin ny.$$

6. The boundary and initial conditions are

$$u(0, y, z) = 0, \qquad u(a, y, z) = 0,$$

$$u(x, 0, z) = 0, \qquad u(x, b, z) = 0,$$

$$u(x, y, 0) = f(x, y), \qquad u(x, y, c) = 0.$$

The conditions $X(0) = Y(0) = 0$ give $c_1 = c_3 = 0$. The conditions $X(a) = Y(b) = 0$ yield two sets of eigenvalues:

$$\alpha = \frac{m\pi}{a}, \; m = 1, 2, 3, \ldots \qquad \text{and} \qquad \beta = \frac{n\pi}{b}, \; n = 1, 2, 3, \ldots.$$

Let

$$\omega_{mn}^2 = \frac{m^2\pi^2}{a^2} + \frac{n^2\pi^2}{b^2}.$$

Then the boundary condition $Z(c) = 0$ gives

$$c_5 \cosh c\omega_{mn} + c_6 \sinh c\omega_{mn} = 0$$

from which we obtain

$$Z(z) = c_5 \left(\cosh w_{mn}z - \frac{\cosh c\omega_{mn}}{\sinh c\omega_{mn}} \sinh \omega z \right)$$

$$= \frac{c_5}{\sinh c\omega_{mn}} (\sinh c\omega_{mn} \cosh \omega_{mn}z - \cosh c\omega_{mn} \sinh \omega_{mn}z) = c_{mn} \sinh \omega_{mn}(c - z).$$

By the Superposition Principle

$$u(x, y, t) = \sum_{m=1}^{\infty} \sum_{n=1}^{\infty} A_{mn} \sinh \omega_{mn}(c - z) \sin \frac{m\pi}{a}x \sin \frac{n\pi}{b}y$$

where

$$A_{mn} = \frac{4}{ab \sinh c\omega_{mn}} \int_0^b \int_0^a f(x, y) \sin \frac{m\pi}{a}x \sin \frac{n\pi}{b}y \, dx \, dy.$$

12.R Chapter 12 in Review

3. Substituting $u(x, t) = v(x, t) + \psi(x)$ into the partial differential equation we obtain

$$k \frac{\partial^2 v}{\partial x^2} + k\psi''(x) = \frac{\partial v}{\partial t}.$$

This equation will be homogeneous provided ψ satisfies

$$k\psi'' = 0 \quad \text{or} \quad \psi = c_1 x + c_2.$$

Considering

$$u(0, t) = v(0, t) + \psi(0) = u_0$$

we set $\psi(0) = u_0$ so that $\psi(x) = c_1 x + u_0$. Now

$$-\frac{\partial u}{\partial x}\bigg|_{x=\pi} = -\frac{\partial v}{\partial x}\bigg|_{x=\pi} - \psi'(x) = v(\pi, t) + \psi(\pi) - u_1$$

is equivalent to

$$\frac{\partial v}{\partial x}\bigg|_{x=\pi} + v(\pi, t) = u_1 - \psi'(x) - \psi(\pi) = u_1 - c_1 - (c_1\pi + u_0),$$

which will be homogeneous when

$$u_1 - c_1 - c_1\pi - u_0 = 0 \quad \text{or} \quad c_1 = \frac{u_1 - u_0}{1 + \pi}.$$

The steady-state solution is

$$\psi(x) = \left(\frac{u_1 - u_0}{1 + \pi}\right) x + u_0.$$

6. The boundary-value problem is

$$\frac{\partial^2 u}{\partial x^2} + x^2 = \frac{\partial^2 u}{\partial t^2}, \quad 0 < x < 1, \quad t > 0,$$

$$u(0, t) = 1, \quad u(1, t) = 0, \quad t > 0,$$

$$u(x, 0) = f(x), \quad u_t(x, 0) = 0, \quad 0 < x < 1.$$

Substituting $u(x, t) = v(x, t) + \psi(x)$ into the partial differential equation gives

$$\frac{\partial^2 v}{\partial x^2} + \psi''(x) + x^2 = \frac{\partial^2 v}{\partial t^2}.$$

This equation will be homogeneous provided $\psi''(x) + x^2 = 0$ or

$$\psi(x) = -\frac{1}{12}x^4 + c_1 x + c_2.$$

From $\psi(0) = 1$ and $\psi(1) = 0$ we obtain $c_1 = -11/12$ and $c_2 = 1$. The new problem is

$$\frac{\partial^2 v}{\partial x^2} = \frac{\partial^2 v}{\partial t^2}, \quad 0 < x < 1, \quad t > 0,$$

$$v(0, t) = 0, \quad v(1, t) = 0, \quad t > 0,$$

$$v(x, 0) = f(x) - \psi(x), \quad v_t(x, 0) = 0, \quad 0 < x < 1.$$

From Section 12.4 in the text we see that $B_n = 0$,

$$A_n = 2 \int_0^1 [f(x) - \psi(x)] \sin n\pi x \, dx = 2 \int_0^1 \left[f(x) + \frac{1}{12} x^4 + \frac{11}{12} x - 1 \right] \sin n\pi x \, dx,$$

and

$$v(x, t) = \sum_{n=1}^{\infty} A_n \cos n\pi t \sin n\pi x.$$

Thus

$$u(x, t) = v(x, t) + \psi(x) = -\frac{1}{12} x^4 - \frac{11}{12} x + 1 + \sum_{n=1}^{\infty} A_n \cos n\pi t \sin n\pi x.$$

9. Using $u = XY$ and $-\lambda$ as a separation constant leads to

$$X'' - \lambda X = 0,$$

and

$$Y'' + \lambda Y = 0,$$

$$Y(0) = 0,$$

$$Y(\pi) = 0.$$

Then

$$X = c_1 e^{nx} + c_2 e^{-nx} \quad \text{and} \quad Y = c_3 \cos ny + c_4 \sin ny$$

for $n = 1, 2, 3, \ldots$. Since u must be bounded as $x \to \infty$) we define $c_1 = 0$. Also $Y(0) = 0$ implies $c_3 = 0$ so

$$u = \sum_{n=1}^{\infty} A_n e^{-nx} \sin ny.$$

Imposing

$$u(0, y) = 50 = \sum_{n=1}^{\infty} A_n \sin ny$$

gives

$$A_n = \frac{2}{\pi} \int_0^{\pi} 50 \sin ny \, dy = \frac{100}{n\pi} [1 - (-1)^n]$$

so that

$$u(x, y) = \sum_{n=1}^{\infty} \frac{100}{n\pi} [1 - (-1)^n] e^{-nx} \sin ny.$$

12. Substituting $u(x, t) = v(x, t) + \psi(x)$ into the partial differential equation results in $\psi'' = -\sin x$ and $\psi(x) = c_1 x + c_2 + \sin x$. The boundary conditions $\psi(0) = 400$ and $\psi(\pi) = 200$ imply $c_1 = -200/\pi$ and $c_2 = 400$ so

$$\psi(x) = -\frac{200}{\pi} x + 400 + \sin x.$$

Solving

$$\frac{\partial^2 v}{\partial x^2} = \frac{\partial v}{\partial t}, \qquad 0 < x < \pi, \quad t > 0$$

$$v(0, t) = 0, \qquad v(\pi, t) = 0, \quad t > 0$$

$$u(x, 0) = 400 + \sin x - \left(-\frac{200}{\pi} x + 400 + \sin x \right) = \frac{200}{\pi} x, \qquad 0 < x < \pi$$

using separation of variables with separation constant $-\lambda$, where $\lambda = \alpha^2$, gives

$$X'' + \alpha^2 X = 0 \qquad \text{and} \qquad T' + \alpha^2 T = 0.$$

Using $X(0) = 0$ and $X(\pi) = 0$ we determine $\alpha^2 = n^2$, $X(x) = c_2 \sin nx$, and $T(t) = c_3 e^{-n^2 t}$. Then

$$v(x, t) = \sum_{n=1}^{\infty} A_n e^{-n^2 t} \sin nx$$

and

$$v(x, 0) = \frac{200}{\pi} x = \sum_{n=1}^{\infty} A_n \sin nx$$

so

$$A_n = \frac{400}{\pi^2} \int_0^{\pi} x \sin nx \, dx = \frac{400}{n\pi} (-1)^{n+1}.$$

Thus

$$u(x, t) = -\frac{200}{\pi} x + 400 + \sin x + \frac{400}{\pi} \sum_{n=1}^{\infty} \frac{(-1)^{n+1}}{n} e^{-n^2 t} \sin nx.$$

15. The boundary conditions are nonhomogeneous so we try to find a solution of the form $u(x, t) = v(x, t) + \psi(x)$. Substituting into the partial differential equation and using the boundary conditions we find that $\psi''(x) = 0$, $\psi(0) = u_0$, $\psi'(1) + \psi(1) = u_1$ so $\psi(x) = ax + b$, $b = u_0$, and $2a + b = u_1$. Solving gives $a = \frac{1}{2}(u_1 - u_0)$, so

$$\psi(x) = \frac{1}{2}(u_1 - u_0)x + u_0.$$

The boundary-value problem for the function $\nu(x, t)$ is given by

$$\frac{\partial^2 \nu}{\partial x^2} = \frac{\partial \nu}{\partial t}, \quad 0 < x < 1, \quad t > 0$$

$$\nu(0, t) = 0, \quad \left.\frac{\partial u}{\partial x}\right|_{x=1} = -\nu(1, t), \quad t > 0$$

$$\nu(x, 0) = \frac{1}{2}(u_0 - u_1), \quad 0 < x < 1.$$

We now proceed as in Example 1 in Section 12.7, using $k = 1$ and $h = 1$ to obtain

$$\nu(x, t) = \sum_{n=1}^{\infty} A_n e^{-\alpha_n^2 t} \sin \alpha_n x,$$

where α_n, $n = 1, 2, \ldots$ are the consecutive positive roots of the equation $\tan \alpha = -\alpha$. The coefficients are determined from the initial condition

$$\frac{1}{2}(u_0 - u_1)x = \sum_{n=1}^{\infty} A_n \sin \alpha_n x.$$

Thus

$$A_n = \frac{\dfrac{1}{2}(u_0 - u_1) \displaystyle\int_0^1 x \sin \alpha_n x \, dx}{\displaystyle\int_0^1 \sin^2 \alpha_n x}$$

$$= \frac{1}{2}(u_0 - u_1) \frac{\dfrac{\sin \alpha_n - \alpha_n \cos \alpha_n}{\alpha_n^2}}{\dfrac{1}{2}\left(1 + \cos^2 \alpha_n\right)}$$

$$= (u_0 - u_1) \frac{-2\alpha_n \cos \alpha_n}{\alpha_n^2 \left(1 + \cos^2 \alpha_n\right)} \quad \leftarrow \begin{array}{l} \tan \alpha_n = -\alpha_n \text{ so} \\ \sin \alpha_n = -\alpha_n \cos \alpha_n \end{array}$$

$$= 2(u_1 - u_o) \frac{\cos \alpha_n}{\alpha_n \left(1 + \cos^2 \alpha_n\right)}.$$

The solution is then

$$\nu(x, t) = 2(u_1 - u_0) \sum_{n=1}^{\infty} \frac{\cos \alpha_n}{\alpha_n \left(1 + \cos^2 \alpha_n\right)} e^{-\alpha_n^2 t} \sin \alpha_n x$$

so

$$u(x, t) = u_0 + \frac{1}{2}(u_1 - u_0)x + 2(u_1 - u_0) \sum_{n=1}^{\infty} \frac{\cos \alpha_n}{\alpha_n \left(1 + \cos^2 \alpha_n\right)} e^{-\alpha_n^2 t} \sin \alpha_n x.$$

13 Boundary-Value Problems in Other Coordinate Systems

13.1 Polar Coordinates

The terminology and concepts listed below provide an outline of the main ideas encountered in this section. These can be useful when preparing for a quiz or test.

Terminology and Concepts

- Laplacian in polar coordinates
- double eigenvalue

The basic skills listed below summarize the more mechanical types of problems encountered in the exercise set for this section.

Basic Skills

- find the steady-state of a plate that is described in terms of polar coordinates

3. We have

$$A_0 = \frac{1}{2\pi} \int_0^{2\pi} \left(2\pi\theta - \theta^2\right) d\theta = \frac{2\pi^2}{3}$$

$$A_n = \frac{1}{\pi} \int_0^{2\pi} \left(2\pi\theta - \theta^2\right) \cos n\theta \, d\theta = -\frac{4}{n^2}$$

$$B_n = \frac{1}{\pi} \int_0^{2\pi} \left(2\pi\theta - \theta^2\right) \sin n\theta \, d\theta = 0$$

and so

$$u(r, \theta) = \frac{2\pi^2}{3} - 4 \sum_{n=1}^{\infty} \frac{r^n}{n^2} \cos n\theta.$$

6. We solve

$$\frac{\partial^2 u}{\partial r^2} + \frac{1}{r} \frac{\partial u}{\partial r} + \frac{1}{r^2} \frac{\partial^2 u}{\partial \theta^2} = 0, \quad 0 < \theta < \frac{\pi}{2}, \quad 0 < r < c,$$

$$u(c, \theta) = f(\theta), \quad 0 < \theta < \frac{\pi}{2},$$

$$u(r, 0) = 0, \quad u(r, \pi/2) = 0, \quad 0 < r < c.$$

Proceeding as in Example 1 in the text we obtain the separated differential equations

$$r^2 R'' + r R' - \lambda R = 0$$

$$\Theta'' + \lambda \Theta = 0.$$

Taking $\lambda = \alpha^2$ the solutions are

$$\Theta(\theta) = c_1 \cos \alpha\theta + c_2 \sin \alpha\theta$$

$$R(r) = c_3 r^\alpha + c_4 r^{-\alpha}.$$

Since we want $R(r)$ to be bounded as $r \to 0$ we require $c_4 = 0$. Applying the boundary conditions $\Theta(0) = 0$ and $\Theta(\pi/2) = 0$ we find that $c_1 = 0$ and $\alpha = 2n$ for $n = 1, 2, 3, \ldots$. Therefore

$$u(r, \theta) = \sum_{n=1}^{\infty} A_n r^{2n} \sin 2n\theta.$$

From

$$u(c, \theta) = f(\theta) = \sum_{n=1}^{\infty} A_n c^n \sin 2n\theta$$

we find

$$A_n = \frac{4}{\pi c^{2n}} \int_0^{\pi/2} f(\theta) \sin 2n\theta \, d\theta.$$

9. Proceeding as in Example 1 in the text and again using the periodicity of $u(r, \theta)$, we have

$$\Theta(\theta) = c_1 \cos \alpha\theta + c_2 \sin \alpha\theta$$

where $\alpha = n$ for $n = 0, 1, 2, \ldots$. Then

$$R(r) = c_3 r^n + c_4 r^{-n}.$$

[We do not have $c_4 = 0$ in this case since $0 < a \le r$.] Since $u(b, \theta) = 0$ we have

$$u(r, \theta) = A_0 \ln \frac{r}{b} + \sum_{n=1}^{\infty} \left[\left(\frac{b}{r} \right)^n - \left(\frac{r}{b} \right)^n \right] \left[A_n \cos n\theta + B_n \sin n\theta \right].$$

From

$$u(a, \theta) = f(\theta) = A_0 \ln \frac{a}{b} + \sum_{n=1}^{\infty} \left[\left(\frac{b}{a} \right)^n - \left(\frac{a}{b} \right)^n \right] \left[A_n \cos n\theta + B_n \sin n\theta \right]$$

we find

$$A_0 \ln \frac{a}{b} = \frac{1}{2\pi} \int_0^{2\pi} f(\theta) \, d\theta,$$

$$\left[\left(\frac{b}{a}\right)^n - \left(\frac{a}{b}\right)^n\right] A_n = \frac{1}{\pi}\int_0^{2\pi} f(\theta)\cos n\theta\, d\theta,$$

and

$$\left[\left(\frac{b}{a}\right)^n - \left(\frac{a}{b}\right)^n\right] B_n = \frac{1}{\pi}\int_0^{2\pi} f(\theta)\sin n\theta\, d\theta.$$

12. We solve

$$\frac{\partial^2 u}{\partial r^2} + \frac{1}{r}\frac{\partial u}{\partial r} + \frac{1}{r^2}\frac{\partial^2 u}{\partial\theta^2} = 0, \quad 0 < \theta < \pi, \quad a < r < b,$$

$$u(a,\theta) = \theta(\pi - \theta), \quad u(b,\theta) = 0, \quad 0 < \theta < \pi,$$

$$u(r,0) = 0, \quad u(r,\pi) = 0, \quad a < r < b.$$

Proceeding as in Example 1 in the text we obtain the separated differential equations

$$r^2 R'' + rR' - \lambda R = 0$$

$$\Theta'' + \lambda\Theta = 0.$$

Taking $\lambda = \alpha^2$ the solutions are

$$\Theta(\theta) = c_1\cos\alpha\theta + c_2\sin\alpha\theta$$

$$R(r) = c_3 r^\alpha + c_4 r^{-\alpha}.$$

Applying the boundary conditions $\Theta(0) = 0$ and $\Theta(\pi) = 0$ we find that $c_1 = 0$ and $\alpha = n$ for $n = 1, 2, 3, \ldots$. The boundary condition $R(b) = 0$ gives

$$c_3 b^n + c_4 b^{-n} = 0 \quad \text{and} \quad c_4 = -c_3 b^{2n}.$$

Then

$$R(r) = c_3\left(r^n - \frac{b^{2n}}{r^n}\right) = c_3\left(\frac{r^{2n} - b^{2n}}{r^n}\right)$$

and

$$u(r,\theta) = \sum_{n=1}^{\infty} A_n\left(\frac{r^{2n} - b^{2n}}{r^n}\right)\sin n\theta.$$

From

$$u(a,\theta) = \theta(\pi - \theta) = \sum_{n=1}^{\infty} A_n\left(\frac{a^{2n} - b^{2n}}{a^n}\right)\sin n\theta$$

we find

$$A_n \left(\frac{a^{2n} - b^{2n}}{a^n} \right) = \frac{2}{\pi} \int_0^\pi (\theta\pi - \theta^2) \sin n\theta \, d\theta = \frac{4}{n^3\pi} [1 - (-1)^n].$$

Thus

$$u(r, \theta) = \frac{4}{\pi} \sum_{n=1}^\infty \frac{1 - (-1)^n}{n^3} \frac{r^{2n} - b^{2n}}{a^{2n} - b^{2n}} \left(\frac{a}{r} \right)^n \sin n\theta.$$

15. We solve

$$\frac{\partial^2 u}{\partial r^2} + \frac{1}{r} \frac{\partial u}{\partial r} + \frac{1}{r^2} \frac{\partial^2 u}{\partial \theta^2} = 0, \quad 0 < \theta < \pi, \quad 0 < r < 2,$$

$$u(2, \theta) = \begin{cases} u_0, & 0 < \theta < \pi/2 \\ 0, & \pi/2 < \theta < \pi \end{cases}$$

$$\frac{\partial u}{\partial \theta} \bigg|_{\theta=0} = 0, \quad \frac{\partial u}{\partial \theta} \bigg|_{\theta=\pi} = 0, \quad 0 < r < 2.$$

Proceeding as in Example 1 in the text we obtain the separated differential equations

$$r^2 R'' + r R' - \lambda R = 0$$

$$\Theta'' + \lambda \Theta = 0.$$

Taking $\lambda = \alpha^2$ the solutions are

$$\Theta(\theta) = c_1 \cos \alpha\theta + c_2 \sin \alpha\theta$$

$$R(r) = c_3 r^\alpha + c_4 r^{-\alpha}.$$

Applying the boundary conditions $\Theta'(0) = 0$ and $\Theta'(\pi) = 0$ we find that $c_2 = 0$ and $\alpha = n$ for $n = 0, 1, 2, \dots$. Since we want $R(r)$ to be bounded as $r \to 0$ we require $c_4 = 0$. Thus

$$u(r, \theta) = A_0 + \sum_{n=1}^\infty A_n r^n \cos n\theta.$$

From

$$u(2, \theta) = \begin{cases} u_0, & 0 < \theta < \pi/2 \\ 0, & \pi/2 < \theta < \pi \end{cases} = A_0 + \sum_{n=1}^\infty A_n 2^n \cos n\theta$$

we find

$$A_0 = \frac{1}{2} \frac{2}{\pi} \int_0^{\pi/2} u_0 \, d\theta = \frac{u_0}{2}$$

and

$$2^n A_n = \frac{2u_0}{\pi} \int_0^{\pi/2} \cos n\theta \, d\theta = \frac{2u_0}{\pi} \frac{\sin n\pi/2}{n}.$$

Therefore

$$u(r, \theta) = \frac{u_0}{2} + \frac{2u_0}{\pi} \sum_{n=1}^\infty \frac{1}{n} \left(\sin \frac{n\pi}{2} \right) \left(\frac{r}{2} \right)^n \cos n\theta.$$

13.2 Polar and Cylindrical Coordinates

The terminology and concepts listed below provide an outline of the main ideas encountered in this section. These can be useful when preparing for a quiz or test.

Terminology and Concepts

- radial symmetry
- radial vibrations
- standing waves
- nodal lines
- Laplacian in cylindrical coordinates

The basic skills listed below summarize the more mechanical types of problems encountered in the exercise set for this section.

Basic Skills

- find the displacement of a circular membrane that is clamped along its circumference and given an initial velocity possessing radial symmetry
- find the steady-state temperature in a cylinder subject to given boundary conditions

3. Referring to Example 2 in the text we have

$$R(r) = c_1 J_0(\alpha r) + c_2 Y_0(\alpha r)$$

$$Z(z) = c_3 \cosh \alpha z + c_4 \sinh \alpha z$$

where $c_2 = 0$ and $J_0(2\alpha) = 0$ defines the positive eigenvalues $\lambda_n = \alpha_n^2$. From $Z(4) = 0$ we obtain

$$c_3 \cosh 4\alpha_n + c_4 \sinh 4\alpha_n = 0 \qquad \text{or} \qquad c_4 = -c_3 \frac{\cosh 4\alpha_n}{\sinh 4\alpha_n}.$$

Then

$$Z(z) = c_3 \left[\cosh \alpha_n z - \frac{\cosh 4\alpha_n}{\sinh 4\alpha_n} \sinh \alpha_n z \right] = c_3 \frac{\sinh 4\alpha_n \cosh \alpha_n z - \cosh 4\alpha_n \sinh \alpha_n z}{\sinh 4\alpha_n}$$

$$= c_3 \frac{\sinh \alpha_n (4 - z)}{\sinh 4\alpha_n}$$

and

$$u(r,z) = \sum_{n=1}^{\infty} A_n \frac{\sinh \alpha_n(4-z)}{\sinh 4\alpha_n} J_0(\alpha_n r).$$

From

$$u(r,0) = u_0 = \sum_{n=1}^{\infty} A_n J_0(\alpha_n r)$$

we obtain

$$A_n = \frac{2u_0}{4J_1^2(2\alpha_n)} \int_0^2 r J_0(\alpha_n r)\, dr = \frac{u_0}{\alpha_n J_1(2\alpha_n)}.$$

Thus the temperature in the cylinder is

$$u(r,z) = u_0 \sum_{n=1}^{\infty} \frac{\sinh \alpha_n(4-z) J_0(\alpha_n r)}{\alpha_n \sinh 4\alpha_n J_1(2\alpha_n)}.$$

6. The boundary-value problem in this case is

$$\frac{\partial^2 u}{\partial r^2} + \frac{1}{r}\frac{\partial u}{\partial r} + \frac{\partial^2 u}{\partial z^2} = 0, \quad 0 < r < 2, \quad 0 < z < 4$$

$$u(2,z) = 50, \quad 0 < z < 4$$

$$\left.\frac{\partial u}{\partial z}\right|_{z=0} = 0, \quad \left.\frac{\partial u}{\partial z}\right|_{z=4} = 0, \quad 0 < r < 2.$$

We have $u(r,z) = v(r,z) + 50$, so

$$\frac{\partial^2 v}{\partial r^2} + \frac{1}{r}\frac{\partial v}{\partial r} + \frac{\partial^2 v}{\partial z^2} = 0, \quad 0 < r < 2, \quad 0 < z < 4$$

$$v(2,z) = 0, \quad 0 < z < 4$$

$$\left.\frac{\partial v}{\partial z}\right|_{z=0} = 0, \quad \left.\frac{\partial v}{\partial z}\right|_{z=4} = 0, \quad 0 < r < 2,$$

which implies that $v(r,z) = 0$ and so $u(r,z) = 50$.

9. Letting $u(r,t) = R(r)T(t)$ and separating variables we obtain

$$\frac{R'' + \frac{1}{r}R'}{R} = \frac{T'}{kT} = -\lambda \quad \text{and} \quad R'' + \frac{1}{r}R' + \lambda R = 0, \quad T' + \lambda kT = 0.$$

From the last equation we find $T(t) = e^{-\lambda kt}$. If $\lambda < 0$, $T(t)$ increases without bound as $t \to \infty$. Thus we assume $\lambda = \alpha^2 > 0$. Now

$$R'' + \frac{1}{r}R' + \alpha^2 R = 0$$

is a parametric Bessel equation with solution

$$R(r) = c_1 J_0(\alpha r) + c_2 Y_0(\alpha r).$$

Since Y_0 is unbounded as $r \to 0$ we take $c_2 = 0$. Then $R(r) = c_1 J_0(\alpha r)$ and the boundary condition $u(c, t) = R(c)T(t) = 0$ implies $J_0(\alpha c) = 0$. This latter equation defines the positive eigenvalues $\lambda_n = \alpha_n^2$. Thus

$$u(r, t) = \sum_{n=1}^{\infty} A_n J_0(\alpha_n r) e^{-\alpha_n^2 kt}.$$

From

$$u(r, 0) = f(r) = \sum_{n=1}^{\infty} A_n J_0(\alpha_n r)$$

we find

$$A_n = \frac{2}{c^2 J_1^2(\alpha_n c)} \int_0^c r J_0(\alpha_n r) f(r)\, dr, \quad n = 1, 2, 3, \ldots.$$

12. We solve

$$\frac{\partial^2 u}{\partial r^2} + \frac{1}{r}\frac{\partial u}{\partial r} + \frac{\partial^2 u}{\partial z^2} = 0, \quad 0 < r < 1, \quad z > 0$$

$$\left.\frac{\partial u}{\partial r}\right|_{r=1} = -hu(1, z), \quad z > 0$$

$$u(r, 0) = u_0, \quad 0 < r < 1.$$

assuming $u = RZ$ we get

$$\frac{R'' + \frac{1}{r}R'}{R} = -\frac{Z''}{Z} = -\lambda$$

and so

$$rR'' + R' + \lambda^2 rR = 0 \quad \text{and} \quad Z'' - \lambda Z = 0.$$

Letting $\lambda = \alpha^2$ we then have

$$R(r) = c_1 J_0(\alpha r) + c_2 Y_0(\alpha r) \quad \text{and} \quad Z(z) = c_3 e^{-\alpha z} + c_4 e^{\alpha z}.$$

We use the exponential form of the solution of $Z'' - \alpha^2 Z = 0$ since the domain of the variable z is a semi-infinite interval. As usual we define $c_2 = 0$ since the temperature is surely bounded as $r \to 0$. Hence $R(r) = c_1 J_0(\alpha r)$. Now the boundary-condition $u_r(1, z) + hu(1, z) = 0$ is equivalent to

$$\alpha J_0'(\alpha) + h J_0(\alpha) = 0. \tag{4}$$

The eigenvalues α_n are the positive roots of (4) above. Finally, we must now define $c_4 = 0$ since the temperature is also expected to be bounded as $z \to \infty$. A product solution of the partial differential equation that satisfies the first boundary condition is given by

$$u_n(r, z) = A_n e^{-\alpha_n z} J_0(\alpha_n r).$$

Therefore

$$u(r, z) = \sum_{n=1}^{\infty} A_n e^{-\alpha_n z} J_0(\alpha_n r)$$

is another formal solution. At $z = 0$ we have $u_0 = A_n J_0(\alpha_n r)$. In view of (4) above we use equations (17) and (18) of Section 11.5 in the text with the identification $b = 1$:

$$A_n = \frac{2\alpha_n^2}{(\alpha_n^2 + h^2) J_0^2(\alpha_n)} \int_0^1 r J_0(\alpha_n r) u_0 \, dr$$

$$= \frac{2\alpha_n^2 u_0}{(\alpha_n^2 + h^2) J_0^2(\alpha_n) \alpha_n^2} t J_1(t) \Big|_0^{\alpha_n} = \frac{2\alpha_n u_0 J_1(\alpha_n)}{(\alpha_n^2 + h^2) J_0^2(\alpha_n)}. \tag{5}$$

Since $J_0' = -J_1$ [see equation (6) of Section 11.5 in the text] it follows from (4) above that $\alpha_n J_1(\alpha_n) = h J_0(\alpha_n)$. Thus (5) above simplifies to

$$A_n = \frac{2u_0 h}{(\alpha_n^2 + h^2) J_0(\alpha_n)}.$$

A solution to the boundary-value problem is then

$$u(r, z) = 2u_0 h \sum_{n=1}^{\infty} \frac{e^{-\alpha_n z}}{(\alpha_n^2 + h^2) J_0(\alpha_n)} J_0(\alpha_n r).$$

15. (a) Writing the partial differential equation in the form

$$g\left(x \frac{\partial^2 u}{\partial x^2} + \frac{\partial u}{\partial x}\right) = \frac{\partial^2 u}{\partial t^2}$$

and separating variables we obtain

$$\frac{xX'' + X'}{X} = \frac{T''}{gT} = -\lambda.$$

Letting $\lambda = \alpha^2$ we obtain

$$xX'' + X' + \alpha^2 X = 0 \qquad \text{and} \qquad T'' + g\alpha^2 T = 0.$$

Letting $x = \tau^2/4$ in the first equation we obtain $dx/d\tau = \tau/2$ or $d\tau/dx = 2\tau$. Then

$$\frac{dX}{dx} = \frac{dX}{d\tau}\frac{d\tau}{dx} = \frac{2}{\tau}\frac{dX}{d\tau}$$

and

$$\frac{d^2 X}{dx^2} = \frac{d}{dx}\left(\frac{2}{\tau}\frac{dX}{d\tau}\right) = \frac{2}{\tau}\frac{d}{dx}\left(\frac{dX}{d\tau}\right) + \frac{dX}{d\tau}\frac{d}{dx}\left(\frac{2}{\tau}\right)$$

$$= \frac{2}{\tau}\frac{d}{d\tau}\left(\frac{dX}{d\tau}\right)\frac{d\tau}{dx} + \frac{dX}{d\tau}\frac{d}{d\tau}\left(\frac{2}{\tau}\right)\frac{d\tau}{dx} = \frac{4}{\tau^2}\frac{d^2 X}{d\tau^2} - \frac{4}{\tau^3}\frac{dX}{d\tau}.$$

Thus

$$xX'' + X' + \alpha^2 X = \frac{\tau^2}{4}\left(\frac{4}{\tau^2}\frac{d^2 X}{d\tau^2} - \frac{4}{\tau^3}\frac{dX}{d\tau}\right) + \frac{2}{\tau}\frac{dX}{d\tau} + \alpha^2 X = \frac{d^2 X}{d\tau^2} + \frac{1}{\tau}\frac{dX}{d\tau} + \alpha^2 X = 0.$$

This is a parametric Bessel equation with solution

$$X(\tau) = c_1 J_0(\alpha\tau) + c_2 Y_0(\alpha\tau).$$

(b) To insure a finite solution at $x = 0$ (and thus $\tau = 0$) we set $c_2 = 0$. The condition $u(L, t) = X(L)T(t) = 0$ implies $X\big|_{x=L} = X\big|_{\tau=2\sqrt{L}} = c_1 J_0(2\alpha\sqrt{L}) = 0$, which defines positive eigenvalues $\lambda_n = \alpha_n^2$. The solution of $T'' + g\alpha^2 T = 0$ is

$$T(t) = c_3 \cos(\alpha_n\sqrt{g}\,t) + c_4 \sin(\alpha_n\sqrt{g}\,t).$$

The boundary condition $u_t(x, 0) = X(x)T'(0) = 0$ implies $c_4 = 0$. Thus

$$u(\tau, t) = \sum_{n=1}^{\infty} A_n \cos(\alpha_n\sqrt{g}\,t) J_0(\alpha_n\tau).$$

From

$$u(\tau, 0) = f(\tau^2/4) = \sum_{n=1}^{\infty} A_n J_0(\alpha_n\tau)$$

we find

$$A_n = \frac{2}{(2\sqrt{L})^2 J_1^2(2\alpha_n\sqrt{L})} \int_0^{2\sqrt{L}} \tau J_0(\alpha_n\tau) f(\tau^2/4)\,d\tau \qquad \boxed{v = \tau/2,\ dv = d\tau/2}$$

$$= \frac{1}{2L J_1^2(2\alpha_n\sqrt{L})} \int_0^{\sqrt{L}} 2v J_0(2\alpha_n v) f(v^2) 2\,dv$$

$$= \frac{2}{LJ_1^2(2\alpha_n\sqrt{L})} \int_0^{\sqrt{L}} v J_0(2\alpha_n v) f(v^2)\, dv.$$

The solution of the boundary-value problem is

$$u(x,t) = \sum_{n=1}^{\infty} A_n \cos(\alpha_n\sqrt{g}\,t) J_0(2\alpha_n\sqrt{x}\,).$$

18. (a) With $c = 10$ in Example 1 in the text the eigenvalues are $\lambda_n = \alpha_n^2 = x_n^2/100$ where x_n is a positive root of $J_0(x) = 0$. From a CAS we find that $x_1 = 2.4048$, $x_2 = 5.5201$, and $x_3 = 8.6537$, so that the first three eigenvalues are $\lambda_1 = 0.0578$, $\lambda_2 = 0.3047$, and $\lambda_3 = 0.7489$. The corresponding coefficients are

$$A_1 = \frac{2}{100 J_1^2(x_1)} \int_0^{10} r J_0(x_1 r/10)(1 - r/10)\, dr = 0.7845,$$

$$A_2 = \frac{2}{100 J_1^2(x_2)} \int_0^{10} r J_0(x_2 r/10)(1 - r/10)\, dr = 0.0687,$$

and

$$A_3 = \frac{2}{100 J_1^2(x_3)} \int_0^{10} r J_0(x_3 r/10)(1 - r/10)\, dr = 0.0531.$$

Since $g(r) = 0$, $B_n = 0$, $n = 1, 2, 3, \ldots$, and the third partial sum of the series solution is

$$S_3(r,t) = \sum_{n=1}^{\infty} A_n \cos(x_n t/10) J_0(x_n r/10)$$

$$= 0.7845\cos(0.2405t) J_0(0.2405r) + 0.0687\cos(0.5520t) J_0(0.5520r)$$

$$+ 0.0531\cos(0.8654t) J_0(0.8654r).$$

(b)

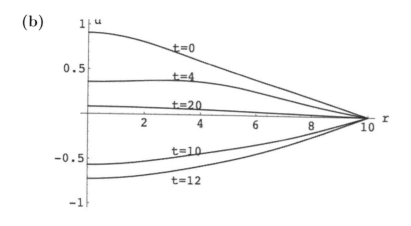

13.3 Spherical Coordinates

The terminology and concepts listed below provide an outline of the main ideas encountered in this section. These can be useful when preparing for a quiz or test.

Terminology and Concepts

- Laplacian in spherical coordinates

The basic skills listed below summarize the more mechanical types of problems encountered in the exercise set for this section.

Basic Skills

- find the steady-state temperature in a sphere or a portion of a sphere subject to given boundary conditions

3. The coefficients are given by

$$A_n = \frac{2n+1}{2c^n} \int_0^\pi \cos\theta \, P_n(\cos\theta) \sin\theta \, d\theta = \frac{2n+1}{2c^n} \int_0^\pi P_1(\cos\theta) P_n(\cos\theta) \sin\theta \, d\theta$$

$$\boxed{x = \cos\theta, \ dx = -\sin\theta \, d\theta}$$

$$= \frac{2n+1}{2c^n} \int_{-1}^1 P_1(x) P_n(x) \, dx.$$

Since $P_n(x)$ and $P_m(x)$ are orthogonal for $m \neq n$, $A_n = 0$ for $n \neq 1$ and

$$A_1 = \frac{2(1)+1}{2c^1} \int_{-1}^1 P_1(x) P_1(x) \, dx = \frac{3}{2c} \int_{-1}^1 x^2 dx = \frac{1}{c}.$$

Thus

$$u(r,\theta) = \frac{r}{c} P_1(\cos\theta) = \frac{r}{c} \cos\theta.$$

6. Referring to Example 1 in the text we have

$$R(r) = c_1 r^n \qquad \text{and} \qquad \Theta(\theta) = P_n(\cos\theta).$$

Now $\Theta(\pi/2) = 0$ implies that n is odd, so

$$u(r,\theta) = \sum_{n=0}^{\infty} A_{2n+1} r^{2n+1} P_{2n+1}(\cos\theta).$$

From

$$u(c,\theta) = f(\theta) = \sum_{n=0}^{\infty} A_{2n+1} c^{2n+1} P_{2n+1}(\cos\theta).$$

we see that

$$A_{2n+1} c^{2n+1} = (4n+3) \int_0^{\pi/2} f(\theta) \sin\theta \, P_{2n+1}(\cos\theta) \, d\theta.$$

Thus

$$u(r, \theta) = \sum_{n=0}^{\infty} A_{2n+1} r^{2n+1} P_{2n+1}(\cos\theta)$$

where

$$A_{2n+1} = \frac{4n+3}{c^{2n+1}} \int_0^{\pi/2} f(\theta) \sin\theta \, P_{2n+1}(\cos\theta) \, d\theta.$$

9. Checking the hint, we find

$$\frac{1}{r} \frac{\partial^2}{\partial r^2}(ru) = \frac{1}{r} \frac{\partial}{\partial r}\left[r\frac{\partial u}{\partial r} + u \right] = \frac{1}{r}\left[r\frac{\partial^2 u}{\partial r^2} + \frac{\partial u}{\partial r} + \frac{\partial u}{\partial r} \right] = \frac{\partial^2 u}{\partial r^2} + \frac{2}{r}\frac{\partial u}{\partial r}.$$

The partial differential equation then becomes

$$\frac{\partial^2}{\partial r^2}(ru) = r\frac{\partial u}{\partial t}.$$

Now, letting $ru(r,t) = v(r,t) + \psi(r)$, since the boundary condition is nonhomogeneous, we obtain

$$\frac{\partial^2}{\partial r^2}\left[v(r,t) + \psi(r) \right] = r\frac{\partial}{\partial t}\left[\frac{1}{r}v(r,t) + \psi(r) \right]$$

or

$$\frac{\partial^2 v}{\partial r^2} + \psi''(r) = \frac{\partial v}{\partial t}.$$

This differential equation will be homogeneous if $\psi''(r) = 0$ or $\psi(r) = c_1 r + c_2$. Now

$$u(r,t) = \frac{1}{r}v(r,t) + \frac{1}{r}\psi(r) \qquad \text{and} \qquad \frac{1}{r}\psi(r) = c_1 + \frac{c_2}{r}.$$

Since we want $u(r,t)$ to be bounded as r approaches 0, we require $c_2 = 0$. Then $\psi(r) = c_1 r$. When $r = 1$

$$u(1,t) = v(1,t) + \psi(1) = v(1,t) + c_1 = 100,$$

and we will have the homogeneous boundary condition $v(1,t) = 0$ when $c_1 = 100$. Consequently, $\psi(r) = 100r$. The initial condition

$$u(r,0) = \frac{1}{r}v(r,0) + \frac{1}{r}\psi(r) = \frac{1}{r}v(r,0) + 100 = 0$$

implies $v(r,0) = -100r$. We are thus led to solve the new boundary-value problem

$$\frac{\partial^2 v}{\partial r^2} = \frac{\partial v}{\partial t}, \quad 0 < r < 1, \quad t > 0,$$

$$v(1,t) = 0, \quad \lim_{r \to 0} \frac{1}{r} v(r,t) < \infty,$$

$$v(r,0) = -100r.$$

Letting $v(r,t) = R(r)T(t)$ and using the separation constant $-\lambda$ we obtain

$$R'' + \lambda R = 0 \quad \text{and} \quad T' + \lambda T = 0.$$

Using $\lambda = \alpha^2 > 0$ we then have

$$R(r) = c_3 \cos \alpha r + c_4 \sin \alpha r \quad \text{and} \quad T(t) = c_5 e^{-\alpha^2 t}.$$

The boundary conditions are equivalent to $R(1) = 0$ and $\lim_{r \to 0} R(r)/r < \infty$. Since

$$\lim_{r \to 0} \frac{\cos \alpha r}{r}$$

does not exist we must have $c_3 = 0$. Then $R(r) = c_4 \sin \alpha r$, and $R(1) = 0$ implies $\alpha = n\pi$ for $n = 1$, $2, 3, \ldots$. Thus

$$v_n(r,t) = A_n e^{-n^2 \pi^2 t} \sin n\pi r$$

for $n = 1, 2, 3, \ldots$. Using the condition $\lim_{r \to 0} R(r)/r < \infty$ it is easily shown that there are no eigenvalues for $\lambda = 0$, nor does setting the common constant to $-\lambda = \alpha^2$ when separating variables lead to any solutions. Now, by the Superposition Principle,

$$v(r,t) = \sum_{n=1}^{\infty} A_n e^{-n^2 \pi^2 t} \sin n\pi r.$$

The initial condition $v(r,0) = -100r$ implies

$$-100r = \sum_{n=1}^{\infty} A_n \sin n\pi r.$$

This is a Fourier sine series and so

$$A_n = 2 \int_0^1 (-100r \sin n\pi r)\, dr = -200 \left[-\frac{r}{n\pi} \cos n\pi r \Big|_0^1 + \int_0^1 \frac{1}{n\pi} \cos n\pi r\, dr \right]$$

$$= -200 \left[-\frac{\cos n\pi}{n\pi} + \frac{1}{n^2 \pi^2} \sin n\pi r \Big|_0^1 \right] = -200 \left[-\frac{(-1)^n}{n\pi} \right] = \frac{(-1)^n 200}{n\pi}.$$

A solution of the problem is thus

$$u(r,t) = \frac{1}{r} v(r,t) + \frac{1}{r} \psi(r) = \frac{1}{r} \sum_{n=1}^{\infty} (-1)^n \frac{20}{n\pi} e^{-n^2 \pi^2 t} \sin n\pi r + \frac{1}{r}(100r)$$

$$= \frac{200}{\pi r} \sum_{n=1}^{\infty} \frac{(-1)^n}{n} e^{-n^2 \pi^2 t} \sin n\pi r + 100.$$

12. Proceeding as in Example 1 we obtain

$$\Theta(\theta) = P_n(\cos\theta) \qquad \text{and} \qquad R(r) = c_1 r^n + c_2 r^{-(n+1)}$$

so that

$$u(r, \theta) = \sum_{n=0}^{\infty} (A_n r^n + B_n r^{-(n+1)}) P_n(\cos\theta).$$

To satisfy $\lim_{r\to\infty} u(r, \theta) = -Er\cos\theta$ we must have $A_n = 0$ for $n = 2, 3, 4, \ldots$. Then

$$\lim_{r\to\infty} u(r, \theta) = -Er\cos\theta = A_0 \cdot 1 + A_1 r\cos\theta,$$

so $A_0 = 0$ and $A_1 = -E$. Thus

$$u(r, \theta) = -Er\cos\theta + \sum_{n=0}^{\infty} B_n r^{-(n+1)} P_n(\cos\theta).$$

Now

$$u(c, \theta) = 0 = -Ec\cos\theta + \sum_{n=0}^{\infty} B_n c^{-(n+1)} P_n(\cos\theta)$$

so

$$\sum_{n=0}^{\infty} B_n c^{-(n+1)} P_n(\cos\theta) = Ec\cos\theta$$

and

$$B_n c^{-(n+1)} = \frac{2n+1}{2} \int_0^\pi Ec\cos\theta\, P_n(\cos\theta) \sin\theta\, d\theta.$$

Now $\cos\theta = P_1(\cos\theta)$ so, for $n \neq 1$,

$$\int_0^\pi \cos\theta\, P_n(\cos\theta) \sin\theta\, d\theta = 0$$

by orthogonality. Thus $B_n = 0$ for $n \neq 1$ and

$$B_1 = \frac{3}{2} Ec^3 \int_0^\pi \cos^2\theta \sin\theta\, d\theta = Ec^3.$$

Therefore,

$$u(r, \theta) = -Er\cos\theta + Ec^3 r^{-2}\cos\theta.$$

13.R | Chapter 13 in Review

3. The conditions $\Theta(0) = 0$ and $\Theta(\pi) = 0$ applied to $\Theta = c_1 \cos \alpha\theta + c_2 \sin \alpha\theta$ give $c_1 = 0$ and $\alpha = n$, $n = 1, 2, 3, \ldots$, respectively. Thus we have the Fourier sine-series coefficients

$$A_n = \frac{2}{\pi} \int_0^\pi u_0(\pi\theta - \theta^2) \sin n\theta \, d\theta = \frac{4u_0}{n^3\pi}[1 - (-1)^n].$$

Thus

$$u(r,\theta) = \frac{4u_0}{\pi} \sum_{n=1}^\infty \frac{1 - (-1)^n}{n^3} r^n \sin n\theta.$$

6. We solve

$$\frac{\partial^2 u}{\partial r^2} + \frac{1}{r} \frac{\partial u}{\partial r} + \frac{1}{r^2} \frac{\partial^2 u}{\partial \theta^2} = 0, \quad r > 1, \quad 0 < \theta < \pi,$$

$$u(r, 0) = 0, \quad u(r, \pi) = 0, \quad r > 1,$$

$$u(1, \theta) = f(\theta), \quad 0 < \theta < \pi.$$

Separating variables we obtain

$$\Theta(\theta) = c_1 \cos \alpha\theta + c_2 \sin \alpha\theta$$

$$R(r) = c_3 r^\alpha + c_4 r^{-\alpha}.$$

Applying the boundary conditions $\Theta(0) = 0$, and $\Theta(\pi) = 0$ gives $c_1 = 0$ and $\alpha = n$ for $n = 1, 2, 3, \ldots$. Assuming $f(\theta)$ to be bounded, we expect the solution $u(r, \theta)$ to also be bounded as $r \to \infty$. This requires that $c_3 = 0$. Therefore

$$u(r, \theta) = \sum_{n=1}^\infty A_n r^{-n} \sin n\theta.$$

From

$$u(1, \theta) = f(\theta) = \sum_{n=1}^\infty A_n \sin n\theta$$

we obtain

$$A_n = \frac{2}{\pi} \int_0^\pi f(\theta) \sin n\theta \, d\theta.$$

9. The boundary-value problem is

$$\frac{\partial^2 u}{\partial r^2} + \frac{1}{r}\frac{\partial u}{\partial r} + \frac{\partial^2 u}{\partial z^2} = 0, \quad 0 < r < 2, \quad 0 < z < 4$$

$$u(2, z) = 50, \quad 0 < z < 4$$

$$\left.\frac{\partial u}{\partial z}\right|_{z=0} = 0, \quad u(r, 4) = 0, \quad 0 < r < 2.$$

We have $u(r, z) = v(r, z) + \psi(r)$ so $\psi'' + (1/r)\psi' = 0$ and $\psi = c_1 \ln r + c_2$. Now, boundedness at $r = 0$ implies $c_2 = 0$. Also, $\psi = c_2$ and $u(r, z) = v(r, z) + \psi(r)$, which implies

$$u(2, z) = v(2, z) + \psi(2) = 50 = c_2,$$

and $\psi(r) = 50$, all of which implies

$$\frac{\partial^2 v}{\partial r^2} + \frac{1}{r}\frac{\partial v}{\partial r} + \frac{\partial^2 v}{\partial z^2} = 0, \quad 0 < r < 2, \quad 0 < z < 4$$

$$v(2, z) = 0, \quad 0 < z < 4$$

$$\left.\frac{\partial v}{\partial z}\right|_{z=0} = 0, \quad v(r, 4) = -50, \quad 0 < r < 2.$$

(See the solution of Problem 5 in Exercises 13.2.) From

$$v(r, z) = -50 \sum_{n=1}^{\infty} \frac{\cosh(\alpha_n z)}{\alpha_n \cosh(4\alpha_n) J_1(2\alpha_n)} J_0(\alpha_n r)$$

we get

$$u(r, z) = 50 - 50 \sum_{n=1}^{\infty} \frac{\cosh(\alpha_n z)}{\alpha_n \cosh(4\alpha_n) J_1(2\alpha_n)} J_0(\alpha_n r).$$

12. Since

$$\frac{1}{r}\frac{\partial^2}{\partial r^2}(ru) = \frac{1}{r}\frac{\partial}{\partial r}\left[r\frac{\partial u}{\partial r} + u\right] = \frac{1}{r}\left[r\frac{\partial^2 u}{\partial r^2} + \frac{\partial u}{\partial r} + \frac{\partial u}{\partial r}\right] = \frac{\partial^2 u}{\partial r^2} + \frac{2}{r}\frac{\partial u}{r}$$

the differential equation becomes

$$\frac{1}{r}\frac{\partial^2}{\partial r^2}(ru) = \frac{\partial^2 u}{\partial t^2} \quad \text{or} \quad \frac{\partial^2}{\partial r^2}(ru) = r\frac{\partial^2 u}{\partial t^2}.$$

Letting $v(r, t) = ru(r, t)$ we obtain the boundary-value problem

$$\frac{\partial^2 v}{\partial r^2} = \frac{\partial^2 v}{\partial t^2}, \quad 0 < r < 1, \quad t > 0$$

$$\left.\frac{\partial v}{\partial r}\right|_{r=1} - v(1, t) = 0, \quad t > 0$$

$$v(r, 0) = rf(r), \quad \left.\frac{\partial v}{\partial t}\right|_{t=0} = rg(r), \quad 0 < r < 1.$$

If we separate variables using $v(r,t) = R(r)T(t)$ and separation constant $-\lambda$ then we obtain

$$\frac{R''}{R} = \frac{T''}{T} = -\lambda$$

so that

$$R'' + \lambda R = 0$$

$$T'' + \lambda T = 0.$$

Letting $\lambda = \alpha^2 > 0$ and solving the differential equations we get

$$R(r) = c_1 \cos \alpha r + c_2 \sin \alpha r$$

$$T(t) = c_3 \cos \alpha t + c_4 \sin \alpha t.$$

Since $u(r,t) = v(r,t)/r$, in order to insure boundedness at $r = 0$ we define $c_1 = 0$. Then $R(r) = c_2 \sin \alpha r$. Now the boundary condition $R'(1) - R(1) = 0$ implies $\alpha \cos \alpha - \sin \alpha = 0$. Thus, the eigenvalues λ_n are determined by the positive solutions of $\tan \alpha = \alpha$. We now have

$$v_n(r,t) = (A_n \cos \alpha_n t + B_n \sin \alpha_n t) \sin \alpha_n r.$$

For the eigenvalue $\lambda = 0$,

$$R(r) = c_1 r + c_2 \qquad \text{and} \qquad T(t) = c_3 t + c_4,$$

and boundedness at $r = 0$ implies $c_2 = 0$. We then take

$$v_0(r,t) = A_0 t r + B_0 r$$

so that

$$v(r,t) = A_0 t r + B_0 r + \sum_{n=1}^{\infty} (a_n \cos \alpha_n t + B_n \sin \alpha_n t) \sin \alpha_n r.$$

Now

$$v(r,0) = r f(r) = B_0 r + \sum_{n=1}^{\infty} A_n \sin \alpha_n r.$$

Since $\{r, \sin \alpha_n r\}$ is an orthogonal set on $[0, 1]$,

$$\int_0^1 r \sin \alpha_n r \, dr = 0 \qquad \text{and} \qquad \int_0^1 \sin \alpha_n r \sin \alpha_n r \, dr = 0$$

for $m \neq n$. Therefore

$$\int_0^1 r^2 f(r) \, dr = B_0 \int_0^1 r^2 \, dr = \frac{1}{3} B_0$$

and

$$B_0 = 3 \int_0^1 r^2 f(r) \, dr.$$

Also

$$\int_0^1 r f(r) \sin \alpha_n r \, dr = A_n \int_0^1 \sin^2 \alpha_n r \, dr$$

and

$$A_n = \frac{\int_0^1 r f(r) \sin \alpha_n r \, dr}{\int_0^1 \sin^2 \alpha_n r \, dr}.$$

Now

$$\int_0^1 \sin^2 \alpha_n r \, dr = \frac{1}{2} \int_0^1 (1 - \cos 2\alpha_n r) \, dr = \frac{1}{2} \left[1 - \frac{\sin 2\alpha_n}{2\alpha_n} \right] = \frac{1}{2} [1 - \cos^2 \alpha_n].$$

Since $\tan \alpha_n = \alpha_n$,

$$1 + \alpha_n^2 = 1 + \tan^2 \alpha_n = \sec^2 \alpha_n = \frac{1}{\cos^2 \alpha_n}$$

and

$$\cos^2 \alpha_n = \frac{1}{1 + \alpha_n^2}.$$

Then

$$\int_0^1 \sin^2 \alpha_n r \, dr = \frac{1}{2} \left[1 - \frac{1}{1 + \alpha_n^2} \right] = \frac{\alpha_n^2}{2(1 + \alpha_n^2)}$$

and

$$A_n = \frac{2(1 + \alpha_n^2)}{\alpha_n^2} \int_0^1 r f(r) \sin \alpha_n r \, dr.$$

Similarly, setting

$$\frac{\partial v}{\partial t} \bigg|_{t=0} = r g(r) = A_0 r + \sum_{n=1}^{\infty} B_n \alpha_n \sin \alpha_n r$$

we obtain

$$A_0 = 3 \int_0^1 r^2 g(r) \, dr$$

and

$$B_n = \frac{2(1 + \alpha_n^2)}{\alpha_n^3} \int_0^1 r g(r) \sin \alpha_n r \, dr.$$

Therefore, since $v(r, t) = r u(r, t)$ we have

$$u(r, t) = A_0 t + B_0 + \sum_{n=1}^{\infty} (A_n \cos \alpha_n t + B_n \sin \alpha_n t) \frac{\sin \alpha_n r}{r},$$

where the α_n are solutions of $\tan \alpha = \alpha$ and

$$A_0 = 3 \int_0^1 r^2 g(r) \, dr$$

$$B_0 = 3 \int_0^1 r^2 f(r) \, dr$$

$$A_n = \frac{2(1 + \alpha_n^2)}{\alpha_n^2} \int_0^1 r f(r) \sin \alpha_n r \, dr$$

$$B_n = \frac{2(1 + \alpha_n^2)}{\alpha_n^3} \int_0^1 r g(r) \sin \alpha_n r \, dr$$

for $n = 1, 2, 3, \ldots$.

15. We use the Superposition Principle for Laplace's equation discussed in Section 12.5 and shown schematically in Figure 12.5.3 in the text. That is,

Solution u = Solution u_1 of Problem 1 + Solution u_2 of Problem 2,

where in Problem 1 the boundary condition on the top and bottom of the cylinder is $u = 0$, while on the lateral surface $r = c$ it is $u = h(z)$, and in Problem 2 the boundary condition on the top of the cylinder $z = L$ is $u = f(r)$, on the bottom $z = 0$ it is $u = g(r)$, and on the lateral surface $r = c$ it is $u = 0$.

$\boxed{\text{Solution for } u_1(r, z)}$

Using λ as a separation constant we have

$$\frac{R'' + \frac{1}{r} R'}{R} = -\frac{Z''}{Z} = \lambda,$$

so

$$rR'' + R' - \lambda r R = 0 \qquad \text{and} \qquad Z'' + \lambda Z = 0.$$

The differential equation in Z, together with the boundary conditions $Z(0) = 0$ and $Z(L) = 0$ is a Sturm-Liouville problem. Letting $\lambda = \alpha^2 > 0$ we note that the above differential equation in R is a modified parametric Bessel equation which is discussed in Section 6.3 in the text. Also, we have $Z(z) = c_1 \cos \alpha z + c_2 \sin \alpha z$. The boundary conditions imply $c_1 = 0$ and $\sin \alpha L = 0$. Thus, $\alpha_n = n\pi/L$, $n = 1, 2, 3, \ldots$, so $\lambda_n = n^2\pi^2/L^2$ and

$$R(r) = c_3 I_0 \left(\frac{n\pi}{L} r\right) + c_4 K_0 \left(\frac{n\pi}{L} r\right).$$

Now boundedness at $r = 0$ implies $c_4 = 0$, so $R(r) = c_3 I_0(n\pi r/L)$ and

$$u_1(r, z) = \sum_{n=1}^{\infty} A_n I_0 \left(\frac{n\pi}{L} r\right) \sin \left(\frac{n\pi}{L} z\right).$$

At $r = c$ for $0 < z < L$ we have

$$h(z) = u_1(c, z) = \sum_{n=1}^{\infty} A_n I_0\left(\frac{n\pi}{L}c\right) \sin\left(\frac{n\pi}{L}z\right)$$

which gives

$$A_n = \frac{2}{L I_0(n\pi c/L)} \int_0^L h(z) \sin\left(\frac{n\pi}{L}z\right) dz.$$

Solution for $u_2(r, z)$

In this case we use $-\lambda$ as a separation constant which leads to

$$\frac{R'' + \frac{1}{r}R'}{R} = -\frac{Z''}{Z} = -\lambda,$$

so

$$rR'' + R' + \lambda r R = 0 \quad \text{and} \quad Z'' - \lambda Z = 0.$$

The differential equation in R is a parametric Bessel equation. Using $\lambda = \alpha^2$ we find $R(r) = c_1 J_0(\alpha r) + c_2 Y_0(\alpha r)$. Boundedness at $r = 0$ implies $c_2 = 0$ so $R(r) = c_3 J_0(\alpha r)$. The boundary condition $R(c) = 0$ then gives the defining equation for the eigenvalues: $J_0(\alpha c) = 0$. Let $\lambda_n = \alpha_n^2$ where $\alpha_n c = x_n$ are the roots. The solution of the differential equation in Z is $Z(z) = c_4 \cosh \alpha_n z + c_5 \sinh \alpha_n z$, so

$$u_2(r, z) = \sum_{n=1}^{\infty} (B_n \cosh \alpha_n z + C_n \sinh \alpha_n z) J_0(\alpha_n r).$$

At $z = 0$, for $0 < r < c$, we have

$$f(r) = u_2(r, 0) = \sum_{n=1}^{\infty} B_n J_0(\alpha_n r),$$

so

$$B_n = \frac{2}{c^2 J_1^2(\alpha_n c)} \int_0^c r f(r) J_0(\alpha_n r)\, dr.$$

At $z = L$, for $0 < r < c$, we have

$$g(r) = u_2(r, L) = \sum_{n=1}^{\infty} (B_n \cosh \alpha_n L + C_n \sinh \alpha_n L) J_0(\alpha_n r),$$

so

$$B_n \cosh \alpha_n L + C_n \sinh \alpha_n L = \frac{2}{c^2 J_1^2(\alpha_n c)} \int_0^c r g(r) J_0(\alpha_n r)\, dr$$

and

$$C_n = -B_n \frac{\cosh \alpha_n L}{\sinh \alpha_n L} + \frac{2}{c^2 (\sinh \alpha_n L) J_1^2(\alpha_n c)} \int_0^c r g(r) J_0(\alpha_n r)\, dr.$$

By the Superposition Principle the solution of the original problem is

$$u(r, z) = u_1(r, z) + u_2(r, z).$$

---

14 Integral Transform

14.1 Error Function

The terminology and concepts listed below provide an outline of the main ideas encountered in this section. These can be useful when preparing for a quiz or test.

Terminology and Concepts

- error function

- complementary error function

The basic skills listed below summarize the more mechanical types of problems encountered in the exercise set for this section.

Basic Skills

- find the Laplace transforms of the error function and complementary error function

3. By the first translation theorem,

$$\mathscr{L}\left\{e^t \operatorname{erf}(\sqrt{t})\right\} = \mathscr{L}\left\{\operatorname{erf}(\sqrt{t})\right\}\bigg|_{s \to s-1} = \frac{1}{s\sqrt{s+1}}\bigg|_{s \to s-1} = \frac{1}{\sqrt{s}\,(s-1)}.$$

6. To find

$$\mathscr{L}^{-1}\left\{\frac{1}{1+\sqrt{s+1}}\right\}$$

we begin by rationalizing first the denominator and then the numerator:

$$\frac{1}{1+\sqrt{s+1}} = \frac{1}{1+\sqrt{s+1}} \cdot \frac{1-\sqrt{s+1}}{1-\sqrt{s+1}} = \frac{1-\sqrt{s+1}}{-s}$$

$$= -\frac{1}{s} + \frac{\sqrt{s+1}}{s} = -\frac{1}{s} + \frac{\sqrt{s+1}}{s} \cdot \frac{\sqrt{s+1}}{\sqrt{s+1}}$$

$$= -\frac{1}{s} + \frac{s+1}{s\sqrt{s+1}} \qquad \longleftarrow \text{ partial fractions}$$

$$= -\frac{1}{s} + \frac{1}{\sqrt{s+1}} + \frac{1}{s\sqrt{s+1}}.$$

331

Therefore, from the first translation theorem and Problem 1(b) in this section,

$$\mathscr{L}^{-1}\left\{-\frac{1}{s} + \frac{1}{\sqrt{s}+1} + \frac{1}{s\sqrt{s}+1}\right\} = \mathscr{L}^{-1}\left\{\frac{1}{s}\right\} + \mathscr{L}^{-1}\left\{\frac{1}{\sqrt{s}+1}\right\} + \mathscr{L}^{-1}\left\{\frac{1}{s\sqrt{s}+1}\right\}$$

$$= -1 + \mathscr{L}^{-1}\left\{\frac{1}{s}\bigg|_{s \to s+1}\right\} + \mathscr{L}^{-1}\left\{\frac{1}{s\sqrt{s}+1}\right\}$$

$$= -1 + \frac{1}{\sqrt{\pi}}\mathscr{L}^{-1}\left\{\frac{\sqrt{\pi}}{s}\bigg|_{s \to s+1}\right\} + \mathscr{L}^{-1}\left\{\frac{1}{s\sqrt{s}+1}\right\}$$

$$= -1 + \frac{1}{\sqrt{\pi}}\frac{e^{-t}}{\sqrt{t}} + \mathrm{erf}(\sqrt{t}) = \frac{e^{-t}}{\sqrt{\pi t}} - (1 - \mathrm{erf}(\sqrt{t}))$$

$$= \frac{e^{-t}}{\sqrt{\pi t}} - \mathrm{erfc}(\sqrt{t}).$$

9. Taking the Laplace transform of both sides of the equation we obtain

$$\mathscr{L}\{y(t)\} = \mathscr{L}\{1\} - \mathscr{L}\left\{\int_0^t \frac{y(\tau)}{\sqrt{t-\tau}}\, d\tau\right\}$$

$$Y(s) = \frac{1}{s} - Y(s)\frac{\sqrt{\pi}}{\sqrt{s}}$$

$$\frac{\sqrt{s} + \sqrt{\pi}}{\sqrt{s}}Y(s) = \frac{1}{s}$$

$$Y(s) = \frac{1}{\sqrt{s}\,(\sqrt{s} + \sqrt{\pi}\,)}.$$

Thus

$$y(t) = \mathscr{L}^{-1}\left\{\frac{1}{\sqrt{s}\,(\sqrt{s} + \sqrt{\pi}\,)}\right\} = e^{\pi t}\,\mathrm{erfc}(\sqrt{\pi t}\,). \qquad \boxed{\text{By entry 5 in Table 14.1.1}}$$

12. Since $f(x) = e^{-x^2}$ is an even function,

$$\int_{-a}^{a} e^{-u^2}\, du = 2\int_0^{a} e^{-u^2}\, du.$$

Therefore,

$$\int_{-a}^{a} e^{-u^2}\, du = \sqrt{\pi}\,\mathrm{erf}(a).$$

14.2 Laplace Transform

The terminology and concepts listed below provide an outline of the main ideas encountered in this section. These can be useful when preparing for a quiz or test.

Terminology and Concepts

- Laplace transform of a function of two variables
- Laplace transform of partial derivatives

The basic skills listed below summarize the more mechanical types of problems encountered in the exercise set for this section.

Basic Skills

- use the Laplace transform to solve a BVP

3. The solution of

$$a^2 \frac{d^2U}{dx^2} - s^2U = 0$$

is in this case

$$U(x, s) = c_1 e^{-(x/a)s} + c_2 e^{(x/a)s}.$$

Since $\lim_{x \to \infty} u(x, t) = 0$ we have $\lim_{x \to \infty} U(x, s) = 0$. Thus $c_2 = 0$ and

$$U(x, s) = c_1 e^{-(x/a)s}.$$

If $\mathscr{L}\{u(0, t)\} = \mathscr{L}\{f(t)\} = F(s)$ then $U(0, s) = F(s)$. From this we have $c_1 = F(s)$ and

$$U(x, s) = F(s)e^{-(x/a)s}.$$

Hence, by the second translation theorem,

$$u(x, t) = f\left(t - \frac{x}{a}\right) \mathscr{U}\left(t - \frac{x}{a}\right).$$

6. Transforming the partial differential equation gives

$$\frac{d^2U}{dx^2} - s^2U = -\frac{\omega}{s^2 + \omega^2} \sin \pi x.$$

Using undetermined coefficients we obtain

$$U(x, s) = c_1 \cosh sx + c_2 \sinh sx + \frac{\omega}{(s^2 + \pi^2)(s^2 + \omega^2)} \sin \pi x.$$

The transformed boundary conditions $U(0, s) = 0$ and $U(1, s) = 0$ give, in turn, $c_1 = 0$ and $c_2 = 0$. Therefore

$$U(x, s) = \frac{\omega}{(s^2 + \pi^2)(s^2 + \omega^2)} \sin \pi x$$

and

$$u(x, t) = \omega \sin \pi x \, \mathcal{L}^{-1} \left\{ \frac{1}{(s^2 + \pi^2)(s^2 + \omega^2)} \right\}$$

$$= \frac{\omega}{\omega^2 - \pi^2} \sin \pi x \, \mathcal{L}^{-1} \left\{ \frac{1}{\pi} \frac{\pi}{s^2 + \pi^2} - \frac{1}{\omega} \frac{\omega}{s^2 + \omega^2} \right\}$$

$$= \frac{\omega}{\pi(\omega^2 - \pi^2)} \sin \pi t \sin \pi x - \frac{1}{\omega^2 - \pi^2} \sin \omega t \sin \pi x.$$

9. Transforming the partial differential equation gives

$$\frac{d^2 U}{dx^2} - s^2 U = -sxe^{-x}.$$

Using undetermined coefficients we obtain

$$U(x, s) = c_1 e^{-sx} + c_2 e^{sx} - \frac{2s}{(s^2 - 1)^2} e^{-x} + \frac{s}{s^2 - 1} xe^{-x}.$$

The transformed boundary conditions $\lim_{x \to \infty} U(x, s) = 0$ and $U(0, s) = 0$ give, in turn, $c_2 = 0$ and $c_1 = 2s/(s^2 - 1)^2$. Therefore

$$U(x, s) = \frac{2s}{(s^2 - 1)^2} e^{-sx} - \frac{2s}{(s^2 - 1)^2} e^{-x} + \frac{s}{s^2 - 1} xe^{-x}.$$

From entries (13) and (26) in the Table of Laplace transforms we obtain

$$u(x, t) = \mathcal{L}^{-1} \left\{ \frac{2s}{(s^2 - 1)^2} e^{-sx} - \frac{2s}{(s^2 - 1)^2} e^{-x} + \frac{s}{s^2 - 1} xe^{-x} \right\}$$

$$= 2(t - x) \sinh(t - x) \, \mathcal{U}(t - x) - te^{-x} \sinh t + xe^{-x} \cosh t.$$

12. We use

$$U(x, s) = c_1 e^{-\sqrt{s}\,x} + c_2 e^{\sqrt{s}\,x} + \frac{u_1 x}{s}.$$

The condition $\lim_{x \to \infty} u(x, t)/x = u_1$ implies $\lim_{x \to \infty} U(x, s)/x = u_1/s$, so we define $c_2 = 0$. Then

$$U(x, s) = c_1 e^{-\sqrt{s}\,x} + \frac{u_1 x}{s}.$$

From $U(0, s) = u_0/s$ we obtain $c_1 = u_0/s$. Hence

$$U(x, s) = u_0 \frac{e^{-\sqrt{s}\,x}}{s} + \frac{u_1 x}{s}$$

and

$$u(x, t) = u_0 \mathcal{L}^{-1}\left\{\frac{e^{-x\sqrt{s}}}{s}\right\} + u_1 x \, \mathcal{L}^{-1}\left\{\frac{1}{s}\right\} = u_0 \operatorname{erfc}\left(\frac{x}{2\sqrt{t}}\right) + u_1 x.$$

15. We use

$$U(x, s) = c_1 e^{-\sqrt{s}\,x} + c_2 e^{\sqrt{s}\,x}.$$

The condition $\lim_{x \to \infty} u(x, t) = 0$ implies $\lim_{x \to \infty} U(x, s) = 0$, so we define $c_2 = 0$. Hence

$$U(x, s) = c_1 e^{-\sqrt{s}\,x}.$$

The transform of $u(0, t) = f(t)$ is $U(0, s) = F(s)$. Therefore

$$U(x, s) = F(s)e^{-\sqrt{s}\,x}$$

and

$$u(x, t) = \mathcal{L}^{-1}\left\{F(s)e^{-x\sqrt{s}}\right\} = \frac{x}{2\sqrt{\pi}} \int_0^t \frac{f(t - \tau)e^{-x^2/4\tau}}{\tau^{3/2}}\, d\tau.$$

18. The solution of the transformed equation

$$\frac{d^2 U}{dx^2} - sU = -100$$

by undetermined coefficients is

$$U(x, s) = c_1 e^{\sqrt{s}\,x} + c_2 e^{-\sqrt{s}\,x} + \frac{100}{s}.$$

From the fact that $\lim_{x \to \infty} U(x, s) = 100/s$ we see that $c_1 = 0$. Thus

$$U(x, s) = c_2 e^{-\sqrt{s}\,x} + \frac{100}{s}. \tag{1}$$

Now the transform of the boundary condition at $x = 0$ is

$$U(0, s) = 20\left[\frac{1}{s} - \frac{1}{s}e^{-s}\right].$$

It follows from (1) that

$$\frac{20}{s} - \frac{20}{s}e^{-s} = c_2 + \frac{100}{s} \qquad \text{or} \qquad c_2 = -\frac{80}{s} - \frac{20}{s}e^{-s}$$

and so

$$U(x, s) = \left(-\frac{80}{s} - \frac{20}{s}e^{-s}\right)e^{-\sqrt{s}\,x} + \frac{100}{s}$$

$$= \frac{100}{s} - \frac{80}{s}e^{-\sqrt{s}\,x} - \frac{20}{s}e^{-\sqrt{s}\,x}e^{-s}.$$

Thus

$$u(x,t) = 100\,\mathscr{L}^{-1}\left\{\frac{1}{s}\right\} - 80\,\mathscr{L}^{-1}\left\{\frac{e^{-\sqrt{s}\,x}}{s}\right\} - 20\,\mathscr{L}^{-1}\left\{\frac{e^{-\sqrt{s}\,x}}{s}e^{-s}\right\}$$

$$= 100 - 80\,\mathrm{erfc}\left(x/2\sqrt{t}\right) - 20\,\mathrm{erfc}\left(x/2\sqrt{t-1}\right)\mathscr{U}(t-1).$$

21. The solution of

$$\frac{d^2U}{dx^2} - sU = -u_0 - u_0\sin\frac{\pi}{L}x$$

is

$$U(x,s) = c_1\cosh(\sqrt{s}\,x) + c_2\sinh(\sqrt{s}\,x) + \frac{u_0}{s} + \frac{u_0}{s+\pi^2/L^2}\sin\frac{\pi}{L}x.$$

The transformed boundary conditions $U(0,s) = u_0/s$ and $U(L,s) = u_0/s$ give, in turn, $c_1 = 0$ and $c_2 = 0$. Therefore

$$U(x,s) = \frac{u_0}{s} + \frac{u_0}{s+\pi^2/L^2}\sin\frac{\pi}{L}x$$

and

$$u(x,t) = u_0\,\mathscr{L}^{-1}\left\{\frac{1}{s}\right\} + u_0\,\mathscr{L}^{-1}\left\{\frac{1}{s+\pi^2/L^2}\right\}\sin\frac{\pi}{L}x = u_0 + u_0 e^{-\pi^2 t/L^2}\sin\frac{\pi}{L}x.$$

24. Letting $\mathscr{L}\{c(x,t)\} = C(x,t)$ we have

$$C(x,s) = c_1\cosh\sqrt{\frac{s}{D}}\,x + c_2\sinh\sqrt{\frac{s}{D}}\,x.$$

The transform of the two boundary conditions are $C(0,s) = c_0/s$ and $C(1,s) = c_0/s$. From these conditions we obtain $c_1 = c_0/s$ and

$$c_2 = c_0(1 - \cosh\sqrt{s/D})/s\sinh\sqrt{s/D}.$$

Therefore

$$C(x,s) = c_0\left[\frac{\cosh\sqrt{s/D}\,x}{s} + \frac{(1-\cosh\sqrt{s/D})}{s\sinh\sqrt{s/D}}\sinh\sqrt{s/D}\,x\right]$$

$$= c_0\left[\frac{\sinh\sqrt{s/D}\,(1-x)}{s\sinh\sqrt{s/D}} + \frac{\sin\sqrt{s/D}\,x}{s\sinh\sqrt{s/D}}\right]$$

$$= c_0 \left[\frac{e^{\sqrt{s/D}\,(1-x)} - e^{-\sqrt{s/D}\,(1-x)}}{s(e^{\sqrt{s/D}} - e^{-\sqrt{s/D}})} + \frac{e^{\sqrt{s/D}\,x} - e^{-\sqrt{s/D}\,x}}{s(e^{\sqrt{s/D}} - e^{-\sqrt{s/D}})} \right]$$

$$= c_0 \left[\frac{e^{-\sqrt{s/D}\,x} - e^{-\sqrt{s/D}\,(2-x)}}{s(1 - e^{-2\sqrt{s/D}})} + \frac{e^{\sqrt{s/D}\,(x-1)} - e^{-\sqrt{s/D}\,(x+1)}}{s(1 - e^{-2\sqrt{s/D}})} \right]$$

$$= c_0 \frac{(e^{-\sqrt{s/D}\,x} - e^{-\sqrt{s/D}\,(2-x)})}{s} \left(1 + e^{-2\sqrt{s/D}} + e^{-4\sqrt{s/D}} + \cdots \right)$$

$$+ c_0 \frac{(e^{\sqrt{s/D}\,(x-1)} - e^{-\sqrt{s/D}\,(x+1)})}{s} \left(1 + e^{-2\sqrt{s/D}} + e^{-4\sqrt{s/D}} + \cdots \right)$$

$$= c_0 \sum_{n=0}^{\infty} \left[\frac{e^{-(2n+x)\sqrt{s/D}}}{s} - \frac{e^{-(2n+2-x)\sqrt{s/D}}}{s} \right]$$

$$+ c_0 \sum_{n=0}^{\infty} \left[\frac{e^{-(2n+1-x)\sqrt{s/D}}}{s} - \frac{e^{-(2n+1+x)\sqrt{s/D}}}{s} \right]$$

and so

$$c(x,t) = c_0 \sum_{n=0}^{\infty} \left[\mathcal{L}^{-1}\left\{ \frac{e^{-\frac{(2n+x)}{\sqrt{D}}\sqrt{s}}}{s} \right\} - \mathcal{L}^{-1}\left\{ \frac{e^{-\frac{(2n+2-x)}{\sqrt{D}}\sqrt{s}}}{s} \right\} \right]$$

$$+ c_0 \sum_{n=0}^{\infty} \left[\mathcal{L}^{-1}\left\{ \frac{e^{-\frac{(2n+1-x)}{\sqrt{D}}\sqrt{s}}}{s} \right\} - \mathcal{L}^{-1}\left\{ \frac{e^{-\frac{(2n+1+x)}{\sqrt{D}}\sqrt{s}}}{s} \right\} \right]$$

$$= c_0 \sum_{n=0}^{\infty} \left[\operatorname{erfc}\left(\frac{2n+x}{2\sqrt{Dt}} \right) - \operatorname{erfc}\left(\frac{2n+2-x}{2\sqrt{Dt}} \right) \right]$$

$$+ c_0 \sum_{n=0}^{\infty} \left[\operatorname{erfc}\left(\frac{2n+1-x}{2\sqrt{Dt}} \right) - \operatorname{erfc}\left(\frac{2n+1+x}{2\sqrt{Dt}} \right) \right].$$

Now using $\operatorname{erfc}(x) = 1 - \operatorname{erf}(x)$ we get

$$c(x,t) = c_0 \sum_{n=0}^{\infty} \left[\operatorname{erf}\left(\frac{2n+2-x}{2\sqrt{Dt}} \right) - \operatorname{erf}\left(\frac{2n+x}{2\sqrt{Dt}} \right) \right]$$

$$+ c_0 \sum_{n=0}^{\infty} \left[\operatorname{erf}\left(\frac{2n+1+x}{2\sqrt{Dt}} \right) - \operatorname{erf}\left(\frac{2n+1-x}{2\sqrt{Dt}} \right) \right].$$

27. Using the Laplace transform with respect to t of the partial differential equation gives

$$\frac{d^2U}{dr^2} + \frac{2}{r}\frac{dU}{dr} = sU - \overbrace{u(r, 0)}^{0},$$

where $\mathscr{L}\{u(r, t)\} = U(r, s)$. The ordinary differential equation is equivalent to

$$rU'' + 2U' - srU = 0.$$

If we let $v(r, s) = rU(r, s)$ then differentiation with respect to r gives $v'' = rU'' + 2U'$. The transformed ordinary differential equation becomes

$$v'' - sv = 0$$

$$v = c_1 e^{-\sqrt{s}r} + c_2 e^{\sqrt{s}r}$$

$$U(r, s) = c_1 \frac{e^{-\sqrt{s}r}}{r} + c_2 \frac{e^{\sqrt{s}r}}{r}.$$

The usual argument about boundedness as $r \to \infty$ implies $c_2 = 0$. Therefore $U(r, s) = c_1 e^{-\sqrt{s}r}/r$. Now the Laplace transform of $u(1, t) = 100$ is

$$U(1, s) = \frac{100}{s} = c_1 e^{-\sqrt{s}}$$

so

$$c_1 = \frac{100 e^{\sqrt{s}}}{s}$$

and

$$U(r, s) = \frac{100}{r} \frac{e^{-\sqrt{s}(r-1)}}{s}.$$

Thus

$$u(r, t) = \frac{100}{r} \mathscr{L}^{-1}\left\{\frac{e^{-(r-1)\sqrt{s}}}{s}\right\}.$$

From the third entry in Table 14.1.1 we get

$$u(r, t) = \frac{100}{r}\operatorname{erfc}\left(\frac{r-1}{2\sqrt{t}}\right).$$

30. (a) Transforming the partial differential equation and using the initial condition gives

$$k\frac{d^2U}{dx^2} - sU = 0.$$

Since the domain of the variable x is an infinite interval we write the general solution of this differential equation as

$$U(x, s) = c_1 e^{-\sqrt{s/k}\,x} + c_2 e^{\sqrt{s/k}\,x}.$$

Transforming the boundary conditions gives $U'(0,s) = -A/s$ and $\lim\limits_{x\to\infty} U(x,s) = 0$. Hence we find $c_2 = 0$ and $c_1 = A\sqrt{k}/s\sqrt{s}$. From

$$U(x,s) = A\sqrt{k}\,\frac{e^{-\sqrt{s/k}\,x}}{s\sqrt{s}}$$

we see that

$$u(x,t) = A\sqrt{k}\,\mathscr{L}^{-1}\left\{\frac{e^{-\sqrt{s/k}\,x}}{s\sqrt{s}}\right\}.$$

With the identification $a = x/\sqrt{k}$ it follows from (47) in the Table of Laplace transforms that

$$u(x,t) = A\sqrt{k}\left\{2\sqrt{\frac{t}{\pi}}\,e^{-x^2/4kt} - \frac{x}{\sqrt{k}}\,\text{erfc}\left(\frac{x}{2\sqrt{kt}}\right)\right\}$$

$$= 2A\,\sqrt{\frac{kt}{\pi}}\,e^{-x^2/4kt} - Ax\,\text{erfc}\left(x/2\sqrt{kt}\right).$$

Since $\text{erfc}(0) = 1$,

$$\lim_{t\to\infty} u(x,t) = \lim_{t\to\infty}\left(2A\sqrt{\frac{kt}{\pi}}e^{-x^2/4kt} - Ax\,\text{erfc}\left(\frac{x}{2\sqrt{kt}}\right)\right) = \infty.$$

(b)

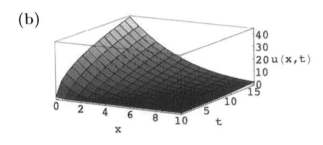
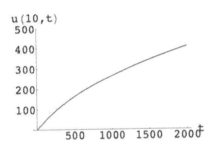

14.3 Fourier Integral

The terminology and concepts listed below provide an outline of the main ideas encountered in this section. These can be useful when preparing for a quiz or test.

Terminology and Concepts

- Fourier integral representation of a function
- convergence conditions for a Fourier integral
- cosine integral representation of an even function

- sine integral representation of an odd function

- complex or exponential form of a Fourier integral

The basic skills listed below summarize the more mechanical types of problems encountered in the exercise set for this section.

Basic Skills

- find the Fourier integral representation of a function defined on $(-\infty, \infty)$

- find the cosine integral representation of an even function

- find the sine integral representation of an odd function

- find both the cosine and sine integral representations of a function defined on $(0, \infty)$

3. From formulas (5) and (6) in the text,

$$A(\alpha) = \int_0^3 x \cos \alpha x \, dx = \frac{x \sin \alpha x}{\alpha} \Big|_0^3 - \frac{1}{\alpha} \int_0^3 \sin \alpha x \, dx$$

$$= \frac{3 \sin 3\alpha}{\alpha} + \frac{\cos \alpha x}{\alpha^2} \Big|_0^3 = \frac{3\alpha \sin 3\alpha + \cos 3\alpha - 1}{\alpha^2}$$

and

$$B(\alpha) = \int_0^3 x \sin \alpha x \, dx = -\frac{x \cos \alpha x}{\alpha} \Big|_0^3 + \frac{1}{\alpha} \int_0^3 \cos \alpha x \, dx$$

$$= -\frac{3 \cos 3\alpha}{\alpha} + \frac{\sin \alpha x}{\alpha^2} \Big|_0^3 = \frac{\sin 3\alpha - 3\alpha \cos 3\alpha}{\alpha^2}.$$

Hence

$$f(x) = \frac{1}{\pi} \int_0^\infty \frac{(3\alpha \sin 3\alpha + \cos 3\alpha - 1) \cos \alpha x + (\sin 3\alpha - 3\alpha \cos 3\alpha) \sin \alpha x}{\alpha^2} \, d\alpha$$

$$= \frac{1}{\pi} \int_0^\infty \frac{3\alpha(\sin 3\alpha \cos \alpha x - \cos 3\alpha \sin \alpha x) + \cos 3\alpha \cos \alpha x + \sin 3\alpha \sin \alpha x - \cos \alpha x}{\alpha^2} \, d\alpha$$

$$= \frac{1}{\pi} \int_0^\infty \frac{3\alpha \sin \alpha(3 - x) + \cos \alpha(3 - x) - \cos \alpha x}{\alpha^2} \, d\alpha.$$

6. From formulas (5) and (6) in the text,

$$A(\alpha) = \int_{-1}^1 e^x \cos \alpha x \, dx$$

$$= \frac{e(\cos \alpha + \alpha \sin \alpha) - e^{-1}(\cos \alpha - \alpha \sin \alpha)}{1 + \alpha^2}$$

$$= \frac{2(\sinh 1)\cos\alpha - 2\alpha(\cosh 1)\sin\alpha}{1+\alpha^2}$$

and

$$B(\alpha) = \int_{-1}^{1} e^x \sin\alpha x\, dx$$

$$= \frac{e(\sin\alpha - \alpha\cos\alpha) - e^{-1}(-\sin\alpha - \alpha\cos\alpha)}{1+\alpha^2}$$

$$= \frac{2(\cosh 1)\sin\alpha - 2\alpha(\sinh 1)\cos\alpha}{1+\alpha^2}.$$

Hence

$$f(x) = \frac{1}{\pi}\int_0^\infty [A(\alpha)\cos\alpha x + B(\alpha)\sin\alpha x]\, d\alpha.$$

9. The function is even. Thus from formula (9) in the text

$$A(\alpha) = \int_0^\pi x\cos\alpha x\, dx = \frac{x\sin\alpha x}{\alpha}\Big|_0^\pi - \frac{1}{\alpha}\int_0^\pi \sin\alpha x\, dx$$

$$= \frac{\pi\sin\pi\alpha}{\alpha} + \frac{1}{\alpha^2}\cos\alpha x\Big|_0^\pi = \frac{\pi\alpha\sin\pi\alpha + \cos\pi\alpha - 1}{\alpha^2}.$$

Hence from formula (8) in the text

$$f(x) = \frac{2}{\pi}\int_0^\infty \frac{(\pi\alpha\sin\pi\alpha + \cos\pi\alpha - 1)\cos\alpha x}{\alpha^2}\, d\alpha.$$

12. The function is odd. Thus from formula (11) in the text

$$B(\alpha) = \int_0^\infty xe^{-x}\sin\alpha x\, dx.$$

Now recall

$$\mathcal{L}\{t\sin kt\} = -\frac{d}{ds}\mathcal{L}\{\sin kt\} = 2ks/(s^2+k^2)^2.$$

If we set $s = 1$ and $k = \alpha$ we obtain

$$B(\alpha) = \frac{2\alpha}{(1+\alpha^2)^2}.$$

Hence from formula (10) in the text

$$f(x) = \frac{4}{\pi}\int_0^\infty \frac{\alpha\sin\alpha x}{(1+\alpha^2)^2}\, d\alpha.$$

15. For the cosine integral,

$$A(\alpha) = \int_0^\infty xe^{-2x}\cos\alpha x\, dx.$$

But we know

$$\mathcal{L}\{t\cos kt\} = -\frac{d}{ds}\frac{s}{(s^2+k^2)} = \frac{(s^2-k^2)}{(s^2+k^2)^2}.$$

If we set $s = 2$ and $k = \alpha$ we obtain

$$A(\alpha) = \frac{4 - \alpha^2}{(4 + \alpha^2)^2}.$$

Hence

$$f(x) = \frac{2}{\pi} \int_0^\infty \frac{(4 - \alpha^2) \cos \alpha x}{(4 + \alpha^2)^2} \, d\alpha.$$

For the sine integral,

$$B(\alpha) = \int_0^\infty x e^{-2x} \sin \alpha x \, dx.$$

From Problem 12, we know

$$\mathcal{L}\{t \sin kt\} = \frac{2ks}{(s^2 + k^2)^2}.$$

If we set $s = 2$ and $k = \alpha$ we obtain

$$B(\alpha) = \frac{4\alpha}{(4 + \alpha^2)^2}.$$

Hence

$$f(x) = \frac{8}{\pi} \int_0^\infty \frac{\alpha \sin \alpha x}{(4 + \alpha^2)^2} \, d\alpha.$$

18. From the formula for sine integral of $f(x)$ we have

$$f(x) = \frac{2}{\pi} \int_0^\infty \left(\int_0^\infty f(x) \sin \alpha x \, dx \right) \sin \alpha x \, dx$$

$$= \frac{2}{\pi} \left[\int_0^1 1 \cdot \sin \alpha x \, d\alpha + \int_1^\infty 0 \cdot \sin \alpha x \, d\alpha \right]$$

$$= \frac{2}{\pi} \frac{(-\cos \alpha x)}{x} \Big|_0^1 = \frac{2}{\pi} \frac{1 - \cos x}{x}.$$

21. (a) From the identity

$$\sin A \cos B = \frac{1}{2}[\sin(A + B) + \sin(A - B)]$$

we have

$$\sin \alpha \cos \alpha x = \frac{1}{2}[\sin(\alpha + \alpha x) + \sin(\alpha - \alpha x)]$$

$$= \frac{1}{2}[\sin \alpha(1+x) + \sin \alpha(1-x)]$$

$$= \frac{1}{2}[\sin \alpha(x+1) - \sin \alpha(x-1)].$$

Then

$$\frac{2}{\pi} \int_0^\infty \frac{\sin \alpha \cos \alpha x}{\alpha} \, d\alpha = \frac{1}{\pi} \int_0^\infty \frac{\sin \alpha(x+1) - \sin \alpha(x-1)}{\alpha} \, d\alpha.$$

(b) Noting that

$$F_b = \frac{1}{\pi} \int_0^b \frac{\sin \alpha(x+1) - \sin \alpha(x-1)}{\alpha} \, d\alpha$$

$$= \frac{1}{\pi} \left[\int_0^b \frac{\sin \alpha(x+1)}{\alpha} \, d\alpha - \int_0^b \frac{\sin \alpha(x-1)}{\alpha} \, d\alpha \right]$$

and letting $t = \alpha(x+1)$ so that $dt = (x+1) \, d\alpha$ in the first integral and $t = \alpha(x-1)$ so that $dt = (x-1) \, d\alpha$ in the second integral we have

$$F_b = \frac{1}{\pi} \left[\int_0^{b(x+1)} \frac{\sin t}{t} \, dt - \int_0^{b(x-1)} \frac{\sin t}{t} \, dt \right].$$

Since $\text{Si}(x) = \int_0^x [(\sin t)/t] \, dt$, this becomes

$$F_b = \frac{1}{\pi}[\text{Si}(b(x+1)) - \text{Si}(b(x-1))].$$

(c) In *Mathematica* we define $\mathbf{f[b\_] := (1/Pi)(SinIntegral[b(x+1)] - SinIntegral[b(x-1)]}.$
Graphs of $F_b(x)$ for $b = 4, 6, 15,$ and 75 are shown below.

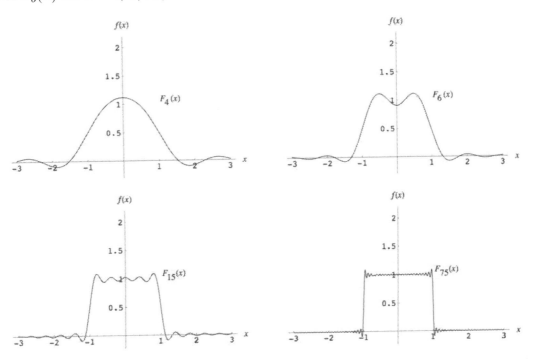

14.4 | Fourier Transform

The terminology and concepts listed below provide an outline of the main ideas encountered in this section. These can be useful when preparing for a quiz or test.

Terminology and Concepts

- integral transform of a function
- inverse transform of a function
- transform pairs
- kernel of a transform
- Fourier transform
- inverse Fourier transform
- Fourier sine transform
- inverse Fourier sine transform
- Fourier cosine transform
- inverse Fourier cosine transform

The basic skills listed below summarize the more mechanical types of problems encountered in the exercise set for this section.

Basic Skills

- use Fourier integral transforms to solve BVPs

For the boundary-value problems in this section it is sometimes useful to note that the identities

$$e^{i\alpha} = \cos\alpha + i\sin\alpha \qquad and \qquad e^{-i\alpha} = \cos\alpha - i\sin\alpha$$

imply

$$e^{i\alpha} + e^{-i\alpha} = 2\cos\alpha \qquad and \qquad e^{i\alpha} - e^{-i\alpha} = 2i\sin\alpha.$$

3. Using the Fourier sine transform, the partial differential equation becomes

$$\frac{dU}{dt} + k\alpha^2 U = k\alpha u_0.$$

The general solution of this linear equation is

$$U(\alpha, t) = ce^{-k\alpha^2 t} + \frac{u_0}{\alpha}.$$

But $U(\alpha, 0) = 0$ implies $c = -u_0/\alpha$ and so

$$U(\alpha, t) = u_0 \frac{1 - e^{-k\alpha^2 t}}{\alpha}$$

and

$$u(x, t) = \frac{2u_0}{\pi} \int_0^\infty \frac{1 - e^{-k\alpha^2 t}}{\alpha} \sin\alpha x \, d\alpha.$$

6. Since the domain of x is $(0, \infty)$ and the condition at $x = 0$ involves $\partial u / \partial x$ we use the Fourier cosine transform:

$$-k\alpha^2 U(\alpha, t) - k u_x(0, t) = \frac{dU}{dt}$$

$$\frac{dU}{dt} + k\alpha^2 U = kA$$

$$U(\alpha, t) = ce^{-k\alpha^2 t} + \frac{A}{\alpha^2}.$$

Since

$$\mathscr{F}_C\{u(x, 0)\} = U(\alpha, 0) = 0$$

we find $c = -A/\alpha^2$, so that

$$U(\alpha, t) = A\frac{1 - e^{-k\alpha^2 t}}{\alpha^2}.$$

Applying the inverse Fourier cosine transform we obtain

$$u(x, t) = \mathscr{F}_C^{-1}\{U(\alpha, t)\} = \frac{2A}{\pi} \int_0^\infty \frac{1 - e^{-k\alpha^2 t}}{\alpha^2} \cos \alpha x \, d\alpha.$$

9. (a) Using the Fourier transform we obtain

$$U(\alpha, t) = c_1 \cos \alpha a t + c_2 \sin \alpha a t.$$

If we write

$$\mathscr{F}\{u(x, 0)\} = \mathscr{F}\{f(x)\} = F(\alpha)$$

and

$$\mathscr{F}\{u_t(x, 0)\} = \mathscr{F}\{g(x)\} = G(\alpha)$$

we first obtain $c_1 = F(\alpha)$ from $U(\alpha, 0) = F(\alpha)$ and then $c_2 = G(\alpha)/\alpha a$ from $dU/dt \,|_{t=0} = G(\alpha)$. Thus

$$U(\alpha, t) = F(\alpha) \cos \alpha a t + \frac{G(\alpha)}{\alpha a} \sin \alpha a t$$

and

$$u(x, t) = \frac{1}{2\pi} \int_{-\infty}^{\infty} \left(F(\alpha) \cos \alpha a t + \frac{G(\alpha)}{\alpha a} \sin \alpha a t \right) e^{-i\alpha x} d\alpha.$$

(b) If $g(x) = 0$ then $c_2 = 0$ and

$$u(x, t) = \frac{1}{2\pi} \int_{-\infty}^{\infty} F(\alpha) \cos \alpha at \, e^{-i\alpha x} d\alpha$$

$$= \frac{1}{2\pi} \int_{-\infty}^{\infty} F(\alpha) \left(\frac{e^{\alpha at i} + e^{-\alpha at i}}{2} \right) e^{-i\alpha x} d\alpha$$

$$= \frac{1}{2} \left[\frac{1}{2\pi} \int_{-\infty}^{\infty} F(\alpha) e^{-i(x-at)\alpha} d\alpha + \frac{1}{2\pi} \int_{-\infty}^{\infty} F(\alpha) e^{-i(x+at)\alpha} d\alpha \right]$$

$$= \frac{1}{2} [f(x - at) + f(x + at)].$$

12. Since the boundary condition at $y = 0$ now involves $u(x, 0)$ rather than $u'(x, 0)$, we use the Fourier sine transform. The transform of the partial differential equation is then

$$\frac{d^2 U}{dx^2} - \alpha^2 U + \alpha u(x, 0) = 0 \qquad \text{or} \qquad \frac{d^2 U}{dx^2} - \alpha^2 U = -\alpha.$$

The solution of this differential equation is

$$U(x, \alpha) = c_1 \cosh \alpha x + c_2 \sinh \alpha x + \frac{1}{\alpha}.$$

The transforms of the boundary conditions at $x = 0$ and $x = \pi$ in turn imply that $c_1 = 1/\alpha$ and

$$c_2 = \frac{\cosh \alpha \pi}{\alpha \sinh \alpha \pi} - \frac{1}{\alpha \sinh \alpha \pi} + \frac{\alpha}{(1 + \alpha^2) \sinh \alpha \pi}.$$

Hence

$$U(x, \alpha) = \frac{1}{\alpha} - \frac{\cosh \alpha x}{\alpha} + \frac{\cosh \alpha \pi}{\alpha \sinh \alpha \pi} \sinh \alpha x - \frac{\sinh \alpha x}{\alpha \sinh \alpha \pi} + \frac{\alpha \sinh \alpha x}{(1 + \alpha^2) \sinh \alpha \pi}$$

$$= \frac{1}{\alpha} - \frac{\sinh \alpha (\pi - x)}{\alpha \sinh \alpha \pi} - \frac{\sinh \alpha x}{\alpha (1 + \alpha^2) \sinh \alpha \pi}.$$

Taking the inverse transform it follows that

$$u(x, y) = \frac{2}{\pi} \int_0^{\infty} \left(\frac{1}{\alpha} - \frac{\sinh \alpha (\pi - x)}{\alpha \sinh \alpha \pi} - \frac{\sinh \alpha x}{\alpha (1 + \alpha^2) \sinh \alpha \pi} \right) \sin \alpha y \, d\alpha.$$

15. We use the Fourier sine transform with respect to x to obtain

$$U(\alpha, y) = c_1 \cosh \alpha y + c_2 \sinh \alpha y.$$

The transforms of $u(x, 0) = f(x)$ and $u(x, 2) = 0$ give, in turn, $U(\alpha, 0) = F(\alpha)$ and $U(\alpha, 2) = 0$. The first condition gives $c_1 = F(\alpha)$ and the second condition then yields

$$c_2 = -\frac{F(\alpha) \cosh 2\alpha}{\sinh 2\alpha}.$$

Hence

$$U(\alpha, y) = F(\alpha) \cosh \alpha y - \frac{F(\alpha) \cosh 2\alpha \sinh \alpha y}{\sinh 2\alpha}$$

$$= F(\alpha) \frac{\sinh 2\alpha \cosh \alpha y - \cosh 2\alpha \sinh \alpha y}{\sinh 2\alpha}$$

$$= F(\alpha) \frac{\sinh \alpha (2 - y)}{\sinh 2\alpha}$$

and

$$u(x, y) = \frac{2}{\pi} \int_0^\infty F(\alpha) \frac{\sinh \alpha (2 - y)}{\sinh 2\alpha} \sin \alpha x \, d\alpha.$$

18. We solve the three boundary-value problems:

Using separation of variables we find the solution of the first problem is

$$u_1(x, y) = \sum_{n=1}^\infty A_n e^{-ny} \sin nx \quad \text{where} \quad A_n = \frac{2}{\pi} \int_0^\pi f(x) \sin nx \, dx.$$

Using the Fourier sine transform with respect to y gives the solution of the second problem:

$$u_2(x, y) = \frac{200}{\pi} \int_0^\infty \frac{(1 - \cos \alpha) \sinh \alpha (\pi - x)}{\alpha \sinh \alpha \pi} \sin \alpha y \, d\alpha.$$

Also, the Fourier sine transform with respect to y gives the solution of the third problem:

$$u_3(x, y) = \frac{2}{\pi} \int_0^\infty \frac{\alpha \sinh \alpha x}{(1 + \alpha^2) \sinh \alpha \pi} \sin \alpha y \, d\alpha.$$

The solution of the original problem is

$$u(x, y) = u_1(x, y) + u_2(x, y) + u_3(x, y).$$

21. Using the Fourier transform with respect to x gives

$$U(\alpha, y) = c_1 \cosh \alpha y + c_2 \sinh \alpha y.$$

The transform of the boundary condition $\partial u/\partial y\,|_{y=0} = 0$ is $dU/dy\,|_{y=0} = 0$. This condition gives $c_2 = 0$. Hence

$$U(\alpha, y) = c_1 \cosh \alpha y.$$

Now by the given information the transform of the boundary condition $u(x, 1) = e^{-x^2}$ is $U(\alpha, 1) = \sqrt{\pi}\, e^{-\alpha^2/4}$. This condition then gives $c_1 = \sqrt{\pi}\, e^{-\alpha^2/4} \cosh \alpha$. Therefore

$$U(\alpha, y) = \sqrt{\pi}\, \frac{e^{-\alpha^2/4} \cosh \alpha y}{\cosh \alpha}$$

and

$$U(x, y) = \frac{1}{2\sqrt{\pi}} \int_{-\infty}^{\infty} \frac{e^{-\alpha^2/4} \cosh \alpha y}{\cosh \alpha} e^{-i\alpha x}\, d\alpha$$

$$= \frac{1}{2\sqrt{\pi}} \int_{-\infty}^{\infty} \frac{e^{-\alpha^2/4} \cosh \alpha y}{\cosh \alpha} \cos \alpha x\, d\alpha$$

$$= \frac{1}{\sqrt{\pi}} \int_{0}^{\infty} \frac{e^{-\alpha^2/4} \cosh \alpha y}{\cosh \alpha} \cos \alpha x\, d\alpha.$$

24. We use the Fourier sine transform with respect to the variable z. The transform of the partial differential equation is

$$\frac{d^2 U}{dr^2} + \frac{1}{r}\frac{dU}{dr} - \alpha^2 U + \alpha U(r, 0) = 0 \qquad \text{or} \qquad \frac{d^2 U}{dr^2} + \frac{1}{r}\frac{dU}{dr} - \alpha^2 U = -\alpha u_0.$$

Thus

$$U(r, \alpha) = c_1 I_0(\alpha r) + c_2 K_0(\alpha r) + \frac{u_0}{\alpha},$$

where $U(r, \alpha) = \mathscr{F}_s\{u(r, z)\}$. The usual argument about u being bounded at $r = 0$ implies $c_2 = 0$ so $U(r, \alpha) = c_1 I_0(\alpha r) + u_0/\alpha$. Now the transform of the boundary condition $u(1, z) = 0$ is $U(1, \alpha) = 0$ and so $c_1 I_0(\alpha) + u_0/\alpha = 0$ or $c_1 = -u_0/\alpha I_0(\alpha)$. Thus

$$U(r, \alpha) = \frac{-u_0 I_0(\alpha r)}{\alpha I_0(\alpha)} + \frac{u_0}{\alpha}.$$

The inverse transform is

$$u(r, z) = \frac{2}{\pi} \int_0^{\infty} \left(-\frac{u_0 I_0(\alpha r)}{\alpha I_0(\alpha)} + \frac{u_0}{\alpha} \right) \sin \alpha z\, d\alpha$$

$$= \frac{2u_0}{\pi} \int_0^{\infty} \frac{\sin \alpha z}{\alpha}\, d\alpha - \frac{2u_0}{\pi} \int_0^{\infty} \frac{I_0(\alpha r)}{\alpha I_0(\alpha)} \sin \alpha z\, d\alpha. \qquad \longleftarrow \quad \text{Problem 4, Exercises 15.4}$$

Therefore

$$u(r, z) = u_0 - \frac{2u_0}{\pi} \int_0^\infty \frac{I_0(\alpha r)}{\alpha I_0(\alpha)} \sin \alpha z \, d\alpha.$$

14.R Chapter 14 in Review

3. The Laplace transform gives

$$U(x, s) = c_1 e^{-\sqrt{s+h}\, x} + c_2 e^{\sqrt{s+h}\, x} + \frac{u_0}{s+h}.$$

The condition $\lim_{x\to\infty} \partial u/\partial x = 0$ implies $\lim_{x\to\infty} dU/dx = 0$ and so we define $c_2 = 0$. Thus

$$U(x, s) = c_1 e^{-\sqrt{s+h}\, x} + \frac{u_0}{s+h}.$$

The condition $U(0, s) = 0$ then gives $c_1 = -u_0/(s+h)$ and so

$$U(x, s) = \frac{u_0}{s+h} - u_0 \frac{e^{-\sqrt{s+h}\, x}}{s+h}.$$

With the help of the first translation theorem we then obtain

$$u(x, t) = u_0 \mathcal{L}^{-1}\left\{\frac{1}{s+h}\right\} - u_0 \mathcal{L}^{-1}\left\{\frac{e^{-\sqrt{s+h}\, x}}{s+h}\right\} = u_0 e^{-ht} - u_0 e^{-ht} \operatorname{erfc}\left(\frac{x}{2\sqrt{t}}\right)$$

$$= u_0 e^{-ht}\left[1 - \operatorname{erfc}\left(\frac{x}{2\sqrt{t}}\right)\right] = u_0 e^{-ht} \operatorname{erf}\left(\frac{x}{2\sqrt{t}}\right).$$

6. The Laplace transform and undetermined coefficients give

$$U(x, s) = c_1 \cosh sx + c_2 \sinh sx + \frac{s-1}{s^2 + \pi^2} \sin \pi x.$$

The conditions $U(0, s) = 0$ and $U(1, s) = 0$ give, in turn, $c_1 = 0$ and $c_2 = 0$. Thus

$$U(x, s) = \frac{s-1}{s^2 + \pi^2} \sin \pi x$$

and

$$u(x, t) = \sin \pi x \, \mathcal{L}^{-1}\left\{\frac{s}{s^2 + \pi^2}\right\} - \frac{1}{\pi} \sin \pi x \, \mathcal{L}^{-1}\left\{\frac{\pi}{s^2 + \pi^2}\right\}$$

$$= (\sin \pi x) \cos \pi t - \frac{1}{\pi}(\sin \pi x) \sin \pi t.$$

9. We solve the two problems

$$\frac{\partial^2 u_1}{\partial x^2} + \frac{\partial^2 u_1}{\partial y^2} = 0, \quad x > 0, \quad y > 0,$$

$$u_1(0, y) = 0, \quad y > 0,$$

$$u_1(x, 0) = \begin{cases} 100, & 0 < x < 1 \\ 0, & x > 1 \end{cases}$$

and

$$\frac{\partial^2 u_2}{\partial x^2} + \frac{\partial^2 u_2}{\partial y^2} = 0, \quad x > 0, \quad y > 0,$$

$$u_2(0, y) = \begin{cases} 50, & 0 < y < 1 \\ 0, & y > 1 \end{cases}$$

$$u_2(x, 0) = 0.$$

Using the Fourier sine transform with respect to x we find

$$u_1(x, y) = \frac{200}{\pi} \int_0^\infty \left(\frac{1 - \cos \alpha}{\alpha} \right) e^{-\alpha y} \sin \alpha x \, d\alpha.$$

Using the Fourier sine transform with respect to y we find

$$u_2(x, y) = \frac{100}{\pi} \int_0^\infty \left(\frac{1 - \cos \alpha}{\alpha} \right) e^{-\alpha x} \sin \alpha y \, d\alpha.$$

The solution of the problem is then

$$u(x, y) = u_1(x, y) + u_2(x, y).$$

12. Using the Laplace transform gives

$$U(x, s) = c_1 \cosh \sqrt{s}\, x + c_2 \sinh \sqrt{s}\, x.$$

The condition $u(0, t) = u_0$ transforms into $U(0, s) = u_0/s$. This gives $c_1 = u_0/s$. The condition $u(1, t) = u_0$ transforms into $U(1, s) = u_0/s$. This implies that $c_2 = u_0(1 - \cosh \sqrt{s})/s \sinh \sqrt{s}$. Hence

$$U(x, s) = \frac{u_0}{s} \cosh \sqrt{s}\, x + u_0 \left[\frac{1 - \cosh \sqrt{s}}{s \sinh \sqrt{s}} \right] \sinh \sqrt{s}\, x$$

$$= u_0 \left[\frac{\sinh \sqrt{s} \cosh \sqrt{s}\, x - \cosh \sqrt{s} \sinh \sqrt{s}\, x + \sinh \sqrt{s}\, x}{s \sinh \sqrt{s}} \right]$$

$$= u_0 \left[\frac{\sinh \sqrt{s}\,(1-x) + \sinh \sqrt{s}\,x}{s \sinh \sqrt{s}} \right]$$

$$= u_0 \left[\frac{\sinh \sqrt{s}\,(1-x)}{s \sinh \sqrt{s}} + \frac{\sinh \sqrt{s}\,x}{s \sinh \sqrt{s}} \right]$$

and

$$u(x,t) = u_0 \left[\mathscr{L}^{-1}\left\{ \frac{\sinh \sqrt{s}\,(1-x)}{s \sinh \sqrt{s}} \right\} + \mathscr{L}^{-1}\left\{ \frac{\sinh \sqrt{s}\,x}{s \sinh \sqrt{s}} \right\} \right]$$

$$= u_0 \sum_{n=0}^{\infty} \left[\mathrm{erf}\left(\frac{2n+2-x}{2\sqrt{t}} \right) - \mathrm{erf}\left(\frac{2n+x}{2\sqrt{t}} \right) \right]$$

$$+ u_0 \sum_{n=0}^{\infty} \left[\mathrm{erf}\left(\frac{2n+1+x}{2\sqrt{t}} \right) - \mathrm{erf}\left(\frac{2n+1-x}{2\sqrt{t}} \right) \right].$$

15. The Fourier cosine transform of the partial differential equation gives

$$\frac{dU}{dt} + k\alpha^2 U = 0 \qquad \text{so} \qquad U(\alpha, t) = c_1 e^{-k\alpha^2 t}.$$

The transform of the initial condition gives

$$U(\alpha, 0) = \frac{\alpha}{\alpha^2 + 1} = c_1$$

$$U(\alpha, t) = \frac{\alpha}{\alpha^2 + 1} e^{-k\alpha^2 t}$$

and so

$$u(x,t) = \frac{2}{\pi} \int_0^\infty \frac{\alpha e^{-k\alpha^2 t}}{\alpha^2 + 1} \cos \alpha x \, d\alpha.$$

15 Numerical Solutions of Partial Differential Equations

15.1 | Laplace's Equation

The terminology and concepts listed below provide an outline of the main ideas encountered in this section. These can be useful when preparing for a quiz or test.

Terminology and Concepts

- difference equation replacement
- mesh size
- interior points
- boundary points
- sparse matrix
- banded matrix
- Gauss-Seidel iteration

The basic skills listed below summarize the more mechanical types of problems encountered in the exercise set for this section.

Basic Skills

- use difference equation replacement to approximate the solution of Laplace's equation with a given mesh size subject to given boundary conditions
- use Gauss-Seidel iteration to approximate the solution of Laplace's equation on the interior of a unit square subject to given boundary conditions

3. The figure shows the values of $u(x, y)$ along the boundary. We need to determine u_{11}, u_{21}, u_{12}, and u_{22}. By symmetry $u_{11} = u_{21}$ and $u_{12} = u_{22}$. The system is

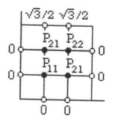

$$u_{21} + u_{12} + 0 + 0 - 4u_{11} = 0$$

$$0 + u_{22} + u_{11} + 0 - 4u_{21} = 0$$

$$u_{22} + \sqrt{3}/2 + 0 + u_{11} - 4u_{12} = 0$$

$$0 + \sqrt{3}/2 + u_{12} + u_{21} - 4u_{22} = 0$$

or

$$3u_{11} + u_{12} = 0$$

$$u_{11} - 3u_{12} = -\frac{\sqrt{3}}{2}.$$

Solving we obtain $u_{11} = u_{21} = \sqrt{3}/16$ and $u_{12} = u_{22} = 3\sqrt{3}/16$.

6. The coefficients of the unknowns are the same as shown above in Problem 5. The constant terms are 7.5, 5, 20, 10, 0, 15, 17.5, 5, 27.5. We use 32.5 as the initial guess for each variable. Then $u_{11} = 21.92$, $u_{21} = 28.30$, $u_{31} = 38.17$, $u_{12} = 29.38$, $u_{22} = 33.13$, $u_{32} = 44.38$, $u_{13} = 22.46$, $u_{23} = 30.45$, and $u_{33} = 46.21$.

15.2 | Heat Equation

The terminology and concepts listed below provide an outline of the main ideas encountered in this section. These can be useful when preparing for a quiz or test.

Terminology and Concepts

- explicit finite difference method
- unstable numerical method
- Crank-Nicholson method
- tridiagonal matrix

The basic skills listed below summarize the more mechanical types of problems encountered in the exercise set for this section.

Basic Skills

- solve the heat equation subject to given boundary and initial conditions using the finite difference method
- solve the heat equation subject to given boundary and initial conditions using the Crank-Nicholson method

3. We identify $c = 1$, $a = 2$, $T = 1$, $n = 8$, and $m = 40$. Then $h = 2/8 = 0.25$, $k = 1/40 = 0.025$, and $\lambda = 2/5 = 0.4$.

TIME	X=0.25	X=0.50	X=0.75	X=1.00	X=1.25	X=1.50	X=1.75
0.000	1.0000	1.0000	1.0000	1.0000	0.0000	0.0000	0.0000
0.025	0.7074	0.9520	0.9566	0.7444	0.2545	0.0371	0.0053
0.050	0.5606	0.8499	0.8685	0.6633	0.3303	0.1034	0.0223
0.075	0.4684	0.7473	0.7836	0.6191	0.3614	0.1529	0.0462
0.100	0.4015	0.6577	0.7084	0.5837	0.3753	0.1871	0.0684
0.125	0.3492	0.5821	0.6428	0.5510	0.3797	0.2101	0.0861
0.150	0.3069	0.5187	0.5857	0.5199	0.3778	0.2247	0.0990
0.175	0.2721	0.4652	0.5359	0.4901	0.3716	0.2329	0.1078
0.200	0.2430	0.4198	0.4921	0.4617	0.3622	0.2362	0.1132
0.225	0.2186	0.3809	0.4533	0.4348	0.3507	0.2358	0.1160
0.250	0.1977	0.3473	0.4189	0.4093	0.3378	0.2327	0.1166
0.275	0.1798	0.3181	0.3881	0.3853	0.3240	0.2275	0.1157
0.300	0.1643	0.2924	0.3604	0.3626	0.3097	0.2208	0.1136
0.325	0.1507	0.2697	0.3353	0.3412	0.2953	0.2131	0.1107
0.350	0.1387	0.2495	0.3125	0.3211	0.2808	0.2047	0.1071
0.375	0.1281	0.2313	0.2916	0.3021	0.2666	0.1960	0.1032
0.400	0.1187	0.2150	0.2725	0.2843	0.2528	0.1871	0.0989
0.425	0.1102	0.2002	0.2549	0.2675	0.2393	0.1781	0.0946
0.450	0.1025	0.1867	0.2387	0.2517	0.2263	0.1692	0.0902
0.475	0.0955	0.1743	0.2236	0.2368	0.2139	0.1606	0.0858
0.500	0.0891	0.1630	0.2097	0.2228	0.2020	0.1521	0.0814
0.525	0.0833	0.1525	0.1967	0.2096	0.1906	0.1439	0.0772
0.550	0.0779	0.1429	0.1846	0.1973	0.1798	0.1361	0.0731
0.575	0.0729	0.1339	0.1734	0.1856	0.1696	0.1285	0.0691
0.600	0.0683	0.1256	0.1628	0.1746	0.1598	0.1214	0.0653
0.625	0.0641	0.1179	0.1530	0.1643	0.1506	0.1145	0.0617
0.650	0.0601	0.1106	0.1438	0.1546	0.1419	0.1080	0.0582
0.675	0.0564	0.1039	0.1351	0.1455	0.1336	0.1018	0.0549
0.700	0.0530	0.0976	0.1270	0.1369	0.1259	0.0959	0.0518
0.725	0.0497	0.0917	0.1194	0.1288	0.1185	0.0904	0.0488
0.750	0.0467	0.0862	0.1123	0.1212	0.1116	0.0852	0.0460
0.775	0.0439	0.0810	0.1056	0.1140	0.1050	0.0802	0.0433
0.800	0.0413	0.0762	0.0993	0.1073	0.0989	0.0755	0.0408
0.825	0.0388	0.0716	0.0934	0.1009	0.0931	0.0711	0.0384
0.850	0.0365	0.0674	0.0879	0.0950	0.0876	0.0669	0.0362
0.875	0.0343	0.0633	0.0827	0.0894	0.0824	0.0630	0.0341
0.900	0.0323	0.0596	0.0778	0.0841	0.0776	0.0593	0.0321
0.925	0.0303	0.0560	0.0732	0.0791	0.0730	0.0558	0.0302
0.950	0.0285	0.0527	0.0688	0.0744	0.0687	0.0526	0.0284
0.975	0.0268	0.0496	0.0647	0.0700	0.0647	0.0495	0.0268
1.000	0.0253	0.0466	0.0609	0.0659	0.0608	0.0465	0.0252

(x,y)	exact	approx	abs error
(0.25,0.1)	0.3794	0.4015	0.0221
(1,0.5)	0.1854	0.2228	0.0374
(1.5,0.8)	0.0623	0.0755	0.0132

6. (a) We identify $c = 15/88 \approx 0.1705$, $a = 20$, $T = 10$, $n = 10$, and $m = 10$. Then $h = 2$, $k = 1$, and $\lambda = 15/352 \approx 0.0426$.

TIME	X=2	X=4	X=6	X=8	X=10	X=12	X=14	X=16	X=18
0	30.0000	30.0000	30.0000	30.0000	30.0000	30.0000	30.0000	30.0000	30.0000
1	28.7216	30.0000	30.0000	30.0000	30.0000	30.0000	30.0000	30.0000	28.7216
2	27.5521	29.9455	30.0000	30.0000	30.0000	30.0000	30.0000	29.9455	27.5521
3	26.4800	29.8459	29.9977	30.0000	30.0000	30.0000	29.9977	29.8459	26.4800
4	25.4951	29.7089	29.9913	29.9999	30.0000	29.9999	29.9913	29.7089	25.4951
5	24.5882	29.5414	29.9796	29.9995	30.0000	29.9995	29.9796	29.5414	24.5882
6	23.7515	29.3490	29.9618	29.9987	30.0000	29.9987	29.9618	29.3490	23.7515
7	22.9779	29.1365	29.9373	29.9972	29.9998	29.9972	29.9373	29.1365	22.9779
8	22.2611	28.9082	29.9057	29.9948	29.9996	29.9948	29.9057	28.9082	22.2611
9	21.5958	28.6675	29.8670	29.9912	29.9992	29.9912	29.8670	28.6675	21.5958
10	20.9768	28.4172	29.8212	29.9862	29.9985	29.9862	29.8212	28.4172	20.9768

(b) We identify $c = 15/88 \approx 0.1705$, $a = 50$, $T = 10$, $n = 10$, and $m = 10$. Then $h = 5$, $k = 1$, and $\lambda = 3/440 \approx 0.0068$.

TIME	X=5	X=10	X=15	X=20	X=25	X=30	X=35	X=40	X=45
0	30.0000	30.0000	30.0000	30.0000	30.0000	30.0000	30.0000	30.0000	30.0000
1	29.7955	30.0000	30.0000	30.0000	30.0000	30.0000	30.0000	30.0000	29.7955
2	29.5937	29.9986	30.0000	30.0000	30.0000	30.0000	30.0000	29.9986	29.5937
3	29.3947	29.9959	30.0000	30.0000	30.0000	30.0000	30.0000	29.9959	29.3947
4	29.1984	29.9918	30.0000	30.0000	30.0000	30.0000	30.0000	29.9918	29.1984
5	29.0047	29.9864	29.9999	30.0000	30.0000	30.0000	29.9999	29.9864	29.0047
6	28.8136	29.9798	29.9998	30.0000	30.0000	30.0000	29.9998	29.9798	28.8136
7	28.6251	29.9720	29.9997	30.0000	30.0000	30.0000	29.9997	29.9630	28.4391
8	28.4391	29.9630	29.9995	30.0000	30.0000	30.0000	29.9995	29.9529	28.2556
9	28.2556	29.9529	29.9992	30.0000	30.0000	30.0000	29.9992	29.9416	28.0745
10	28.0745	29.9416	29.9989	30.0000	30.0000	30.0000	29.9989	29.9416	28.0745

(c) We identify $c = 50/27 \approx 1.8519$, $a = 20$, $T = 10$, $n = 10$, and $m = 10$. Then $h = 2$, $k = 1$, and $\lambda = 25/54 \approx 0.4630$.

TIME	X=2	X=4	X=6	X=8	X=10	X=12	X=14	X=16	X=18
0	18.0000	32.0000	42.0000	48.0000	50.0000	48.0000	42.0000	32.0000	18.0000
1	16.1481	30.1481	40.1481	46.1481	48.1481	46.1481	40.1481	30.1481	16.1481
2	15.1536	28.2963	38.2963	44.2963	46.2963	44.2963	38.2963	28.2963	15.1536
3	14.2226	26.8414	36.4444	42.4444	44.4444	42.4444	36.4444	26.8414	14.2226
4	13.4801	25.4452	34.7764	40.5926	42.5926	40.5926	34.7764	25.4452	13.4801
5	12.7787	24.2258	33.1491	38.8258	40.7407	38.8258	33.1491	24.2258	12.7787
6	12.1622	23.0574	31.6460	37.0842	38.9677	37.0842	31.6460	23.0574	12.1622
7	11.5756	21.9895	30.1875	35.4385	37.2238	35.4385	30.1875	21.9895	11.5756
8	11.0378	20.9636	28.8232	33.8340	35.5707	33.8340	28.8232	20.9636	11.0378
9	10.5230	20.0070	27.5043	32.3182	33.9626	32.3182	27.5043	20.0070	10.5230
10	10.0420	19.0872	26.2620	30.8509	32.4400	30.8509	26.2620	19.0872	10.0420

(d) We identify $c = 260/159 \approx 1.6352$, $a = 100$, $T = 10$, $n = 10$, and $m = 10$. Then $h = 10$, $k = 1$, and $\lambda = 13/795 \approx 00164$.

TIME	X=10	X=20	X=30	X=40	X=50	X=60	X=70	X=80	X=90
0	8.0000	16.0000	24.0000	32.0000	40.0000	32.0000	24.0000	16.0000	8.0000
1	8.0000	16.0000	23.6075	31.3459	39.2151	31.6075	23.7384	15.8692	8.0000
2	8.0000	15.9936	23.2279	30.7068	38.4452	31.2151	23.4789	15.7384	7.9979
3	7.9999	15.9812	22.8606	30.0824	37.6900	30.8229	23.2214	15.6076	7.9937
4	7.9996	15.9631	22.5050	29.4724	36.9492	30.4312	22.9660	15.4769	7.9874
5	7.9990	15.9399	22.1606	28.8765	36.2228	30.0401	22.7125	15.3463	7.9793
6	7.9981	15.9118	21.8270	28.2945	35.5103	29.6500	22.4610	15.2158	7.9693
7	7.9967	15.8791	21.5037	27.7261	34.8117	29.2610	22.2112	15.0854	7.9575
8	7.9948	15.8422	21.1902	27.1709	34.1266	28.8733	21.9633	14.9553	7.9439
9	7.9924	15.8013	20.8861	26.6288	33.4548	28.4870	21.7172	14.8253	7.9287
10	7.9894	15.7568	20.5911	26.0995	32.7961	28.1024	21.4727	14.6956	7.9118

9. (a) We identify $c = 15/88 \approx 0.1705$, $a = 20$, $T = 10$, $n = 10$, and $m = 10$. Then $h = 2$, $k = 1$, and $\lambda = 15/352 \approx 0.0426$.

TIME	X=2.00	X=4.00	X=6.00	X=8.00	X=10.00	X=12.00	X=14.00	X=16.00	X=18.00
0.00	30.0000	30.0000	30.0000	30.0000	30.0000	30.0000	30.0000	30.0000	30.0000
1.00	28.7733	29.9749	29.9995	30.0000	30.0000	30.0000	29.9998	29.9916	29.5911
2.00	27.6450	29.9037	29.9970	29.9999	30.0000	30.0000	29.9990	29.9679	29.2150
3.00	26.6051	29.7938	29.9911	29.9997	30.0000	29.9999	29.9970	29.9313	28.8684
4.00	25.6452	29.6517	29.9805	29.9991	30.0000	29.9997	29.9935	29.8839	28.5484
5.00	24.7573	29.4829	29.9643	29.9981	29.9999	29.9994	29.9881	29.8276	28.2524
6.00	23.9347	29.2922	29.9421	29.9963	29.9997	29.9988	29.9807	29.7641	27.9782
7.00	23.1711	29.0836	29.9134	29.9936	29.9995	29.9979	29.9711	29.6945	27.7237
8.00	22.4612	28.8606	29.8782	29.9899	29.9991	29.9966	29.9594	29.6202	27.4870
9.00	21.7999	28.6263	29.8362	29.9848	29.9985	29.9949	29.9454	29.5421	27.2666
10.00	21.1829	28.3831	29.7878	29.9783	29.9976	29.9927	29.9293	29.4610	27.0610

12. We identify $c = 1$, $a = 1$, $T = 1$, $n = 5$, and $m = 20$. Then $h = 0.2$, $k = 0.04$, and $\lambda = 1$. The values below were obtained using *Excel*, which carries more than 12 significant digits. In order to see evidence of instability use $0 \le t \le 2$.

TIME	X=0.2	X=0.4	X=0.6	X=0.8
0.00	0.5878	0.9511	0.9511	0.5878
0.04	0.3633	0.5878	0.5878	0.3633
0.08	0.2245	0.3633	0.3633	0.2245
0.12	0.1388	0.2245	0.2245	0.1388
0.16	0.0858	0.1388	0.1388	0.0858
0.20	0.0530	0.0858	0.0858	0.0530
0.24	0.0328	0.0530	0.0530	0.0328
0.28	0.0202	0.0328	0.0328	0.0202
0.32	0.0125	0.0202	0.0202	0.0125
0.36	0.0077	0.0125	0.0125	0.0077
0.40	0.0048	0.0077	0.0077	0.0048
0.44	0.0030	0.0048	0.0048	0.0030
0.48	0.0018	0.0030	0.0030	0.0018
0.52	0.0011	0.0018	0.0018	0.0011
0.56	0.0007	0.0011	0.0011	0.0007
0.60	0.0004	0.0007	0.0007	0.0004
0.64	0.0003	0.0004	0.0004	0.0003
0.68	0.0002	0.0003	0.0003	0.0002
0.72	0.0001	0.0002	0.0002	0.0001
0.76	0.0001	0.0001	0.0001	0.0001
0.80	0.0000	0.0001	0.0001	0.0000
0.84	0.0000	0.0000	0.0000	0.0000
0.88	0.0000	0.0000	0.0000	0.0000
0.92	0.0000	0.0000	0.0000	0.0000
0.96	0.0000	0.0000	0.0000	0.0000
1.00	0.0000	0.0000	0.0000	0.0000

TIME	X=0.2	X=0.4	X=0.6	X=0.8
1.04	0.0000	0.0000	0.0000	0.0000
1.08	0.0000	0.0000	0.0000	0.0000
1.12	0.0000	0.0000	0.0000	0.0000
1.16	0.0000	0.0000	0.0000	0.0000
1.20	-0.0001	0.0001	-0.0001	0.0001
1.24	0.0001	-0.0002	0.0002	-0.0001
1.28	-0.0004	0.0006	-0.0006	0.0004
1.32	0.0010	-0.0015	0.0015	-0.0010
1.36	-0.0025	0.0040	-0.0040	0.0025
1.40	0.0065	-0.0106	0.0106	-0.0065
1.44	-0.0171	0.0277	-0.0277	0.0171
1.48	0.0448	-0.0724	0.0724	-0.0448
1.52	-0.1172	0.1897	-0.1897	0.1172
1.56	0.3069	-0.4965	0.4965	-0.3069
1.60	-0.8034	1.2999	-1.2999	0.8034
1.64	2.1033	-3.4032	3.4032	-2.1033
1.68	-5.5064	8.9096	-8.9096	5.5064
1.72	14.416	-23.326	23.326	-14.416
1.76	-37.742	61.067	-61.067	37.742
1.80	98.809	-159.88	159.88	-98.809
1.84	-258.68	418.56	-418.56	258.685
1.88	677.24	-1095.8	1095.8	-677.245
1.92	-1773.1	2868.9	-2868.9	1773.1
1.96	4641.9	-7510.8	7510.8	-4641.9
2.00	-12153	19663	-19663	12153

15.3 Wave Equation

The basic skills listed below summarize the more mechanical types of problems encountered in the exercise set for this section.

Basic Skills

- solve the wave equation subject to given boundary and initial conditions using the finite difference method

3. (a) Identifying $h = 1/5$ and $k = 0.5/10 = 0.05$ we see that $\lambda = 0.25$.

TIME	X=0.2	X=0.4	X=0.6	X=0.8
0.00	0.5878	0.9511	0.9511	0.5878
0.05	0.5808	0.9397	0.9397	0.5808
0.10	0.5599	0.9059	0.9059	0.5599
0.15	0.5256	0.8505	0.8505	0.5256
0.20	0.4788	0.7748	0.7748	0.4788
0.25	0.4206	0.6806	0.6806	0.4206
0.30	0.3524	0.5701	0.5701	0.3524
0.35	0.2757	0.4460	0.4460	0.2757
0.40	0.1924	0.3113	0.3113	0.1924
0.45	0.1046	0.1692	0.1692	0.1046
0.50	0.0142	0.0230	0.0230	0.0142

(b) Identifying $h = 1/5$ and $k = 0.5/20 = 0.025$ we see that $\lambda = 0.125$.

TIME	X=0.2	X=0.4	X=0.6	X=0.8
0.00	0.5878	0.9511	0.9511	0.5878
0.03	0.5860	0.9482	0.9482	0.5860
0.05	0.5808	0.9397	0.9397	0.5808
0.08	0.5721	0.9256	0.9256	0.5721
0.10	0.5599	0.9060	0.9060	0.5599
0.13	0.5445	0.8809	0.8809	0.5445
0.15	0.5257	0.8507	0.8507	0.5257
0.18	0.5039	0.8153	0.8153	0.5039
0.20	0.4790	0.7750	0.7750	0.4790
0.23	0.4513	0.7302	0.7302	0.4513
0.25	0.4209	0.6810	0.6810	0.4209
0.28	0.3879	0.6277	0.6277	0.3879
0.30	0.3527	0.5706	0.5706	0.3527
0.33	0.3153	0.5102	0.5102	0.3153
0.35	0.2761	0.4467	0.4467	0.2761
0.38	0.2352	0.3806	0.3806	0.2352
0.40	0.1929	0.3122	0.3122	0.1929
0.43	0.1495	0.2419	0.2419	0.1495
0.45	0.1052	0.1701	0.1701	0.1052
0.48	0.0602	0.0974	0.0974	0.0602
0.50	0.0149	0.0241	0.0241	0.0149

6. We identify $c = 24944.4$, $k = 0.00010022$ seconds $= 0.10022$ milliseconds, and $\lambda = 0.25$. Time in the table is expressed in milliseconds.

TIME	X=10	X=20	X=30	X=40	X=50
0.00000	0.2000	0.2667	0.2000	0.1333	0.0667
0.10022	0.1958	0.2625	0.2000	0.1333	0.0667
0.20045	0.1836	0.2503	0.1997	0.1333	0.0667
0.30067	0.1640	0.2307	0.1985	0.1333	0.0667
0.40089	0.1384	0.2050	0.1952	0.1332	0.0667
0.50111	0.1083	0.1744	0.1886	0.1328	0.0667
0.60134	0.0755	0.1407	0.1777	0.1318	0.0666
0.70156	0.0421	0.1052	0.1615	0.1295	0.0665
0.80178	0.0100	0.0692	0.1399	0.1253	0.0661
0.90201	-0.0190	0.0340	0.1129	0.1184	0.0654
1.00223	-0.0435	0.0004	0.0813	0.1077	0.0638
1.10245	-0.0626	-0.0309	0.0464	0.0927	0.0610
1.20268	-0.0758	-0.0593	0.0095	0.0728	0.0564
1.30290	-0.0832	-0.0845	-0.0278	0.0479	0.0493
1.40312	-0.0855	-0.1060	-0.0639	0.0184	0.0390
1.50334	-0.0837	-0.1237	-0.0974	-0.0150	0.0250
1.60357	-0.0792	-0.1371	-0.1275	-0.0511	0.0069
1.70379	-0.0734	-0.1464	-0.1533	-0.0882	-0.0152
1.80401	-0.0675	-0.1515	-0.1747	-0.1249	-0.0410
1.90424	-0.0627	-0.1528	-0.1915	-0.1595	-0.0694
2.00446	-0.0596	-0.1509	-0.2039	-0.1904	-0.0991
2.10468	-0.0585	-0.1467	-0.2122	-0.2165	-0.1283
2.20491	-0.0592	-0.1410	-0.2166	-0.2368	-0.1551
2.30513	-0.0614	-0.1349	-0.2175	-0.2507	-0.1772
2.40535	-0.0643	-0.1294	-0.2154	-0.2579	-0.1929
2.50557	-0.0672	-0.1251	-0.2105	-0.2585	-0.2005
2.60580	-0.0696	-0.1227	-0.2033	-0.2524	-0.1993
2.70602	-0.0709	-0.1219	-0.1942	-0.2399	-0.1889
2.80624	-0.0710	-0.1225	-0.1833	-0.2214	-0.1699
2.90647	-0.0699	-0.1236	-0.1711	-0.1972	-0.1435
3.00669	-0.0678	-0.1244	-0.1575	-0.1681	-0.1115
3.10691	-0.0649	-0.1237	-0.1425	-0.1348	-0.0761
3.20713	-0.0617	-0.1205	-0.1258	-0.0983	-0.0395
3.30736	-0.0583	-0.1139	-0.1071	-0.0598	-0.0042
3.40758	-0.0547	-0.1035	-0.0859	-0.0209	0.0279
3.50780	-0.0508	-0.0889	-0.0617	0.0171	0.0552
3.60803	-0.0460	-0.0702	-0.0343	0.0525	0.0767
3.70825	-0.0399	-0.0478	-0.0037	0.0840	0.0919
3.80847	-0.0318	-0.0221	0.0297	0.1106	0.1008
3.90870	-0.0211	0.0062	0.0648	0.1314	0.1041
4.00892	-0.0074	0.0365	0.1005	0.1464	0.1025
4.10914	0.0095	0.0680	0.1350	0.1558	0.0973
4.20936	0.0295	0.1000	0.1666	0.1602	0.0897
4.30959	0.0521	0.1318	0.1937	0.1606	0.0808
4.40981	0.0764	0.1625	0.2148	0.1581	0.0719
4.51003	0.1013	0.1911	0.2291	0.1538	0.0639
4.61026	0.1254	0.2164	0.2364	0.1485	0.0575
4.71048	0.1475	0.2373	0.2369	0.1431	0.0532
4.81070	0.1659	0.2526	0.2315	0.1379	0.0512
4.91093	0.1794	0.2611	0.2217	0.1331	0.0514
5.01115	0.1867	0.2620	0.2087	0.1288	0.0535

15.R   Chapter 15 in Review

3. (a)

TIME	X=0.0	X=0.2	X=0.4	X=0.6	X=0.8	X=1.0
0.00	0.0000	0.2000	0.4000	0.6000	0.8000	0.0000
0.01	0.0000	0.2000	0.4000	0.6000	0.5500	0.0000
0.02	0.0000	0.2000	0.4000	0.5375	0.4250	0.0000
0.03	0.0000	0.2000	0.3844	0.4750	0.3469	0.0000
0.04	0.0000	0.1961	0.3609	0.4203	0.2922	0.0000
0.05	0.0000	0.1883	0.3346	0.3734	0.2512	0.0000

(b)

TIME	X=0.0	X=0.2	X=0.4	X=0.6	X=0.8	X=1.0
0.00	0.0000	0.2000	0.4000	0.6000	0.8000	0.0000
0.01	0.0000	0.2000	0.4000	0.6000	0.8000	0.0000
0.02	0.0000	0.2000	0.4000	0.6000	0.5500	0.0000
0.03	0.0000	0.2000	0.4000	0.5375	0.4250	0.0000
0.04	0.0000	0.2000	0.3844	0.4750	0.3469	0.0000
0.05	0.0000	0.1961	0.3609	0.4203	0.2922	0.0000

(c) The table in part (b) is the same as the table in part (a) shifted downward one row.

3. If $t = x^3$, then $dt = 3x^2\,dx$ and $x^4\,dx = \frac{1}{3}t^{2/3}\,dt$. Now

$$\int_0^\infty x^4 e^{-x^3}\,dx = \int_0^\infty \frac{1}{3}t^{2/3}e^{-t}\,dt = \frac{1}{3}\int_0^\infty t^{2/3}e^{-t}\,dt$$

$$= \frac{1}{3}\Gamma\left(\frac{5}{3}\right) = \frac{1}{3}(0.89) \approx 0.297.$$

6. For $x > 0$ we integrate by parts:

$$\Gamma(x+1) = \int_0^\infty t^x e^{-t}\,dt$$

$u = t^x$	$dv = e^{-t}\,dt$
$du = xt^{x-1}\,dt$	$v = -e^{-t}$

$$= -t^x e^{-t}\Big|_0^\infty - \int_0^\infty xt^{x-1}\left(-e^{-t}\right)dt$$

$$= x\int_0^\infty t^{x-1}e^{-t}\,dt = x\,\Gamma(x).$$

A II | Matrices

3. (a) $\mathbf{AB} = \begin{pmatrix} -2-9 & 12-6 \\ 5+12 & -30+8 \end{pmatrix} = \begin{pmatrix} -11 & 6 \\ 17 & -22 \end{pmatrix}$

(b) $\mathbf{BA} = \begin{pmatrix} -2-30 & 3+24 \\ 6-10 & -9+8 \end{pmatrix} = \begin{pmatrix} -32 & 27 \\ -4 & -1 \end{pmatrix}$

(c) $\mathbf{A}^2 = \begin{pmatrix} 4+15 & -6-12 \\ -10-20 & 15+16 \end{pmatrix} = \begin{pmatrix} 19 & -18 \\ -30 & 31 \end{pmatrix}$

(d) $\mathbf{B}^2 = \begin{pmatrix} 1+18 & -6+12 \\ -3+6 & 18+4 \end{pmatrix} = \begin{pmatrix} 19 & 6 \\ 3 & 22 \end{pmatrix}$

6. (a) $\mathbf{AB} = \begin{pmatrix} 5 & -6 & 7 \end{pmatrix} \begin{pmatrix} 3 \\ 4 \\ -1 \end{pmatrix} = (-16)$

(b) $\mathbf{BA} = \begin{pmatrix} 3 \\ 4 \\ -1 \end{pmatrix} \begin{pmatrix} 5 & -6 & 7 \end{pmatrix} = \begin{pmatrix} 15 & -18 & 21 \\ 20 & -24 & 28 \\ -5 & 6 & -7 \end{pmatrix}$

(c) $(\mathbf{BA})\mathbf{C} = \begin{pmatrix} 15 & -18 & 21 \\ 20 & -24 & 28 \\ -5 & 6 & -7 \end{pmatrix} \begin{pmatrix} 1 & 2 & 4 \\ 0 & 1 & -1 \\ 3 & 2 & 1 \end{pmatrix} = \begin{pmatrix} 78 & 54 & 99 \\ 104 & 72 & 132 \\ -26 & -18 & -33 \end{pmatrix}$

(d) Since $\mathbf{AB}$ is 1×1 and $\mathbf{C}$ is 3×3 the product $(\mathbf{AB})\mathbf{C}$ is not defined.

9. (a) $(\mathbf{AB})^T = \begin{pmatrix} 7 & 10 \\ 38 & 75 \end{pmatrix}^T = \begin{pmatrix} 7 & 38 \\ 10 & 75 \end{pmatrix}$

(b) $\mathbf{B}^T\mathbf{A}^T = \begin{pmatrix} 5 & -2 \\ 10 & -5 \end{pmatrix}\begin{pmatrix} 3 & 8 \\ 4 & 1 \end{pmatrix} = \begin{pmatrix} 7 & 38 \\ 10 & 75 \end{pmatrix}$

12. $\begin{pmatrix} 6t \\ 3t^2 \\ -3t \end{pmatrix} + \begin{pmatrix} -t+1 \\ -t^2+t \\ 3t-3 \end{pmatrix} - \begin{pmatrix} 6t \\ 8 \\ -10t \end{pmatrix} = \begin{pmatrix} -t+1 \\ 2t^2+t-8 \\ 10t-3 \end{pmatrix}$

15. Since $\det \mathbf{A} = 0$, $\mathbf{A}$ is singular.

18. Since $\det \mathbf{A} = -6$, $\mathbf{A}$ is nonsingular.

$$\mathbf{A}^{-1} = -\frac{1}{6}\begin{pmatrix} 2 & -10 \\ -2 & 7 \end{pmatrix}$$

21. Since $\det \mathbf{A} = -9$, $\mathbf{A}$ is nonsingular. The cofactors are

$$\begin{array}{lll} A_{11} = -2 & A_{12} = -13 & A_{13} = 8 \\ A_{21} = -2 & A_{22} = 5 & A_{23} = -1 \\ A_{31} = -1 & A_{32} = 7 & A_{33} = -5. \end{array}$$

Then

$$\mathbf{A}^{-1} = -\frac{1}{9}\begin{pmatrix} -2 & -13 & 8 \\ -2 & 5 & -1 \\ -1 & 7 & -5 \end{pmatrix}^T = -\frac{1}{9}\begin{pmatrix} -2 & -2 & -1 \\ -13 & 5 & 7 \\ 8 & -1 & -5 \end{pmatrix}.$$

24. Since $\det \mathbf{A}(t) = 2e^{2t} \neq 0$, $\mathbf{A}$ is nonsingular.

$$\mathbf{A}^{-1} = \frac{1}{2}e^{-2t}\begin{pmatrix} e^t \sin t & 2e^t \cos t \\ -e^t \cos t & 2e^t \sin t \end{pmatrix}$$

27. $\mathbf{X} = \begin{pmatrix} 2e^{2t} + 8e^{-3t} \\ -2e^{2t} + 4e^{-3t} \end{pmatrix}$ so that $\dfrac{d\mathbf{X}}{dt} = \begin{pmatrix} 4e^{2t} - 24e^{-3t} \\ -4e^{2t} - 12e^{-3t} \end{pmatrix}.$

30. (a) $\dfrac{d\mathbf{A}}{dt} = \begin{pmatrix} -2t/(t^2+1)^2 & 3 \\ 2t & 1 \end{pmatrix}$

(b) $\dfrac{d\mathbf{B}}{dt} = \begin{pmatrix} 6 & 0 \\ -1/t^2 & 4 \end{pmatrix}$

(c) $\displaystyle\int_0^1 \mathbf{A}(t)\,dt = \begin{pmatrix} \tan^{-1} t & \frac{3}{2}t^2 \\ \frac{1}{3}t^3 & \frac{1}{2}t^2 \end{pmatrix}\Bigg|_{t=0}^{t=1} = \begin{pmatrix} \frac{\pi}{4} & \frac{3}{2} \\ \frac{1}{3} & \frac{1}{2} \end{pmatrix}$

(d) $\displaystyle\int_1^2 \mathbf{B}(t)\,dt = \begin{pmatrix} 3t^2 & 2t \\ \ln t & 2t^2 \end{pmatrix}\Bigg|_{t=1}^{t=2} = \begin{pmatrix} 9 & 2 \\ \ln 2 & 6 \end{pmatrix}$

(e) $\mathbf{A}(t)\mathbf{B}(t) = \begin{pmatrix} 6t/(t^2+1)+3 & 2/(t^2+1)+12t^2 \\ 6t^3+1 & 2t^2+4t^2 \end{pmatrix}$

(f) $\dfrac{d}{dt}\mathbf{A}(t)\mathbf{B}(t) = \begin{pmatrix} (6-6t^2)/(t^2+1)^2 & -4t/(t^2+1)^2+24t \\ 18t^2 & 12t \end{pmatrix}$

(g) $\displaystyle\int_1^t \mathbf{A}(s)\mathbf{B}(s)\,ds = \int_1^t \begin{pmatrix} 6s/(s^2+1)+3 & 2/(s^2+1)+12s^2 \\ 6s^3+1 & 6s^2 \end{pmatrix} ds$

$$= \begin{pmatrix} 3\ln(t^2+1)+3t-3\ln 2-3 & 2\tan^{-1} t+4t^3-\pi/2-4 \\ 3t^4/2+t-5/2 & 2t^3-2 \end{pmatrix}$$

33. $\left(\begin{array}{ccc|c} 1 & -1 & -5 & 7 \\ 5 & 4 & -16 & -10 \\ 0 & 1 & 1 & -5 \end{array} \right) \implies \left(\begin{array}{ccc|c} 1 & -1 & -5 & 7 \\ 0 & 1 & 1 & -5 \\ 0 & 9 & 9 & -45 \end{array} \right) \implies \left(\begin{array}{ccc|c} 1 & 0 & -4 & 2 \\ 0 & 1 & 1 & -5 \\ 0 & 0 & 0 & 0 \end{array} \right)$

Letting $z = t$ we find $y = -5 - t$, and $x = 2 + 4t$.

36. $\begin{pmatrix} 1 & 0 & 2 & | & 8 \\ 1 & 2 & -2 & | & 4 \\ 2 & 5 & -6 & | & 6 \end{pmatrix} \implies \begin{pmatrix} 1 & 0 & 2 & | & 8 \\ 0 & 2 & -4 & | & -4 \\ 0 & 5 & -10 & | & -10 \end{pmatrix} \implies \begin{pmatrix} 1 & 0 & 2 & | & 8 \\ 0 & 1 & -2 & | & -2 \\ 0 & 0 & 0 & | & 0 \end{pmatrix}$

Letting $z = t$ we find $y = -2 + 2t$, and $x = 8 - 2t$.

39. $\begin{pmatrix} 1 & 2 & 4 & | & 2 \\ 2 & 4 & 3 & | & 1 \\ 1 & 2 & -1 & | & 7 \end{pmatrix} \implies \begin{pmatrix} 1 & 2 & 4 & | & 2 \\ 0 & 0 & -5 & | & -3 \\ 0 & 0 & -5 & | & 5 \end{pmatrix} \implies \begin{pmatrix} 1 & 2 & 0 & | & -2/5 \\ 0 & 0 & 1 & | & 3/5 \\ 0 & 0 & 0 & | & 8 \end{pmatrix}$

There is no solution.

42. $\begin{bmatrix} 2 & 4 & -2 & | & 1 & 0 & 0 \\ 4 & 2 & -2 & | & 0 & 1 & 0 \\ 8 & 10 & -6 & | & 0 & 0 & 1 \end{bmatrix} \xrightarrow[\text{operations}]{\text{row}} \begin{bmatrix} 1 & 2 & -1 & | & \frac{1}{2} & 0 & 0 \\ 0 & 1 & -\frac{1}{3} & | & \frac{1}{3} & -\frac{1}{6} & 0 \\ 0 & 0 & 0 & | & -2 & -1 & 1 \end{bmatrix}$; $\mathbf{A}$ is singular.

45. $\begin{bmatrix} 1 & 2 & 3 & 1 & | & 1 & 0 & 0 & 0 \\ -1 & 0 & 2 & 1 & | & 0 & 1 & 0 & 0 \\ 2 & 1 & -3 & 0 & | & 0 & 0 & 1 & 0 \\ 1 & 1 & 2 & 1 & | & 0 & 0 & 0 & 1 \end{bmatrix} \xrightarrow[\text{operations}]{\text{row}} \begin{bmatrix} 1 & 2 & 3 & 1 & | & 1 & 0 & 0 & 0 \\ 0 & 1 & \frac{5}{2} & 1 & | & \frac{1}{2} & \frac{1}{2} & 0 & 0 \\ 0 & 0 & 1 & -\frac{2}{3} & | & \frac{1}{3} & -1 & -\frac{2}{3} & 0 \\ 0 & 0 & 0 & 1 & | & -\frac{1}{2} & 1 & \frac{1}{2} & \frac{1}{2} \end{bmatrix}$

$\xrightarrow[\text{operations}]{\text{row}} \begin{bmatrix} 1 & 0 & 0 & 0 & | & -\frac{1}{2} & -\frac{2}{3} & -\frac{1}{6} & \frac{7}{6} \\ 0 & 1 & 0 & 0 & | & 1 & \frac{1}{3} & \frac{1}{3} & -\frac{4}{3} \\ 0 & 0 & 1 & 0 & | & 0 & -\frac{1}{3} & -\frac{1}{3} & \frac{1}{3} \\ 0 & 0 & 0 & 1 & | & -\frac{1}{2} & 1 & \frac{1}{2} & \frac{1}{2} \end{bmatrix}$; $\mathbf{A}^{-1} = \begin{bmatrix} -\frac{1}{2} & -\frac{2}{3} & -\frac{1}{6} & \frac{7}{6} \\ 1 & \frac{1}{3} & \frac{1}{3} & -\frac{4}{3} \\ 0 & -\frac{1}{3} & -\frac{1}{3} & \frac{1}{3} \\ -\frac{1}{2} & 1 & \frac{1}{2} & \frac{1}{2} \end{bmatrix}$

48. We solve

$$\det(\mathbf{A} - \lambda\mathbf{I}) = \begin{vmatrix} 2 - \lambda & 1 \\ 2 & 1 - \lambda \end{vmatrix} = \lambda(\lambda - 3) = 0.$$

For $\lambda_1 = 0$ we have

$$\begin{pmatrix} 2 & 1 & | & 0 \\ 2 & 1 & | & 0 \end{pmatrix} \implies \begin{pmatrix} 1 & 1/2 & | & 0 \\ 0 & 0 & | & 0 \end{pmatrix}$$

so that $k_1 = -\frac{1}{2}k_2$. If $k_2 = 2$ then

$$\mathbf{K}_1 = \begin{pmatrix} -1 \\ 2 \end{pmatrix}.$$

For $\lambda_2 = 3$ we have

$$\begin{pmatrix} -1 & 1 & | & 0 \\ 2 & -2 & | & 0 \end{pmatrix} \implies \begin{pmatrix} 1 & -1 & | & 0 \\ 0 & 0 & | & 0 \end{pmatrix}$$

so that $k_1 = k_2$. If $k_2 = 1$ then

$$\mathbf{K}_2 = \begin{pmatrix} 1 \\ 1 \end{pmatrix}.$$

51. We solve

$$\det(\mathbf{A} - \lambda\mathbf{I}) = \begin{vmatrix} 5 - \lambda & -1 & 0 \\ 0 & -5 - \lambda & 9 \\ 5 & -1 & -\lambda \end{vmatrix} = \lambda(4 - \lambda)(\lambda + 4) = 0.$$

If $\lambda_1 = 0$ then

$$\begin{pmatrix} 5 & -1 & 0 & | & 0 \\ 0 & -5 & 9 & | & 0 \\ 5 & -1 & 0 & | & 0 \end{pmatrix} \implies \begin{pmatrix} 1 & 0 & -9/25 & | & 0 \\ 0 & 1 & -9/5 & | & 0 \\ 0 & 0 & 0 & | & 0 \end{pmatrix}$$

so that $k_1 = \frac{9}{25}k_3$ and $k_2 = \frac{9}{5}k_3$. If $k_3 = 25$ then

$$\mathbf{K}_1 = \begin{pmatrix} 9 \\ 45 \\ 25 \end{pmatrix}.$$

If $\lambda_2 = 4$ then

$$\begin{pmatrix} 1 & -1 & 0 & | & 0 \\ 0 & -9 & 9 & | & 0 \\ 5 & -1 & -4 & | & 0 \end{pmatrix} \implies \begin{pmatrix} 1 & 0 & -1 & | & 0 \\ 0 & 1 & -1 & | & 0 \\ 0 & 0 & 0 & | & 0 \end{pmatrix}$$

so that $k_1 = k_3$ and $k_2 = k_3$. If $k_3 = 1$ then

$$\mathbf{K}_2 = \begin{pmatrix} 1 \\ 1 \\ 1 \end{pmatrix}.$$

If $\lambda_3 = -4$ then

$$\begin{pmatrix} 9 & -1 & 0 & | & 0 \\ 0 & -1 & 9 & | & 0 \\ 5 & -1 & 4 & | & 0 \end{pmatrix} \implies \begin{pmatrix} 1 & 0 & -1 & | & 0 \\ 0 & 1 & -9 & | & 0 \\ 0 & 0 & 0 & | & 0 \end{pmatrix}$$

so that $k_1 = k_3$ and $k_2 = 9k_3$. If $k_3 = 1$ then

$$\mathbf{K}_3 = \begin{pmatrix} 1 \\ 9 \\ 1 \end{pmatrix}.$$

54. We solve

$$\det(\mathbf{A} - \lambda\mathbf{I}) = \begin{vmatrix} 1-\lambda & 6 & 0 \\ 0 & 2-\lambda & 1 \\ 0 & 1 & 2-\lambda \end{vmatrix} = (3-\lambda)(1-\lambda)^2 = 0.$$

For $\lambda = 3$ we have

$$\begin{pmatrix} -2 & 6 & 0 & | & 0 \\ 0 & 0 & 0 & | & 0 \\ 0 & 1 & -1 & | & 0 \end{pmatrix} \implies \begin{pmatrix} 1 & 0 & -3 & | & 0 \\ 0 & 1 & -1 & | & 0 \\ 0 & 0 & 0 & | & 0 \end{pmatrix}$$

so that $k_1 = 3k_3$ and $k_2 = k_3$. If $k_3 = 1$ then

$$\mathbf{K}_1 = \begin{pmatrix} 3 \\ 1 \\ 1 \end{pmatrix}.$$

For $\lambda_2 = \lambda_3 = 1$ we have

$$\begin{pmatrix} 0 & 6 & 0 & | & 0 \\ 0 & 1 & 1 & | & 0 \\ 0 & 1 & 1 & | & 0 \end{pmatrix} \implies \begin{pmatrix} 0 & 1 & 0 & | & 0 \\ 0 & 0 & 1 & | & 0 \\ 0 & 0 & 0 & | & 0 \end{pmatrix}$$

so that $k_2 = 0$ and $k_3 = 0$. If $k_1 = 1$ then

$$\mathbf{K}_2 = \begin{pmatrix} 1 \\ 0 \\ 0 \end{pmatrix}.$$

57. Let

$$\mathbf{A} = \begin{pmatrix} a_{11} & a_{12} \\ a_{21} & a_{22} \end{pmatrix}.$$

Then

$$\frac{d}{dt}[\mathbf{A}(t)\mathbf{X}(t)] = \frac{d}{dt}\begin{pmatrix} a_1 & a_2 \\ a_3 & a_4 \end{pmatrix}\begin{pmatrix} x_1 \\ x_2 \end{pmatrix} = \frac{d}{dt}\begin{pmatrix} a_1 x_1 + a_2 x_2 \\ a_3 x_1 + a_4 x_2 \end{pmatrix} = \begin{pmatrix} a_1 x_1' + a_1' x_1 + a_2 x_2' + a_2' x_2 \\ a_3 x_1' + a_3' x_1 + a_4 x_2' + a_4' x_2 \end{pmatrix}$$

$$= \begin{pmatrix} a_1 & a_2 \\ a_3 & a_4 \end{pmatrix}\begin{pmatrix} x_1' \\ x_2' \end{pmatrix} + \begin{pmatrix} a_1' & a_2' \\ a_3' & a_4' \end{pmatrix}\begin{pmatrix} x_1 \\ x_2 \end{pmatrix} = \mathbf{A}(t)\mathbf{X}'(t) + \mathbf{A}'(t)\mathbf{X}(t).$$

60. Since

$$(\mathbf{AB})(\mathbf{B}^{-1}\mathbf{A}^{-1}) = \mathbf{A}(\mathbf{BB}^{-1})\mathbf{A}^{-1} = \mathbf{AIA}^{-1} = \mathbf{AA}^{-1} = \mathbf{I}$$

and

$$(\mathbf{B}^{-1}\mathbf{A}^{-1})(\mathbf{AB}) = \mathbf{B}^{-1}(\mathbf{A}^{-1}\mathbf{A})\mathbf{B} = \mathbf{B}^{-1}\mathbf{IB} = \mathbf{B}^{-1}\mathbf{B} = \mathbf{I}$$

we have

$$(\mathbf{AB})^{-1} = \mathbf{B}^{-1}\mathbf{A}^{-1}.$$

Some Basic Mathematica Syntax and Commands

■ *Mathematica* conventions

■ Most implementations of *Mathematica* consist of two parts, the *front end* and the *kernel*. The lines you type into the Front End of *Mathematica* are called inputs and are shown in **boldface type** (unless you specify another convention). The output from *Mathematica* is generated by the kernel and is shown in plainface type. Generally, the output is displayed unless you specifically suppress it by using a semicolon. The construct (* **comment** *) is ignored by the kernel and can be used to annotate inputs in the front end.

In[1]:= **2**

Out[1]= 2

In[2]:= **3;**   (* **The corresponding output of 3 is suppressed, and Out[2] is not shown.** *)

In[3]:= **4; 5**   (* **The 4 is suppressed and the 5 is shown.** *)

Out[3]= 5

■ An equal sign (=) is used for assignment.

In[4]:= **a=5**

Out[4]= 5

In[5]:= **b=3;**

In[6]:= **b**   (* **Just to be sure that b was correctly input.** *)

Out[6]= 3

In[7]:= **a+b**

Out[7]= 8

In[8]:= **ab**   (* **This is the variable ab, not the product of a and b.** *)

Out[8]= ab

In[9]:= **a b**   (* **When a and b are separated by a space we have the product of a and b.** *)

Out[9]= 15

■ Grouping is done with parentheses only, not with square backets [ ] or braces { }.

In[10]:= **a/(a+b)**

Out[10]= $\dfrac{5}{8}$

■ Arguments of functions are enclosed in square brackets.

In[11]:= **Sqrt[9]**

Out[11]= 3

- Braces are used for ordered sets - called *lists* in *Mathematica*. The individual elements of a list are obtained using double square brackets [[ ]].

In[12]:= **m = {2, 7, -8}**

Out[12]= {2, 7, −8}

In[13]:= **m[[2]]**

Out[13]= 7

■ Built-in constants

■ π, E, I, Infinity

In[14]:= **π**   (* This is input using option-p. Equivalently, Pi can be typed. *)

Out[14]= π

In[15]:= **N[π]**   (* N is a function than converts a number to decimal form. *)

Out[15]= 3.14159

In[16]:= **E//N**   (* Using //, a function can be typed following its argument. *)

Out[16]= 2.71828

In[17]:= **I^2**   (* I is the square root of −1. *)

Out[17]= −1

In[18]:= **Infinity**

Out[18]= ∞

■ Built-in mathematical functions

- The following functions are built into *Mathematica* with the syntax shown.

  absolute value: **Abs[x]**
  square root: **Sqrt[x]**
  trigonometric: **Sin[x], Cos[x], Tan[x], Cot[x], Sec[x], Csc[x]**
  inverse trigonometric: **ArcSin[x], ArcCos[x], ArcTan[x], ArcCot[x], ArcSec[x], ArcCsc[x]**
  exponential: **Exp[x]** or **E^x**
  natural logarithm: **Log[x]**
  hyperbolic: **Sinh[x], Cosh[x], Tanh[x], Coth[x], Sech[x], Csch[x]**

  In[19]:= **Sqrt[−1]**

  Out[19]= i

  In[20]:= **Sqrt[90]**

  Out[20]= $3\sqrt{10}$

  In[21]:= **Sqrt[90]//N**

Out[21]= 9.48638

In[22]:= **N[Sin[Pi/3]]**

Out[22]= 0.866025

In[23]:= **Log[5]**

Out[23]= Log[5]

In[24]:= **Log[5]//N**

Out[24]= 1.60944

In[25]:= **Log[Exp[6]]**

Out[25]= 6

In[26]:= **Sinh[Log[3]]**

Out[26]= $\dfrac{4}{3}$

■ Derivatives

■ The syntax for finding the derivative of a function is discussed in Section 1.1 of this manual. Below we show how *Mathematica* can be used to verify that $2x/(3 - x^2)$ is a solution of the differential equation $xy' - y = xy^2$.

In[27]:= **Clear[y]**
lhs = x y′[x] - y[x]
rhs = x y[x]^2

Out[28]= $-y[x] + x\ y'[x]$

Out[29]= $x\ y[x]^2$

In[30]:= **y[x_]:=2x/(3 − x^2)**

y[x]

Out[31]= $\dfrac{2x}{3 - x^2}$

In[32]:= **lhs − rhs//Simplify**

Out[32]= 0

■ Rules

■ We saw above under ***Mathematica* conventions** how to assign a specific value to a variable. For example, if we want the variable x to have the value 2 we input **x=2**. Once this is done, until x is either reassigned or cleared, it will have the value 2. Suppose, on the other hand, we have an expression containing x, such as $x^2 + 2x - 5$, and we want the value of this expression when $x = 2$, but we do not want to permanently assign the value 2 to x. One way to do this is to define the expression as a function **f** and evaluate **f[2]**. However, if we are only going to do this for a single value of x, a faster way is to use the following syntax.

In[33]:= **x^2 + 3x - 1 /. x ->2**

Out[33]= 9

The arrow from x to 2 is obtained by typing a *minus sign* followed by the *greater than* symbol. The expression **x ->2** is called a rule and **/.** is a replacement operator. The complete input expression **x^2 + 3x - 1 /. x ->2** says to replace x with 2 in $x^2 + 3x - 1$, but do not permanently assign the value 2 to x.

In[34]:= **x+3** (* **x has not been permanently assigned the value 2.** *)

Out[34]= 3+x

■ Graphing with color and dashing

■ In Section 2.2 of this manual we saw that the **Plot** command can be supplemented with options like **PlotRange** to modify the appearance of a graph. *Mathematica* also has options for colorizing graphs and drawing graphs with dashed lines or thicker lines. The **RGBColor** option is used to obtain graphs of various colors. Red is **[1,0,0]**, green is **[0,1,0]**, and blue is **[0,0,1]**. Other colors are obtained by using numbers between 0 and 1 as the arguments of **RGBColor**. On a black and white printer the colors will appear as shades of gray as seen below.

In[35]:= **Plot[{ x^2, x, Sin[x], 2.5 - x^2}, {x,-3,3}, PlotStyle – > { RGBColor[1,0,0], RGBColor[0,1,0], RGBColor[0,0,1], RGBColor[1,0.5,0.8]}]**

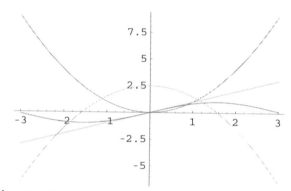

Out[35]= -Graphics-

■ The example below shows how to obtain a dashed graph, and thicker graphs in various shades of gray. The arguments for **Dashing** specify, alternately, the length of the dash and the length of the open space. The argument for **Thickness** is a number from 0 to 1 where 1 represents the width of the entire plot. The argument for **GrayLevel** is also a number from 0 to 1 where 0 is black and 1 is white.

In[36]:= Plot[{ x^2, x, Sin[x], 2.5 - x^2},{ x,-3,3}, PlotStyle- > {{ Dashing[{0.05,0.05]},

      { Dashing[{0.01,0.06]}, { Thickness[0.015], GrayLevel[0.4]},

      { Thickness[0.025], GrayLevel[0.8]}}]

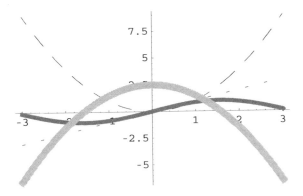

Out[36]= -Graphics-

This is explanatory only and should not be incorporated into the SRSM. The following should go at the very end of this manual immediately after "Out[36]= -Graphics-". The paragraph below starting with "Mathematica contains..." should be completely aligned on the word "Mathematica".

■ Help

■ *Mathematica* contains excellent help features that are easily accessed. If you have a question about a particular command you can select the Help menu at the top of your screen (in the *Mathematica* program) and Find Selected Function (or something comparable). Then use the SEARCH box to type in an appropriate command. For example, if you are interested in plotting a graph you could try typing in "plot" and see where that takes you.

47484515R00214